Industrial Automation: Circuit Design and Components

Industrial Automation: Circuit Design and Components

Contributors

Felipe Padilla, Aurora Torres et al.

www.aurisreference.com

Industrial Automation: Circuit Design and Components

Contributors: Felipe Padilla, Aurora Torres et al.

Published by Auris Reference Limited

www.aurisreference.com

United Kingdom

Industrial Automation: Circuit Design and Components

ISBN: 978-1-78154-936-0

British Library Cataloguing in Publication Data
A CIP record for this book is available from the British Library

Printed in the United Kingdom

Exclusively distributed by CBS Publishers & Distributors Pvt. Ltd.

Sales & Distribution Rights only for India, Pakistan, Bangladesh, Sri Lanka, Nepal and Bhutan.This book is not to be sold outside these territories.

Contents

List of Abbreviations

ACK	Acknowledgements
CAD	computer aided design
CRC	Cyclic redundancy check
DLL	Data Link Layer
DTI	declaration time instantiation
DFS	Depth-first search
DFM	Design for Manufacturability
DHA	discrete hybrid automata
DNP3	Distributed network protocol
EHW	Evolvable hardware
FTA	Fault Tree Analysis
FT	Fault trees
FSM	finite state machine
FB	Function Block
FBD	Function Block Diagram
FBN	Function Block Network
HMI	Human machine interface () 186
MTTF	Mean time to failure
MAC	Medium access control
MPS	Minimal path sets
MLD	Mixed Logical Dynamical
MS	mode selector
MDD	Model Driven Development
MPC	Model Predictive Control
MC	Monte Carlo
OSI	Open systems interconnection
POU	Program Organization Unit
QoS	Quality of Service
RTU	Remote terminal station
RC	Resistance and capacitance
SFC	Sequential Function Chart
ST	Structured Text
TCP	Transport control protocol
TL	Transport layer
TSDU	Transport service data unit
UML	Unified Modeling Language
VLSI	very large scale integration
WSN	Wireless Sensor Networks,

List of Contributors

Felipe Padilla
Aguascalientes University México
École de Technologie Supérieure Canada

Aurora Torres
Aguascalientes University México

Julio Ponce
Aguascalientes University México

María Dolores Torres
Aguascalientes University México

Sylvie Ratté
École de Technologie Supérieure Canada

Eunice Ponce-de-León
Aguascalientes University México

David C. Potts
Fairchild Semiconductor Corporation USA

Reza Hashemian
Northern Illinois University United States

Bruno Apolloni
Department of Computer Science, University of Milan, Via Comelico 39/41, 20135 Milano

Simone Bassis
Department of Computer Science, University of Milan, Via Comelico 39/41, 20135 Milano

Angelo Ciccazzo
STMicroelectronics, Stradale Primo Sole 50, 95121 Catania

Angelo Marotta
STMicroelectronics, Stradale Primo Sole 50, 95121 Catania

Salvatore Rinaudo
STMicroelectronics, Stradale Primo Sole 50, 95121 Catania

Orazio Muscato
Department of Mathematics and Informatics, University of Catania, Viale Andrea Doria 6, 95125 Catania Italy

Martin Coors
Institute for Integrated Signal Processing Systems, Aachen University of Technology, 52056 Aachen, Germany

Holger Keding
Institute for Integrated Signal Processing Systems, Aachen University of Technology, 52056 Aachen, Germany

Olaf Luthje
Institute for Integrated Signal Processing Systems, Aachen University of Technology, 52056 Aachen, Germany

Heinrich Meyr
Institute for Integrated Signal Processing Systems, Aachen University of Technology, 52056 Aachen, Germany

Goran Stojanovski
Department of Automation and System Engineering, Faculty of Electrical Engineering and Information Technologies, Ss. Cyril and Methodius University, Skopje, Macedonia

Mile Stankovski
Department of Automation and System Engineering, Faculty of Electrical Engineering and Information Technologies, Ss. Cyril and Methodius University, Skopje, Macedonia

Kleanthis Thramboulidis
Electrical and Computer Engineering, University of Patras, Patras, Greece
Saarland University, Saarbrucken, Germany

Georg Frey
Saarland University, Saarbrucken, Germany

Aamir Shahzad
Center for Advanced Image and Information Technology, School of Electronics & Information Engineering, Chon Buk National University, 664-14, 1Ga, Deokjin-Dong, Jeonju 561-756, Korea

Malrey Lee
Center for Advanced Image and Information Technology, School of Electronics & Information Engineering, Chon Buk National University, 664-14, 1Ga, Deokjin-Dong, Jeonju 561-756, Korea

Neal Naixue Xiong
School of Information Technology, Jiangxi University of Finance and Economics, Nanchang 330013, China
Department of Business and Computer Science, Southwestern Oklahoma State University, Oklahoma, OK 73096, USA

Gisung Jeong
Department of Fire Service Administration, WonKwang University, Iksan 570-749, Korea

Young-Keun Lee
Department of Orthopedic Surgery, Chonbuk National University Hospital, Jeonju 561-756, Korea

Jae-Young Choi
College of Information and Communication Engineering, Sungkyunkwan University, Suwon 16419, Korea

Abdul Wheed Mahesar
Department of Computer Science, International Islamic University Malaysia, Kuala Lumpur 53100, Malaysia

Iftikhar Ahmad
Department of Software Engineering, College of Computer and Information Sciences, King Saud University, Riyadh 11543, Saudi Arabia

Delphine Christin
Secure Mobile Networking Lab, Center for Advanced Security Research Darmstadt, Department of Computer Science, Technische Universität Darmstadt, Mornewegstr. 32, 64293 Darmstadt, Germany

Parag S. Mogre
Multimedia Communications Lab, Department of Computer Science, Technische Universität Darmstadt, Rundeturmstr. 10, 64283 Darmstadt, Germany

Matthias Hollick
Secure Mobile Networking Lab, Center for Advanced Security Research Darmstadt, Department of Computer Science, Technische Universität Darmstadt, Mornewegstr. 32, 64293 Darmstadt, Germany

Ivanovitch Silva
Department of Computer Engineering and Automation, Federal University of Rio Grande do Norte, Campus Universitário 59078-900, Natal, Brazil

Luiz Affonso Guedes
Department of Computer Engineering and Automation, Federal University of Rio Grande do Norte, Campus Universitário 59078-900, Natal, Brazil

Paulo Portugal
ISR, Department of Electrical and Computer Engineering, University of Porto, Porto 4200-465, Portugal

Francisco Vasques
IDMEC, Department of Mechanical Engineering, University of Porto, Porto 4200-465, Portugal

Preface

Industrial automation is the use of control systems, such as computers or robots, and information technologies for handling different processes and machineries in an industry to replace a human being. The text Industrial Automation: Circuit Design and Components covers methods immediately applicable to industrial problems, showing how to select the most appropriate control method for a given application, then design the necessary circuit. First chapter discusses evolvable metaheuristics on circuit design. Statistical analog circuit simulation has been focused in second chapter. A new approach to biasing design of analog circuits has been presented in third chapter. Advanced statistical methodologies for tolerance analysis in analog circuit design have been proposed in fourth chapter. In fifth chapter, we describe the principles and elements of FRIDGE and outline the seamless design flow as it becomes possible with this design environment. A hybrid system approach for high consumption industrial furnace control has been described in sixth chapter. Seventh chapter discusses possible alternatives to integrate the current but also the emerging specification of IEC 61131 in the model driven development process of automation systems. A secure, intelligent, and smart-sensing approach for industrial system automation and transmission over unsecured wireless networks has been presented in eighth chapter. In ninth chapter, we provide a detailed survey on wireless sensor networks (WSNs) standards dedicated to industrial automation networks. In last chapter, we propose a methodology based on an automatic generation of a fault tree to evaluate the reliability and availability of wireless sensor networks, when permanent faults occur on network devices.

Chapter 1

EVOLVABLE METAHEURISTICS ON CIRCUIT DESIGN

Felipe Padilla[1,2], Aurora Torres[1], Julio Ponce[1], María Dolores Torres[1], Sylvie Ratté[2] and Eunice Ponce-de-León[1]

[1]Aguascalientes University México

[2]École de Technologie Supérieure Canada

INTRODUCTION

Evolutionary computation algorithms are stochastic optimization methods; they are conveniently presented using the metaphor of natural evolution: a randomly initialized population of individuals evolves following a simulation of the Darwinian principle. New individuals are generated using genetic operations such as mutation and crossover. The probability of survival of the newly generated solutions depends on their fitness (Michalewicz et al., 1995). Evolutionary algorithms (EAs) have been successfully used to solve different types of optimization problems (Back, 1996). In the most general terms, evolution can be described as a two-step iterative process, consisting of random variation followed by selection. The structure of any evolutionary computation algorithm is shown in the figure 1.

```
procedure evolutionary algorithm
t ←0
initialize P(t)
evaluate P(t)
while (not termination-condition) do
begin
        t←t + 1
        select P(t) from P(t - 1)
        alter P(t)
        evaluate P(t)
end
```

Figure. 1: Structure of any evolutionary algorithm

The term evolutionary computation is used to describe techniques such as genetic algorithms, evolution strategies, evolutionary programming and genetic programming. The different approaches are distinguished by the genetic structures under adaption and the genetic operators that generate new candidate solutions (Cordon et al., 2001). Evolvable hardware (EHW) is an exquisite combination of evolutionary computation and electronic hardware. While the most common techniques of evolutionary computation are genetic algorithms and genetic programming, electronic hardware implies not only digital but analog circuits also. This field has earned importance since the early 1990's because of the advent of reconfigurable hardware. The ultimate objective of this field is to design and construct intelligent hardware, capable of online adaptation (Yao and Higuchi, 1999). The first classification of evolvable hardware can be found in (De Garis, 1993). In this work De Garis established there are extrinsic and intrinsic EHW. While Extrinsic EHW simulates evolution by software and downloads to hardware only the best configuration; intrinsic EHW simulates evolution directly in hardware. Nowadays the scope of this discipline has grown vastly. According to Zebulum (Zebulum, 1996), evolvable hardware can be classified by several criterion like hardware evaluation, evolvable computation approach, application area and evolvable platform. In regard to its application area EHW in divided in: Circuit design, robotics and control, pattern recognition, fault tolerance and very large scale integration (VLSI). We are interested in discuss about the first one. Circuit design is the art of constructing a sized circuit from user specifications (Das and Vemuri, 2009). This task is divided according to the kind of circuits that are handled in digital and analog circuit design. Nowadays there are different algorithms that can be used to solve problems of optimization of circuits like: Genetic Programming, Genetic Algorithm, Estimation of the Distribution Algorithms, Ant Colony Optimizations, Others. The more amenable nature of digital circuits made researchers like Louis (Louis, 1993) and Koza (Koza, 1992) to focus first on the production of functional logic circuits. Afterwards, the goal was not only to obtain functional circuits, but optimum ones. The work of Louis (Louis, 1993) was pioneer on the use of genetic algorithms on the design of combinational circuits; Thompson et al (Thompson et al., 1996) were the first in coding logic gates and its connections. Other outstanding researches on digital design are Higuchi et al. (Higuchi et al., 1996) specially focused on intrinsic evolution based on neural networks; Hernández and Coello (Hernández and Coello, 2003) first worked with genetic algorithms and later with genetic programming and Information Theory. A very interesting case is the use of ACO on the optimization of combinatorial circuits (Mendoza, 2001). The analog synthesis world also has numerous successful implementations of different metaheuristics like genetic algorithms (Lohn and Colombano, 1998),

(Zebulum et al., 2000), (Goh and Li, 2001), (Das and Vemuri, 2007), (Khalifa et al., 2008), (Torres et al., 2010); genetic programming (Koza et al., 1997), (Hu et al., 2005)(Chang et al., 2006) and estimation of the distribution algorithms (Torres et al., 2009). Analog circuit synthesis is a process composed of two phases: the selection of a suitable topology and the sizing of all its components (Torres et al., 2010). While topology consists on the determination of the type of components and its connections; sizing refers to the selection of the components values. Further on this document, will be discuss some of the mentioned approaches. Others types of evolutionary algorithms are based in biological systems in which complex collective behaviour emerges from the local interaction of simple components. Some examples of these algorithms are Swarm Intelligence, Ant Colony, Bees Algorithm, etc. We will speak of an ant colony, this algorithm is based in the foraging behaviour of some species of ants. Ant colonies are capable of finding the shortest paths between their nest and food sources, through a substance denominated pheromone.

OPTIMIZATION ALGORITHM

Actual trends in VLSI technology are towards integration of mixed analog-digital circuits as a complete system-on-a-chip. Most of the knowledge intensive and challenging design effort spent in such systems design is due to the analog building blocks (Balkir et al., 2004). Analog design has been traditionally a difficult discipline of integrated circuits (IC) design. In circuit design optimization, a circuit and its performance specifications are given and the goal is to automatically determine the device sizes in order to meet the given performance specifications while minimizing a cost function, such as a weighted sum of the active area or power dissipation (Baghini et al., 2007). This is a difficult and critical step for several reasons: 1) most analog circuits require a custom optimized design; 2) the design problem is typically under constrained with many degrees of freedom; and 3) it is common that many (often conflicting) performance requirements must to be taken into account, and tradeoffs must be made that satisfy the designer (Rutenbar et al., 2007). Fuzzy techniques have been successfully applied in a variety of fields such as automatic control data classification, decision analysis, expert systems, computer vision, multi-criteria evaluation, genetic algorithms, ant colony systems, optimization, etc. Works showing the possibility of application of fuzzy logic in computer aided design (CAD) of electronic circuits started to appear in late 1980s and early 1990s. An argument for fuzzy logic application in CAD is derived from the nature of the algorithm used for solving design problems. The majority of algorithms for synthesis use heuristics that are based on human knowledge acquired through experience and understanding of problems. Another important source of knowledge is numerical data. Fuzzy logic systems are appropriate in such situations because they are able to deal simultaneously with both types of information: linguistic and numerical. Also, fuzzy systems being universal appoximators can model any nonlinear functions of arbitrary complexity. This is very useful in modelling complex circuit functions of high accuracy at low cost, necessary in performance evaluation. Design optimization of an electronic circuit is a technique used to find the design parameter values (length and width of MOS transistors, bias current, capacitor values, etc.) in such a way that the final circuit performances (de gain, gain-bandwidth, slew rate, phase margin, etc.) meet as close as possible the design requirements. There is no general design procedure independent of the circuit; also, there is no formal representation to connect the circuit functions on its structure in a consistent manner. The major obstacle consists in the peculiarity of the analog signals: the continuous domain of the signals` amplitude and their continuous time dependency. Hereby the analog circuit design is known like an iterative, multi-phase task that necessitates a large spectrum of knowledge and abilities of designers.

GENETIC ALGORITHMS

Genetic algorithms originally were called "reproductive plans" by John Holland (Holland, 1975), and were the first emulators of the genetic evolution that produced practical results. In 1989, when Goldberg (Goldberg, 1989) published his book, mentioned more than 70 successful applications of this paradigm that continues winning popularity nowadays. According to Coello (Coello, 1996), a good definition of genetic algorithm was established by Koza in his book of 1992 (Koza, 1992), he says the following: "The genetic algorithm is a highly parallel mathematical algorithm that transforms a group (population) of individual mathematical objects (that usually have the form of chains of characters of fixed longitude), each one with an associate aptitude value, in new populations (for example the following generation) using modelling of operations under the Darwinian principle of the reproduction and survival of the "most capable", naturally, after the occurrence of the genetic operators (sexual recombination)". Ponce de León (Ponce de León, 1997) summarizes the mechanism of operation of the simple genetic algorithm in the following way; "it is generated a population of n structures aleatorily (chains, chromosomes or individuals) and then, some operators act transforming the population. The transformation is carried out by means of the application of three operators; once this culminates, it is said that a generational cycle has finished". The three operators Ponce references are: selection, crossover and mutation. The genetic algorithm in the form like Holland illustrates it

(Holland, 1975) has the following characteristic elements: 1. Representation of binary chains. 2. Proportional selection. 3. Crossover like the main method to produce new individuals. After the Holland's proposal, have been carried out different modifications; either by means of the use of different representation outlines, or until certain modifications to the selection operators, crossover, mutation and elitism. The diagram shown in the following figure presents the simplest version in the genetic algorithm, well-known as SGA (for the initials in English of "Simple Genetic Algorithm").

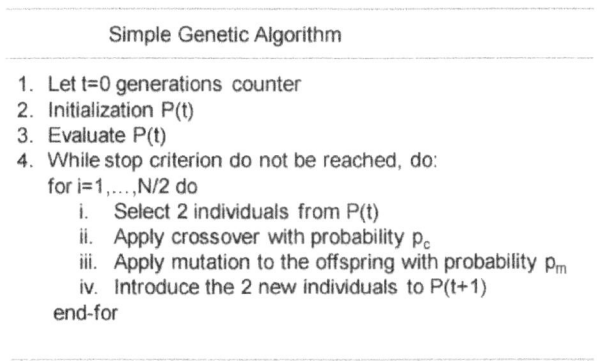

Figure. 2: Pseudocode of SGA

Although the general mechanism of this algorithm is extremely simple, it can be demonstrated by means of Markov's chains that the evolutionary algorithms that use elitist selection mechanisms, will converge to a good global solution of certain functions whose domain can be an arbitrary space (Torres, 2010). Günter Rudolph in 1996, generalized the previous developments in theory of the convergence for binary search spaces and Euclidian ones to general search spaces (Rudolph, 1996).

Genetic Algorithms in Automated Analog Design

Due to the high level of complexity that implies the task of designing and also to the strong dependence that this task has with the knowledge and experience experts; the automatic design of analogical circuits is a challenge and a necessity. Some researchers of the area believe that the automation of the design should be preceded by a change in the process of current design, for example, governed by the execution of the restrictions (Jerke, 2009). The fact is that nowadays, it has not still been possible to automate this process in a complete way. One of the metaheuristics that have shown better benefits in the realization of this task are the genetic algorithm and the genetic programming; this space belongs to the genetic algorithm. Lohn and Colombado (Lohn and

Colombado, 1998) used the genetic algorithm to design two analog filters, one of low complexity and one of medium complexity. The contribution of these researchers resides in that they demonstrated that it was feasible to use a very simple lineal representation. They proposed a code outline in which each element was represented by a fixed number of bytes called bytecodes in which they included an operation code that dictated the connection of each element and three bytes more they used to code its value. Koza on the other hand, continued making use of the genetic programming in the synthesis of computational circuits (Koza, 1997b) and controllers, filters and other kind of circuits (Koza, 1997). According to Ricardo Zebulum and his collaborators (Zebulum et al., 1998), the Evolutionary Electronics is an area that seeks to find new techniques of automatic design based on Darwinian concepts. The authors of the mentioned work, made the comparison of three different methodologies in the design of electronic filters. Their work was put on approval with two cases of study: A low-pass filter discussed in (Koza, 1996) and a filter pass-band with band in passing between 2000 and 3000 Hz and the bands of rejection above 4000 and below 1000Hz. The methodologies on approval were the following: "Outline of representation of variable longitude in combination with an evolutionary algorithm that restricts the topology of the filter (parallel meshes of two elements each one). For the simulation, an own tool was used in C, based on Laplace´s analysis. "Outline of representation of fixed longitude in combination with an evolutionary algorithm that doesn't restrict the topology of the circuit. To analyse the circuits they used Smash and SPICE, obtaining the same results. "Outline of representation of variable longitude in combination with an evolutionary algorithm that doesn't restrict the topology of the circuit. For the simulation of the circuits they used as much Smash as SPICE, obtaining the same results. In this work, Zebulum and his collaborators demonstrated that making use of an evolutionary algorithm based on the "Genetic Algorithm of Adaptation of Species (SAGA) of Harvey (Harvey, 1993), they could be obtained results comparable with those obtained using genetic programming, as for the answer in frequency of the obtained circuits using much smaller populations. This work concludes settling down that as for time, the first methodology was better, however this can explain to you for the rigidity of the used topology that allowed the use of a tool of quicker simulation. In spite of the success of this work, all the methodologies had inducer circuits whose values were so big as a result (2.2H for example) that are not very practical. On the other hand, investigators as Grimbleby and their collaborators (Grimbleby et al., 1995) they were working with mechanisms of numeric optimization in combination with genetic algorithms for the synthesis of analogical circuits using a chromosome of fixed length and a type of null component to fight with the variable size of the real circuits.

The XXI century has also been witness of numerous efforts made toward the automation of the synthesis of the analogical circuits, for example, in the year 2000, Zebulum et al. (Zebulum et al., 2000), established some advantages of variable length representation systems. Among other things, they argued that when using a fixed size, it is not only required expert knowledge of the problem, but the potential of the evolutionary algorithms is also limiting. That same year, they also proved an outline of representation of variable longitude that they understood passive elements, connected nodes and disconnected nodes. The authors emphasize the use of resistances and capacitors with programmable values in their architecture. These investigators intend to work the two phases of the evolution of an electric circuit (topology and adjustment of the parameters) in a sequential way, instead of making it simultaneously. In the year 2001, the investigating Goh and Li (Goh and Li, 2001) they began to outline some of the weaknesses that persisted in the process of design of analogical circuits that they were commented later by investigators as Khalifa and their collaborators (Khalifa et al., 2008), (Das, 2008) among others. The weaknesses that these investigators declare that they should be assisted, the reduction of the enormous computational effort that implies the evaluation of big generations of circuits

Table 1: Relevant research on analog circuit synthesis using Genetic Algorithms (Torres, 2010)

Year	Author	Application
1993	Horrocks and Spittle	Active low-pass filter
1994	Horrocks and Khalifa	Low-pass filter
1995	Grimbleby	High-pass filter
1996	Horrocks and Khalifa	Low-pass filter
1998	Lohn and Colombano	Low-pass filter
1998	Zebulum et al.	Low-pass filter Band- pass filter
1999	Krasnicki et al.	OP-AMP
2000	Ando and Iba	Passive filters
2000	Zebulum et al.	Passive filters
2001	Goh and Li	Low-pass filter High-pass filter
2007	Das and Vemuri	Low-pass filter
2008	Khalifa et al.	Low-pass filter High-pass filter
2008	Das and Vemuri	OP-AMP
2010	Torres et al.	Low-pass filter

that they don't always produce results and the reduction of the breach between the evolved circuits and those that finally are taken to the physical implementation, due to the restrictions of commercial physical devices. Other equally important aspects are related with the elaboration of tools that due to their complexity, they require expert personnel's manipulation or with a considerable level of knowledge (Krasnicki, 2001); as well as the execution in teams whose level of sophistication is outside of the reach of a great number of people.

ESTIMATION OF DISTRIBUTION ALGORITHMS

Estimation of distribution algorithms (EDA's) constitute a relatively new field of the Evolutionary Computation (Larrañaga, 2002) that replaces genetic operators (crossover and mutation) for the estimation of the distribution of the selected individuals and the sampling from the distribution to obtain the new population. The objective of this paradigm is to avoid the use of arbitrary operators as crossover and mutation, to modeling explicitly the most promising solutions for sampling solutions from its distribution. Pseudocode of the algorithm EDA:

- Step 1: Random generation of M individuals (initial Population)
- Step 2: Repeat the
- steps 3-5 for the generation l=1, 2,… until an stop criterion is reached
- Step 3: Select N <= M individuals from Dl-1 according to a selection method
- Step 4: Estimate the distribution of probability pl(x) from the group of selected individuals
- Step 5: Sample M individuals (new population) from pl(x)

EDAs can be classified according to two fundamental approaches. The first is the level of interdependences of variables, and the second is the type of involved variables. With regard to the level of interdependences EDAs are divided in 3, when the variables are independent, when there are bivaluated dependences and when there are multiple dependences. With regard to the type of involved variables, they can be discrete, continuous or mixed. The easiest version of an EDA is the "Univariate Marginal Distribution Algorithm" (UMDA) introduced by Mühlenbein (Mühlenbein and Paad, 1996). This algorithm works on the supposition of complete independence among variables. Pseudocode of this algorithm in presented in figure 3.

UMDA_AC

---

1. Begin
2. $D_0 \leftarrow$ Generate M individuals at random
3. Repeat for $l = 1, 2, \ldots$ until the stopping criteria met
   a) $D_{l-1}^{Se} \leftarrow$ Select $N \le M$ individuals from D_{l-1} according to the selection method
   b) $p_l(x) = P(x \mid D_{l-1}^{Se}) = \Pi_{i=1}^{n} p_l(x_i) = \Pi_{i=1}^{n} \dfrac{\sum_{j=1}^{N} \delta_j (X_i = xi \mid D_{l-1}^{Se})}{N} \leftarrow$
   Estimate the joint probability distribution
   c) $D_{l-1}^{Se} \leftarrow$ Sample M individuals from $p_l(x)$

Figure. 3: Pseudocode for UMDA (Larrañaga, 2002).

Another very common approach for the estimation of the distribution supposing independence among the variables is the algorithm PBIL ("Population-based incremental learning") (Baluja, 1994) that contrary to UMDA, doesn't estimate a new model in each generation, but refines it. The main problem of the distribution of the estimation algorithms, is to estimate the model; because as it gets more complicated, the dependences among the variables are captured in a better way, however, its estimation becomes more expensive (Larrañaga, 2002). Regarding models that consider bivariated dependences (dependences among pairs of variables), the most outstanding methods according its use in the literature are those that use chains like the "MIMIC" algorithm (Mutual Information Maximizing Input Clustering Algorithm) (De Bonet et al., 1996), those that use trees, as the case of the COMIT (Baluja and Davies, 1997) that uses the method of Chow and Liu [Chow 1968] based on the concept of mutual information and the BMDA (Pelikan, 1999), in which Pelikan and Mühlenbein propose a factoring of the distribution of joint probability. This algorithm is based on the construction of an acyclic directed graph of dependences that is not necessarily connected. Finally, the most common n-varied models are those that allow estimating a model in a Bayesian-net form. This approach has originated a great variety of algorithms according to the learning method, according to the nature of the variables (discrete or continuous), according to the imposed restrictions, etc. (Larrañaga, 2002). The great success genetic algorithms (GAs) have shown on several synthesis problems, has motivated some researches to explore the EDA's world in analog circuit synthesis. Next table show some examples.

Table 2: Relevant works on analog circuit synthesis by means of Estimation of the Distribution Algorithms

Year	Author	Application	Used metaheuristic
2002	Mühlenbein et al.	Low-pass	UMDA
2007	Zinchenko et al.	Mixed circuit	UMDA
2009	Torres et al.	Filters	UMDA
2010	Torres et al.	Filters	MITEDA

From table 2 it can be seen UMDA is the most common approach implemented on the analog circuit synthesis, nevertheless, MITEDA represents an effort on exploring the behavior of more complex EDAS. This algorithm was developed inspired by the COMIT and it uses the concept of mutual information used by Baluja and Davies (Baluja, 1997) to build the tree of dependences. Later this tree is sampling in order to create new generations. This algorithm represents the first tool that considers bi-valuated dependencies used in the design of analogical circuits we know until this moment.

ANT COLONY OPTIMIZATION

The Ant Colony Optimization Algorithm is a meta-heuristic bio-inspired in the behavior of real ant colonies. The first algorithm which can be classified within this framework was presented in 1991 by Marco Dorigo. In his PHD thesis with Title: "Optimization, learning, and Natural Algorithms", modeling the way real ants solve problems using pheromones. Real ants are capable of finding the shortest path from a food source to their nest. The ants deposit a concentration of pheromone in theirs paths, and they follows with more probability the way with more concentration of pheromone that it was previously deposited by other ants, the essential trait of ACO algorithms is the combination of a priori information about the structure of a promising solution with a posteriori information about the structure of previously obtained good solutions. In the Ant Colony Algorithms a number of artificial ants (agents) build solutions for an optimization problem and exchange information on their quality via a scheme of global communication that is reminiscent of the one adopted by real ants. When exist paths without any amount of pheromone, the ants explore the neighbourhood area in a totally random way. In presence of an amount of pheromone, the ants follow a path with a probability based in the pheromone concentration. The ants deposit additional pheromone concentrations during his travels. Since the pheromone evaporates, the pheromone concentration in non-used paths tends to disappear slowly. To find the shortest path, a moving ants lay some pheromone on the ground, so an ant encountering a previously trail can detect it and decide with high probability to follow it. As a result, the collective behavior that emerges is a form of a positive feedback loop where the probability with which each ant choose the next path increases with the number of ants that previously chose the same path. The Ant Colony System (ACS) models the behavior of ants, which are able to find the shortest path from their nest to a food source. Although individual ants move in a quasirandom form, performing relatively simple tasks, the entire colony of ants can collectively accomplish sophisticated movement patterns. Ants accomplish this by depositing a substance called a pheromone as they move. This chemical trail can be detected by other ants, which are probabilistically more likely to follow a path rich in pheromone. This trail information can be utilized to adapt to sudden unexpected changes to the terrain, such as when an obstruction blocks a previously used part of the path.

Application of Ant Colony to the Design of Combinatory Logic Circuits

To apply Ant Colony Algorithm to the design of logic circuits, in (Mendoza, 2001) is shown as the design of logic circuits with ACO. In the case of the logic circuits, the treatment of the problem does not seem to be so immediate.

Circuit Representation

The circuits are represent used a bidimensional matrix. Where each element of the matrix is a triplet of the type [Entrance 1, Entrance 2, Type of floodgate] (see figure 5). Was used five types of floodgates: AND, OR, NOT, XOR and WIRE, although this last one is not a floodgate, but rather it is a connection (a wire) that unites an element of certain column with another one of the previous column. Each element of the matrix receives its entrances solely of the exits of the previous column.

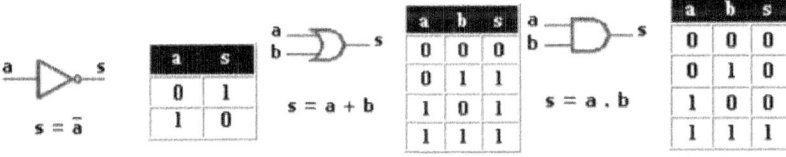

Figure. 4: Basic floodgate Not, Or, And

The first column directly receives its entrances of the table really of the given circuit. The last column provides the exits of the circuit. The first N rows corresponds to the N exits of the circuit. This form to represent a circuit has been used successfully. In the following figure are shown the basic floodgate.

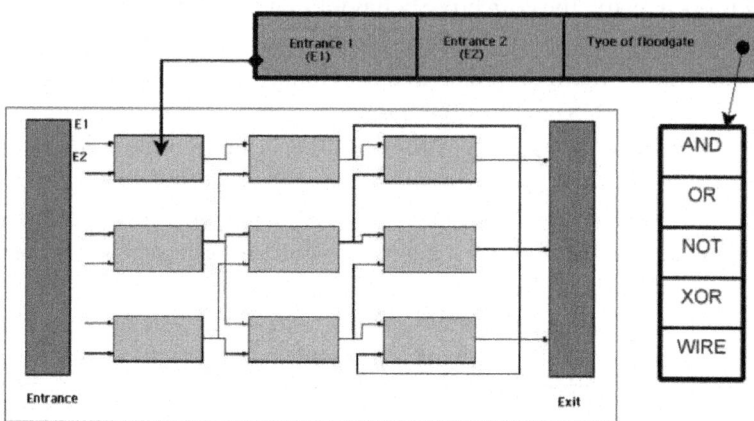

Figure. 5: Matrix used to circuit representation

Implementation

The route of an ant or agent will be a complete circuit. While each ant crosses a route, it constructs a circuit. In the TSP the ants find the route in terms of distance, do it here in terms of the number of floodgates. A state or city is a column, which is made up of several elements to which it is called substates to them, being these each one of the floodgates of a column and the number of combinations of possible entrances of each floodgate of this column. The first N substates (N is the number of exits in the circuit) is chosen with a selection factor P, and the others are chosen randomly. The distance between cities or states is measured as the increase or diminution from the successes to the exits of the circuit when changing from a level to another one. Unlike the problem of the TSP, in a same route (circuit), they do not have to visit all the states. The pheromones keep in a matrix called Trails. The length of this matrix corresponds to the number of exits of the circuit. Each element of Trails is a three-dimensional matrix as well. Next it is explained what they represent each one of the dimensions of the element. The first dimension of this matrix corresponds to the combination of possible entrances to the floodgate and goes from 0 to 6. The possible combinations of entrances, independent of the incoming number of the table really. The second dimension corresponds to the number of floodgate, that is to say, goes of 0 to the number of floodgates except one (NumGates-1). The third dimension corresponds to the number of successes that take until the level (column) previous and really goes of 0 to the number of lines in the table, because the number of successes that can be had in any level is between 0 and the number of lines of the true table.

The Construction of a Solution (Route)

As it was already mentioned before, a state is a column of the matrix, each element of the column is a floodgate with its respective entrances and their exit. Because of that, the election of a state is a process that becomes by parts (floodgate by floodgate), reason why we will call to each floodgate (element of the column) a substate. A state a combination of three elements (floodgate, IN1, IN2). In order to choose a substate of anyone of the first N rows, a value is assigned to him to each one of the possible combinations, call selection factor P, with which it will compete remaining in that position. The distance is a heuristic value and is given by the number of successes that the portion of the circuit constructed until the moment produces with respect to exit 1 of the True table. This is analogous to the distance in the TSP. Once it has assigned a factor of selection to all the combinations, is chosen what of them remains in the position in game. This is repeated with all the substates that belong to one of the rows that represent an exit of the circuit. The other substates, are chosen randomly. This is repeated until arriving at the last state from the circuit or column of the matrix. When all the ants finish their route, the pheromone signs are updated. This becomes in two steps: 1. First the amount is due to update pheromone in the ways, simulating the pheromone evaporation of the ways by the artificial ants to the passage of time. 2. The ways are due to update or to increase according to the routes constructed by each ant in the algorithm. This becomes of the following form: If the circuit result of the route is not valid (that it does not produce all the exits).

MULTIOBJECTIVE OPTIMIZATION

A population based evolutionary multiobjective optimization approach (Coello, 2009) to design combinatorial circuit was proposed for first time by Coello and Hernández in 2000 (Coello and Hernández, 2000). This approach reduced the computational effort required by genetic algorithm to design circuit at gate level. The main motivation was the reduction of fitness function evaluations while keeping the capabilities of the GA to generate novel designs. The main ideas behind MGA algorithm are: 1. Circuit representation as a matrix (originally proposed by Louis in 1991 (Louis and Rawlins, 1991)) and an n-cardinality alphabet. 2. Incremental method to resized of matrix used to fit a circuit. 3. Fitness function in two stages. At the beginning only validity of the circuit outputs is taken into account, and at the ending the fitness function is modified such that any valid designs produced are rewarded for each WIRE gate that they include. (WIRE gate indicates a null operation, that is, the absence of gate) 4. Use a multi-objective optimization technique (Fonseca and Fleming, 1995) (Coello, 1999). In general, it redefines the single-objective optimization of as a multiobjective optimization problem in which we will have $m + 1$ objectives, where Œm is the number of constraints. There is a new vector, $\bar{v} = (f, f_1, ..., f_n))$, where $_\iota$ is the objective function $f_1, ..., f_n$ are the original constraints of the problem. An ideal solution X X would thus have $_\iota f_i(X) = 0$ for $i = 1, ..., m$, and $f(X) \leq f(Y)$ for all feasible ; (assuming minimization). For combinatorial logic circuit design this technique consists on using a population based multiobjective optimization technique such as VEGA (Schaffer, 1984) to handle each of the outputs of the circuit as an objective. At each generation, the population is split in to $m + 1$ sub-populations, $m = 2^0$ (outputs), n: inputs of the circuit. The main mission of each sub-population is to match its corresponding output with the value indicated by the user in the truth table. After one of these objectives is satisfied, its corresponding sub-population is merged with the rest of the individuals in what becomes a joint effort to minimize the total amount of mismatches produced (between the encoded circuit and the truth table). Once a feasible individual is found, all individuals cooperate to minimize its number of gates (Coello and Hernández, 2002). The MGA algorithm outperformance the GA algorithm in quality of solution and decreased the evaluation amount of fitness function. This approach made a path in solving evolutionary design of combinational logic circuits.

Formulation of Multiobjective Optimization Problem

The multiobjective optimization problem can be formulated as follows (Coello and Hernández, 2000): A General Multiobjective Optimization Problem (MOP): Find the vector $\vec{x}^* = [x_1^*, ..., x_n^*]^T$ which will satisfy the m inequality constraints:

$$g_i(\vec{x}) \geq 0, i = 1, ..., m \tag{1}$$

the equality constraints

$$h_i(\vec{x}) = 0, i = 1, ..., p \tag{2}$$

and optimizes the vector function

$$\vec{f}(\vec{x}) = [f_1(\vec{x}), ..., f_k(\vec{x})]^T \tag{3}$$

where $\vec{x} = [x_1, ..., x_n]^T$ is the vector of decision variables.

That is, we wish to determine from among all $\vec{x} = [x_1, ..., x_n]^T$, which satisfy the inequality and equality constraints above, the particular $\vec{x}^* = [x_1^*, ..., x_n^*]^T$ which yields the optimum values of all the objective functions of the problem. Let be the set defined as all vectors $\vec{x} = [x_1, ..., x_n]^T$, that do not violate the constraints.

Pareto   Optimality   Definition: We   say   that
$\vec{x}^* = [x_1^*, ..., x_n^*]^T \in \Omega, \Omega \subseteq \mathbb{R}^n, f_i: \mathbb{R}^n \to \mathbb{R}$, is *Pareto*   is   Pareto   optimal   if   for   every
$\vec{x} = [x_1, ..., x_n]^T$, and $I = \{1, ..., k\}$ either,

$$\underset{i \in I}{\wedge}(f_i(\vec{x}) = f_i(\vec{x}^*))$$ (4)

Or, there is at least one $i \in I$ such that

$$`f_i(\vec{x}) > f_i(\vec{x}^*)$$ (5)

$\vec{x}^* = [x_1^*, ..., x_n^*]^T$ is Pareto optimal if there exists no feasible vector $\vec{x} = [x_1, ..., x_n]^T$ which would decrease some criterion without causing a simultaneous increase in at least one other criterion.

Pareto Dominance Definition: A vector $\vec{u} = (u_1, ..., u_n)$ is said to dominate $\vec{v} = (v_1, ..., v_n)$ (denoted by $\vec{u} \preccurlyeq \vec{v}$) if and only if $\vec{u}$ is partially less than $\vec{v}$, i.e., $\forall i \in \{1, ..., k\}, u_i \leq v_i \wedge \exists i \in \{1, ..., k\}: u_i < v_i$.

Pareto Optimal Set Definition: For a given $\mathcal{MOP}$, $\vec{f}(\vec{x}) = [f_1(\vec{x}), ..., f_k(\vec{x})]^T$, the Pareto optimal set $(\mathcal{P}^*)$ is defined as:

$$\mathcal{P}^* = \{\vec{x} \in \Omega \,/\, \nexists \vec{x}' = [x_1', ..., x_n']^T \in \Omega (\vec{f}(\vec{x}') \preccurlyeq \vec{f}(\vec{x}))\}$$ (6)

Pareto Front Definition: For a given $\mathcal{MOP}, \vec{f}(\vec{x}) = [f_1(\vec{x}), ..., f_k(\vec{x})]^T$ and Pareto optimal set $\mathcal{P}^*$, the Pareto front $(\mathcal{PF}^*)$ is defined as:

$$\mathcal{PF}^* = \{\vec{u} = \vec{f}(\vec{x}) = [f_1(\vec{x}), ..., f_k(\vec{x})]^T \,/\, \vec{x} \in \mathcal{P}^*\}$$ (7)

APPLICATION

Due to the enormous success genetic algorithms has proved on the field of circuit design, this section has the purpose of show how this metaheuristic could be used for the synthesis of analog circuits. In order to implement a genetic algorithm for the artificial evolution of any kind of process, is indispensable to find a way to represent a solution of the given problem, to find the way to generate possible solutions, to be able to evaluate the quality of the solutions and to have a group of operators that let transform one solution into another. Figure 6 shows the general flow used to implement a genetic algorithm in the analog circuit design according to Azizi (Azizi, 2001).

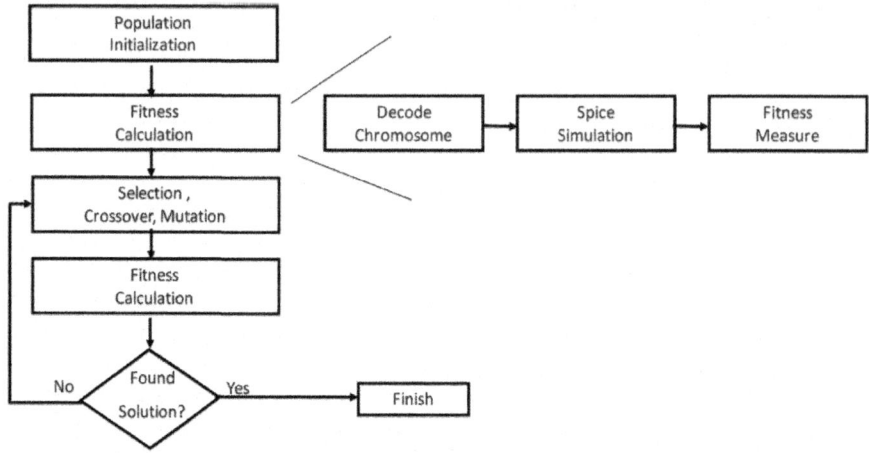

Figure. 6: Genetic Algorithm Flow for Analog Circuit Synthesis.

Representation mechanism In order to initialize the first population, the programmer has to establish how each solution is going to be represented and how the population can be generated.

A genetic encoding for artificial evolution of analog networks must be capable of representing both; the topology and the sizing of the network (Mattiussi and Floreano, 2007). While topology refers to the way each element is going to be connected to each other; sizing refers to the type and dimension of each element on a net. Other important aspects of the representation mechanism are its ability to capture any kind of circuit and the chance to reduce the process and time inverted in translate the circuit into a netlist (net description list). The representation mechanism has also to be flexible enough to be used with a wide range of components values but sufficiently short to be computational handling. (Torres et al., 2009). Torres et al (2009), reported a representation mechanism for passive elements of two terminals. This mechanism uses a gene of six parts to represent an analog element as figure 7 shows. Each circuit is a linked list of several genes.

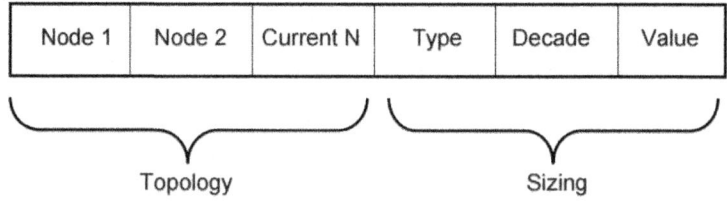

Figure. 7: Gene description.

While node 1 and node 2 refers to the terminals of an electrical device; current N is a pointer that is going to be used to build the network. Type, decade and value are the parameters that completely characterize a specific element (Torres, 2009). These parameters use integer coding according to table 3.

Table 3: Sizing encoding system

Type	Decade	Value	
C(0)	$10^{-6} - 10^{-9}$ (0-3)	E6	(0-5)
R(1)	$10^{+3} - 10^{+6}$ (0-3)	E12	(0-11)
L(2)	$10^{-3} - 10^{-6}$ (0-3)	E12	(0-11)

Next figure shows an element and its corresponding gene. We refers to initial node, that represents the beginning of an analog circuit.

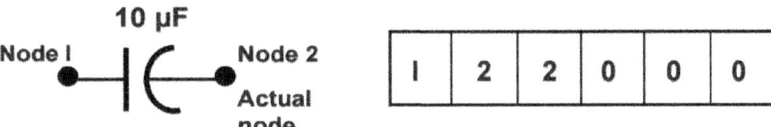

Figure. 8: An element of circuit and its corresponding gene.

Generation mechanism Once, a representation mechanism has been selected, the generation routine need to be established. The generation mechanism proposed by Torres et al. (2009) is based on an operation code randomly generated. The operation code establishes the connection that has to be done in the construction process of an admitted topology. The process begins in "Initial node" and ends when certain termination criterion is reached. This criterion could be one of two possibilities: the connection is done with the "Final Node" or the circuit reaches a preset amount of elements. Next figure describes how the generation mechanism works (CNode refers to the current node, and INode corresponds to Initial Node) (Torres et al., 2009).

Generation mechanism

---

1. begin
2.   CNode <- INode
3.   while(Not meet termination criterion)
    - Node1 = Cnode
    - Generate OP-Code
    - Execute_connection (Update Node2 and Cnode)
    - Generate Type, Decade and Value
4.   end_while
5. end

Figure. 9: Algorithm for the generation of each solution.

The circuit creation process performed by the former algorithm is very flexible. Once the operation code has been chosen and the connection has been done, type, decade and value of each element are generated. All operation codes used and their meaning are depicted in table 4.

Table 4: The operation code of the generation mechanism (Torres et al. 2010)

Op code	Instruction
0	Connect to grown
1	Connect to final node
2	Connect to x node
3	Connect to new node

Evaluation Function

Evolvable process depends on the ability to distinguish good and bad solutions, because it consists in continuously improve solutions from one generation to another. Therefore, a fitness function that describes how close a circuit is from the target is needed. Within the scope of analog circuit design, filters and amplifiers are the most frequently discussed. Fitness function used on the synthesis of low-pass filter will be presented below. Filters are circuits that block certain frequencies or bands of frequencies (Curtis, 2003). A low pass filter is the one that let pass low frequencies while blocks high frequencies. Next figure, illustrates the frequency response of an ideal and a real filter.

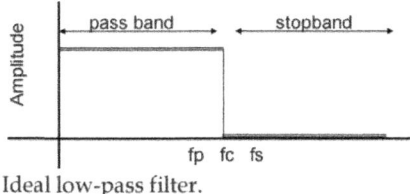 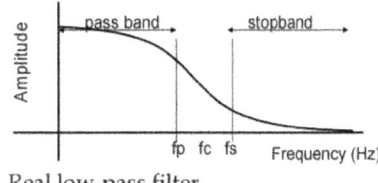

Ideal low-pass filter.   Real low-pass filter.

Figure. 10: Frequency response of an ideal and a real low-pass filter.

The fitness function used by Torres et al, is based on the measurement of the distance between the ideal and the real (evolved) filter. This function is an adaptation of the one used by Koza (Koza et al., 1997) and Hilder and Tyrrell (Hilder and Tyrrell, 2007) among other researchers. This function is the sum of errors between the ideal frequency response and the actual candidate, along N sampling points. Equation 8 describes the fitness measure for filters.

$$F = \frac{1}{1 + \xi} \tag{8}$$

Where :

$$\xi = \sum_{i=1}^{N} \lambda(\varepsilon_i) * \varepsilon_i \tag{9}$$

$$\varepsilon_i = \left| M(f_i)_{T \arg et} - M(f_i)_{Actual} \right| \tag{10}$$

"ξ" represents the error over the N points of frequency. If the deviation from target magnitude is inacceptable according to the frequency band, then a penalty factor "λ" has to be assigned to the error function.

A sample error function "ε" give us the absolute deviation between the actual output response and the target response over the "i" sampling point. $M(f_i)$Target denotes target magnitude at a f_i frequency, $M(f_i)$Actual is the magnitude of the actual evolved circuit at a fi frequency and fi is the sampling frequency.

Transformation of a Solution

Finally, when representation, generation and evaluation of candidate solutions have been solved, the programmer needs to find a group of operators to transform one solution into another. Starting from two parents chosen by any selection routine, an offspring is produce through two possible operators:

crossover and mutation. There are several selection algorithms; one of the more popular is the roulette-wheel. Roulette-wheel selection is an operator used for selecting potentially useful solutions for recombination. The fitness level of each solution is used to associate a probability of selection. If fi is the fitness of individual i in the population, its probability of being selected is $pi = \frac{fi}{\sum_{j=1}^{n} fj}$, where n is the number of individuals in the population. Crossover operation, introduces new solutions into the genetic algorithm starting from previous circuits; this operator is the responsible of changing some parts of a circuit by parts from another one. According to Dastidar et al., (Dastidar et al.,2005) and Das and Vemuri (Das and Vemuri, 2007), the use of some suitable connectivity rules, can reduce the unwanted search space not only for active, but for passive circuit synthesis. The crossover operator proposed by Torres et al., generates topological modifications because it alters the connection order of the offspring. This operator can be applied to one or two crossover points. Next figure shows how this operator can be executed on the condense chromosome of two progenitors, using the representation mechanisms proposed by Torres et al. In the figure "T" refers to ground connection and "F" represents the final node of the analog circuit. This condense representation of each solution only has connection nodes and type of each element.

Figure. 11: Crossover operator (Torres et al., 2010)

Mutation is an operator that traditionally introduces new solutions modifying only one chromosome. There are several ways to implement mutation, Goh and Li (Goh and Li, 2001) show a nice group on operators. The mutation exhibit in this section was proposed by Torres et al. This mutation operator is executed at gene level; and it works by altering a randomly chosen gene with another randomly generated. A mutated gene corresponds to a different type of element with different value, but connected to the same pair of nodes (Torres et al., 2010). Next figure shows an example of the use of this operator.

Figure. 12: Mutation operator.

Using all elements discuss in this section, the interested reader can implement an effective genetic algorithm for the automated synthesis of a passive filter.

CONCLUSION AND THE FUTURE RESEARCH

Nowadays exist applications in real life problems, where is possible used evolvable metaheuristics based on populations to the circuit design process, in this chapter was present some algorithms and applications, like Genetic Algorithms, Estimation of the distribution algorithms, Ant colony optimization. As shown there are multiple metaheuristics that can be used to circuit design trough different representations. We describe how is the representation with Genetic Algorithms. Since avoiding non valid topologies and non simulable networks, implies a very high reduction on time and computational resources in our problem; mainly three algorithms were compared at designing a low pass filter; a genetic algorithm (GA-AC), Ant Colony Systems (ACO-AC) and an estimation of distribution algorithm (UMDA-AC). Experimental results demonstrated that the group of mechanisms used in theses algorithms, worked better with GA-AC than with UMDA-AC and ACO-AC, according to the Pearson's Chisquared tests with respect to the generation of low rate of non spice-simulable circuits. Although UMDA-AC and ACO-AC performed faster the execution, and found a better individual on 200 generations' execution; statistically it cannot be said, the time difference is significant. With respect to the number of fitness evaluations, it can be said with statistical base, that UMDA-AC performs less evaluations than GA-AC per execution. In order to improve the performance of this algorithms, next step is the creation of a tool that blends the strengths of each metaheuristic. The work team is already working on the design of some new operators to be inserted on the EDA-AC and ACO-AC. GA-AC could be improved by enhancing the algorithm with some mechanisms of diversity control, like other kind of operators and another type of selection, in order to improve its exploration and delays its convergence. As future work is to continue working with various tools and algorithms that allow us to improve new circuit design. A new Artificial Intelligence that can be in charge of these systems, continues being distant into the horizon, in the same way that we still lack of methods to understand the original and peculiar things of each form to represent circuits.

REFERENCES

1. Back, T.: (1996). Evolutionary Algorithms in Theory and Practice. Oxford University Press, New York.

2. Baghini, M.; Kanphade, R.; Wakade, P.; Gawande, M.; Changani, M.; Patil, M. (2007). GPbased Design and Optimization of a Floating Voltage Source for Low-Power and Highly Tunable OTA Applications, WSEAS Transactions on Circuits and Systems,Issue 10, Volume 6, October 2007, pp. 588-582.

3. Balkir, S.; Dundar, G.; Alpaydin, G. (2004). Evolution Based Synthesis of Analog Integrated Circuits and Systems, IEEE NASA/DoD Conference on Evolution Hardware,EH`04. Pp 26-29.

4. Baluja, S. (1994). Population-Based Incremental Learning: A Method for Integrating GeneticSearch Based Function Optimization and Competitive Learning. Technical Report TR CMU-CS 94-163, Carnegie Mellon University.

5. Baluja, S. and Davies, S. (1997) Combining Multiple Optimization Runs with Optimal Dependency Trees. Technical Report TR CMU-CS-97-157, Carnegie Mellon University.

6. Chang, S.; Hou, H. and Su, Y. (2006). Automated Passive Filter Synthesis Using an Novel Tree Representation and Genetic Programming. IEEE ransactions on Evolutionary Computation, Vol. 10. No.1, February 2006. Pp. 93-100.

7. Coello, C. (1996). An Empirical Study of Evolutionary Techniques for Multiobjective Optimization in Engineering Design, PhD thesis, Department of Computer Science, Tulane University, New Orleans, Louisiana, USA..

8. Coello, C. A. A Comprehensive Survey of Evolutionary-Based Multiobjective Optimization Techniques. Knowledge and Information Systems. An International Journal, 1(3):269–308, August 1999.

9. Coello, C. A. and Hernández, A. (2002) Design of combinational logic circuits through an evolutionary multi-objective optimization approach. Artificial Intelligence for Engineering, Design, Analysis and Manufacture, 16(1): 39-53.

10. Coello, C. A. Hernández, A. and Buckles, B. P. (2000) Evolutionary

Multiobjective Design of Combinational Logic Circuits, eh, pp.161-172, The Second NASA/DoD Workshop on Evolvable Hardware (EH'00).

11. Coello, C. A., Lamont and, G. B., and Van Veldhuizen, D. A. (2007) Evolutionary Algorithms for Solving Multi-Objective Problems, Second Edition, Springer, New York, ISBN 978-0-387-33254-3.

12. Cordon, O.; Herrera, F.; Homann, F. and Magdalena, L. (2001). Genetic Fuzzy Systems: Evolutionary Tuning and Learning of Fuzzy Knowledge Bases, World Scientic.

13. Das A. and Vemuri R. (2007). An Automated Passive Analog Circuit Synthesis Framework using Genetic Algorithms. IEEE Computer Society Annual Symposium on VLSI ISVL '07. pp. 145-152.

14. Das, A. (2008) Algorithms for Topology Synthesis of Analog Circuits. Doctoral thesis. University of Cincinnati. November.

15. Das, A. and Vemuri, R. (2009). A Graph Grammar Based Approach to Automated MultiObjective Analog Circuit Design. Design, Automation and Test in Europe Conference and Exhibition 2009. IEEE Conferences., pp. 700-705.

16. De Bonet, J.; Isbell, C. and Viola, C. (1996) MIMIC: Finding Optima by Estimating Probability Densities. Proceeding of Neural Information Processing Systems. Pp.424-430.

17. De Garis, H. (1993). Evolvable Hardware: Genetic Programming of a Darwin Machine, in Artificial Neural Nets and Genetic Algorithms, Albretch, R.F., Reeves, C.R., and Steele, N.C., Eds., Springer-Verlag, New York.

18. Dorigo, M. (1991). Positive Feedback as a Search Strategy. Technical Report. No. 91-016. Politecnico Di Milano, Italy.

19. Fonseca, C. M. and Fleming, P. J. Genetic Algorithms for Multiobjective Optimization: Formulation, Discussion and Generalization. In S. Forrest, editor, Proceedings of the Fifth International Conference on Genetic Algorithms, pages 416–423, San Mateo, California, 1993. University of Illinois at Urbana-Champaign, Morgan Kauffman Publishers.

20. Goh, C. and . Li (2001). GA Automated Design and Synthesis of Analog Circuits with Practical Constraints. Proc. IEEE Int. Conf. Evol. Computation. pp. 170-177.

21. Goldberg, D. (1989) Genetic Algorithms in Search Optimization & Machine Learning. Addison-Wesley .

22. Grimbleby, J. (1995) Automatic Analogue Network Synthesis Using Genetic Algorithms, Proceedings of the First IEE/IEEE International

Conference on Genetic Algorithms in Engineering Systems (GALESIAS-95), pp.53-58, UK.

23. Harvey, I. (1993). The Artificial Evolution of Adaptive Behaviour, PhD Thesis, University of Sussex, School of Cognitive and Computing Sciences, September, 1993.

24. Hernandez A. and Coello C. (2003). Evolutionary Synthesis of Logic Circuits Using Information Theory. Artificial Intelligence Review 20: 445–471, Kluwer Academic Publishers. Printed in the Netherlands. Higuchi, T.; Iwata, M.; Kajitani, I.; Murakawa, M.; Yoshizawa, S. and Furuya, T. (1996).

25. Hardware Evolution at Gate and Function Levels. In Proceedings of the International Conference on Biologically Inspired Autonomous Systems: Computation, Cognition and Action, Durham, North Carolina.

26. Holland, J. (1975). Adaptation in Natural and Artificial Systems. University of Michigan Press. Ann Arbor, MI, 1975; MIT Press, Cambridge, MA 1992.

27. Hu, J.; Zhong, X. and Goodman, E. (2005). Open-ended Robust Design of Analog Filters Using Genetic Programming Proceedings of the 2005 conference on Genetic and evolutionary computation, June 25-29, Washington, DC, USA, pp. 1619-1626.

28. Jerke, G. and Lienig, J. (2009) "Constraint-driven Design — The Next Step Towards Analog Design Automation". Proceedings of the 2009 international symposium on Physical design. San Diego, California, USA. Pp. 75-82. 2009.

29. Khalifa, Y.; Khan, B. and Taha, F. (2008). Multi-objective Optimization Tool for A Free Structure Analog Circuits Design Using Genetic Algorithms and Incorporating Parasitics. Hindawi Publishing Corporation. Journal of Artificial Evolution and Applications. Volume 2008, PP. 0-9.

30. Koza, J. (1992) Genetic Programming. On the Programming of Computers by Means of Natural Selection. MIT Press. Cambridge, Massachussetts, 1992.

31. Koza, J.; Bennett, F.; Andre, D. and Keane, M.. (1996) Toward Evolution of Electronic Animals Using Genetic Programming. Artificial Life V: Proceedings of the Fifth International Workshop on the Synthesis and Simulation of Living Systems.Cambridge, MA: The MIT Press.

32. Koza, J.; Bennethh, F.; Lohn, J.; Dunlap, F.; Keane M. and Andre D. (1997b) Automated synthesis of computacional circuits using genetic programming" in Proc. 1997 IEEEConf. Evolutionary Computation. Piscataway, NJ: IEEE Press, pp. 447–452, 1997.

33. Koza, J.; Bennethh, F.; Andre, D. and Keane, M. (1997). Automated Synthesis of Analog Electrical Circuits by Means of Genetic Programming. IEEE Transactions on Evolutionary Computation, Vol 1, No.2, pp. 109-128.

34. Krasnicki, M.; Phelps, R.; Hellums, J.; McClung, M.; Rutenbar, R. and Carley, L. (2001). ASF: a practical simulation based methodology for the synthesis of custom analog circuits, In Proceedings of ICCAD 2001, pp. 350–357.

35. Larrañaga P. and Lozano, J. (2002) Estimation of Distribution Algorithms: A New Tool for Evolutionary Computation". Kluwer Academic Publishers.

36. Lohn, J. and Colombano, S. (1998). Automated Analog Circuit Synthesis using a Linear Representation. Proc. of the Second Int'l Conf on Evolvable Systems: From Biology to Hardware, Springer-Verlag, Berlin, pp. 125-133.

37. Louis, S. (1993). Genetic Algorithms as a Computational Tool for Design. PhD Thesis,Department of Computer Science, Indiana University.

38. Louis, S. J. and Rawlins, G. J. E. (1991) Using Genetic Algorithm to Design Structures. Technical Report 326. Computer Science Department, Indiana University, Bloomington, Indiana.

39. Mendoza, B. (2001). Uso de del Sistema de la Colonia de Hormigas para Optimizar Circuitos Lógicos Combinatorios. Tesis de Maestría en Inteligencia Artificial de la Universidad Veracruzana. México.

40. Michalewicz, Z.; Dasgupta, D.; Le Riche, R. and Schoenauer M. (1995). Evolutionary Algorithms for Constrained Engineering Problems.

41. Muhlenbein, H. and Paad G. (1996). From Recombination of Genes to the Estimation of Distributions I. Binary Parameters, in H.M.Voigt, et al., eds., Lecture Notes in Computer Science 1411: Parallel Problem Solvingfrom Nature - PPSN IV, pp. 178-187.

42. Pelikan M. and Mühlenbein, H. (1999). The Bivariate Marginal Distribution Algorithm. Advances in Soft Computing-Engineering Design and Manufacturing. Pp. 521-535.

43. Ponce de León, E., (1997) Algoritmos Genéticos y su Aplicación a Problemas de Secuenciación". PhD. Tesis. Centro de Inteligencia Artificial. Instituto de Cibernética, Matemática y Física.

44. Rudolph, G. (1996) Convergence of Evolutionary Algorithms in General Search Spaces, In Proceedings of the Third IEEE Conference on Evolutionary Computation .

45. Rutenbar, R.; Gielen, G.; Roychowdhury, J. (2007). Hierarchical Modeling, Optmization, and Synthesis for System-level Anlog and RF Designs, Proc. of the IEEE, Vol. 95, Issue 3, March 2007, pp. 640-669.

46. Schaffer, J. D. (1984) Multiple Objective Optimization with Vector Evaluated Genetic Algorithms. PhD thesis, Vanderbilt University.

47. Shragowitz, E.; Lee,J.; Kang, Q. (1998). Application of Fuzzy Logic in Computer-Aided VLSI Design, IEEE Trans. on Fuzzy Systems, Vol. 6, No 1, February 1998, pp. 163-172.

48. Thompson, A.; Harvey, I. and Husbands, P. (1996). Unconstrained evolution and hard consequences. In E. Sanchez and M. Tomassini, editors, Toward Evolvable Hardware: The Evolutionary Engineering Approach (Lecture Notes in Computer Science, Vol. 1062), pages 136--165, Heidelberg, Germany, Springer-Verlag.

49. Torres, A. (2010). Metaheurísticas Evolutivas en el Diseño de Circuitos Analógicos. Tesis Doctoral. Universidad Autónoma de Aguascalientes.

50. Torres, A.; Ponce de León, E.; Hernández, A.; Torres, M.D. and Díaz, E. (2010). A Robust Evolvable System for the Synthesis of Analog Circuits. Computación y Sistemas, Revista Iberoamericana de Computación. April-June, Vol. 13, No.4, pp. 295-312.

51. Torres, A.; Ponce de León, E.; Torres, M. D.; Díaz E. and Padilla, F. (2009). "Comparison of Two Evolvable Systems in the Automated Analog Circuit Synthesis". Artificial Intelligence, MICAI 2009. Eighth Mexican International Conference on, vol.1, pp 3-8, 8-13.

52. Yao, X. and Higuchi, T. (1999) "Promises and Challenges of Evolvable Hardware", IEEE Transactions on Systems, Man and Cybernetics_Part C. Applications and Reviews, Vol 29, No.1.

53. Zebulum, R.; Pacheco M. and Vellasco M. (1996). Evolvable Systems in Hardware Design: Taxonomy, Survey and Applications, Proceedings of The First International Conference on Evolvable Systems: From Biology to Hardware (ICES'96), Lecture Notes in Computer Science 1259, pp. 344-358, Tsukuba, Japan, October 7-8.

54. Zebulum, R.; Pacheco M. and Vellasco, M. (1998). Comparison of different evolutionary methodologies applied to electronic filter design. In Proc. Of IEEE. Inttl. Conf. On Evolutionary Computation. May.

55. Zebulum, R.; Vellasco, M. and Pacheco, M. (2000) Variable length representation in evolutionary electronics. Evol. Comput., vol. 8, no. 1, pp. 93–120.

Chapter 2

STATISTICAL ANALOG CIRCUIT SIMULATION: MOTIVATION AND IMPLEMENTATION

David C. Potts

Fairchild Semiconductor Corporation USA

INTRODUCTION

New technologies are continually being developed that enable designers to create faster, more complex circuits, packed within a shrinking die. However, along with the promise of speed and density comes the challenge of variability, as intra-die device mismatch looms proportionately greater. Analog designs typically employ multiple core building block circuits, including current mirrors, band gap references, differential pairs and op amps, that are especially sensitive to device mismatch. Understanding the impact and potential interactions of variations between these matched devices can be critical in producing a commercially viable product. The first part of this chapter will provide a background on the statistical nature of the semiconductor manufacturing process, with a particular focus on their implications on device performance. Due to the complexity of interactions coupled with circuit-specific design sensitivities, traditional corner models do not provide the designer with sufficient accuracy and visibility to thoroughly assess and improve the quality of their designs. Corner models also do not account for mismatch, which is a major concern for analog designs. A statistical simulation system that realistically replicates process variability will provide the designer with insights to optimize the design. The second part of the chapter will delve into the extraction and use of statistical models within a statistical simulation system. A properly implemented statistical design tool can become one of the greatest assets available to the designer. Following a discussion of various published statistical model formulations and extraction methodologies from literature, we will consider how they might be incorporated and used within commercially available simulators. We conclude the chapter with a demonstration that systematically evaluates the components of a band gap circuit to isolate matching sensitivities and refine the design for optimized results. With the assistance of statistical design analysis, a designer can make informed choices that will produce better circuit performance and manufacturability.

SEMICONDUCTOR PROCESS VARIATION

Semiconductor device and circuit performance will fluctuate due to the inherent underlying statistical variation in the process itself. This variation can include both random and systematic components. As illustrated in Figure 1, the overall total variance can be partitioned into components reflecting the physical separation of the material during processing.

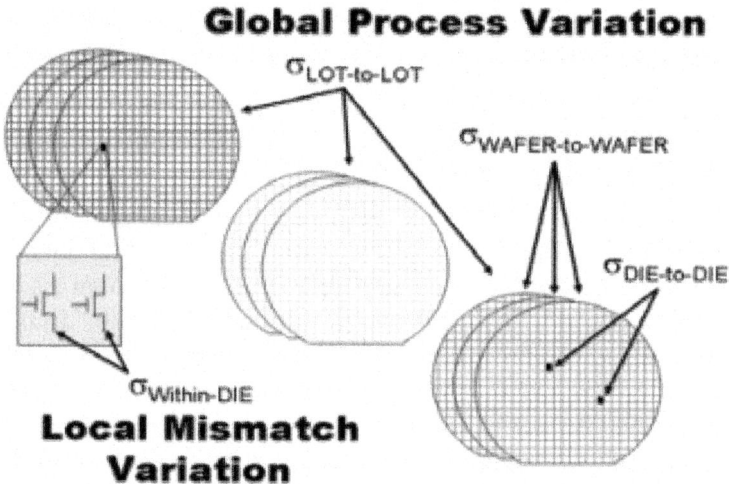

Figure. 1: Classifications of Statistical Variation

Lot-to-lot variance is generally the largest of the components as it reflects significant sources of variation not seen in the other groups, including variation across different tools that may be used at a given process step, variation between batches of raw materials, along with timebased trends and cycles relating to tool aging, preventive maintenance, upgrades and adjustments. Wafer to wafer variance can result from the slight differences experienced between wafers at single wafer processing steps as well as from gradients across batch processed wafers, such as induced by temperature and flow gradients within a furnace tube. Die-to-die variance can be an artifact of differences in exposures in stepper based lithography or gradients or localized disturbances of wafer uniformity. Lot-to-lot, wafer-towafer and die-to-die variance combined are often referred to as Global Variation, because all devices found on any particular die will be simultaneously and equally affected by them in the same way. In other words, in the world of that particular die, this is a global effect. Within-die (device-to-device) variation may include a more localized contribution of some of the wafer uniformity effects driving die-to-die variance, as well as individual device definition effects resulting in slight non-uniformities in film thicknesses and edge definitions, dopant distributions, junction depths, surface roughness, and so on. Within-die variance is generally referred to as Local Variation, because the performance of each individual device on a given die will be affected slightly differently by it. This variation can include both random and systematic components. The designer may have some limited control over certain systematic components relating to device layout, but needs to be aware of and have some means to estimate the effects of variation on circuit performance. Traditionally, this was done using so-called 'corner' models, intended to represent the worst case corners of the process variation.

ISSUES WITH TRADITIONAL CORNER MODELS

In traditional corner methodologies, 'worst case' models were typically created by evaluating the sensitivities of critical model parameters individually and then setting each of them to their worst case values simultaneously. The accuracy of this approach, however, would be highly dependent on the actual physical correlation between the parameters as well as the cumulative probability that all would be worst case at the same time (Nardi et al., 1999). The corner method also assumes a 'one-size-fits-all' solution, when in reality different designs and circuit architectures will exhibit different worst case sensitivities. Finally, fixed corner models do not account for the intra-device variations that can have a major impact on analog circuit performance.

The Issue of Correlation

To demonstrate the impact of correlation, consider two standard normal variables, X and Y, which are summed and scaled to create Z. Figure 2 depicts the results for 3 cases representing negative, zero and positive correlation between X and Y:

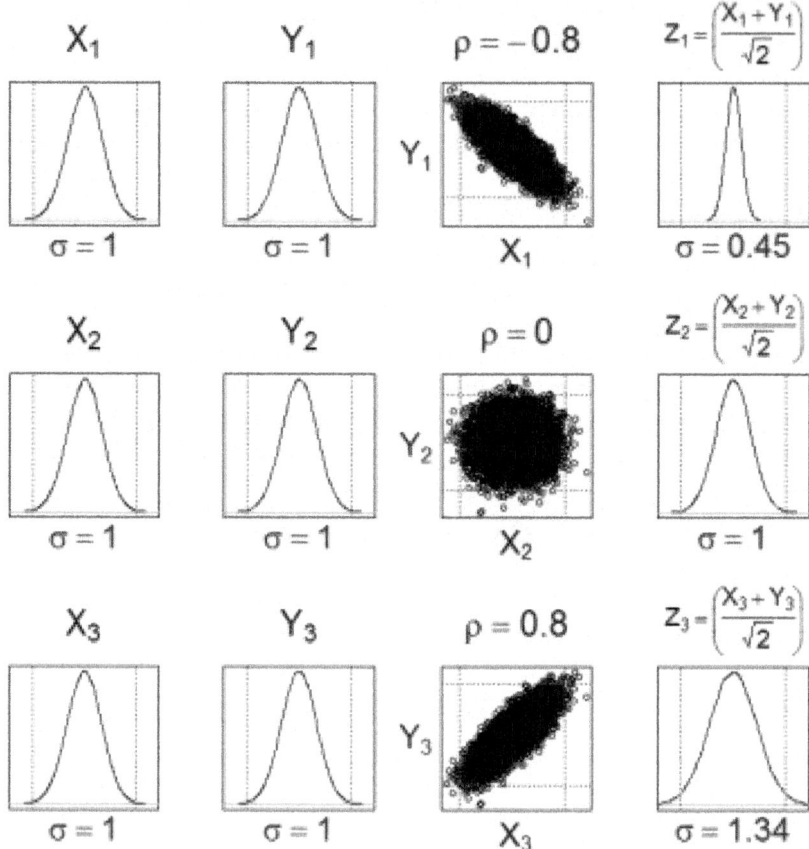

Figure. 2: The Impact of Correlation

In this simple example, it is intuitively obvious that when X and Y are negatively correlated, they would tend to cancel each other out, thus minimizing the resulting variability of Z. Conversely, when they are positively correlated, they would tend to reinforce each other, creating greater variability. Semiconductor processes, of course, are much more complex with a great number of interacting variables. The fact that there are a large number of variables brings in the next problem: how to determine which combinations of these variables best define the corners?

The Issue of Corner Selection

Assume we have a normally distributed process and we want to define a set of worst case corners that encompass an interval of ± 3 standard deviations about its mean ($\mu \pm 3\sigma$). In other words, the probability the process would fall outside of our $\mu \pm 3\sigma$ corners would be about 0.0027. The probability that two different uncorrelated normally distributed variables would both simultaneously fall outside their respective $\mu \pm 3\sigma$ is only $(0.0027)2 = 0.00000729$. As the number of independent variables increases, the probability that they would all simultaneously fall outside their respective $\mu \pm 3\sigma$ windows drops off rapidly, as shown in Figure 3a. Instead of putting all variables at $\pm 3\sigma$, we might prefer to find a $\pm k\sigma$ window such that the probability of falling outside remains constant at 0.0027 (for n variables, this corresponds to the standard normal z score for area of $(0.00271/n)/2$). As the number of independent variables increases, the k value drops, as shown in Figure 3b. Of course, there is nothing that forces us to select a corner that puts each variable at the same k value. Figure 3c show the line that plots possible solutions of k values when there are only 2 variables to consider (for 3 variables, the solution would be a surface and for n variables, it would be an n dimensional space).

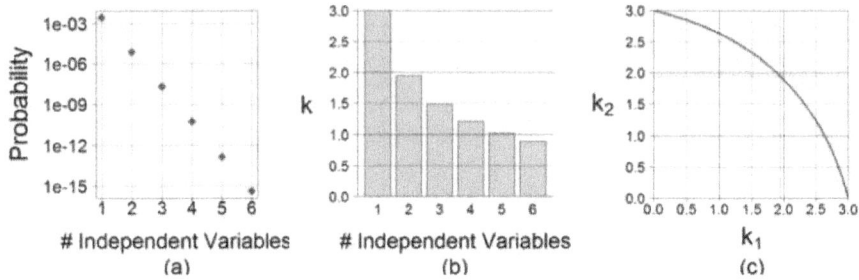

Figure. 3: (a) Probability of Multiple Variables Falling Outside Their Respective $\mu \pm 3\sigma$ Windows (b) k Values vs. # Variables for Cumulative Probability Outside $\mu \pm k\sigma$ = 0.0027 (c) Possible Solutions for k1 and k2 for Constant Probability Outside $\mu n \pm k_n \sigma_n = 0.0027$

The more variables there are in a given process, the less likely that the uncorrelated components within them will all be worst case at the same time. Ideally, a worst case corner would place those parameters that have greatest impact on circuit performance at more extreme values, while letting other less important parameters remain at more nominal levels. In the context of semiconductor device and circuit performance, the relative importance of a given process parameter often depends on the device architecture and operating conditions. Figure 4 depicts the sensitivities of several simulated MOS I_{DS} conditions to SPICE model parameters lint (channel length offset fitting parameter), wint (channel width offset fitting parameter), vth0 (threshold voltage @ Vbs=0), tox (gate oxide thickness) and rdsw (parasitic resistance per unit width). The underlying independent process variables that would contribute to that variation include poly gate lithography, gate oxide deposition and source drain implant and anneal (Mutlu & Rahman, 2005). Being independent, the probability of all of them being worst case at the same time is quite low. Figure 5 further demonstrates this effect, showing the results of a 10000 trial Monte Carlo simulation of the propagation delay of a simple inverter cell. Although the Monte Carlo completely covers the range of values defined by the worst case corner models for the individual model parameters, the resulting propagation delay distribution falls well inside the values predicted by the corners, simply because the occurrence of those simultaneous worst case conditions is so improbable:

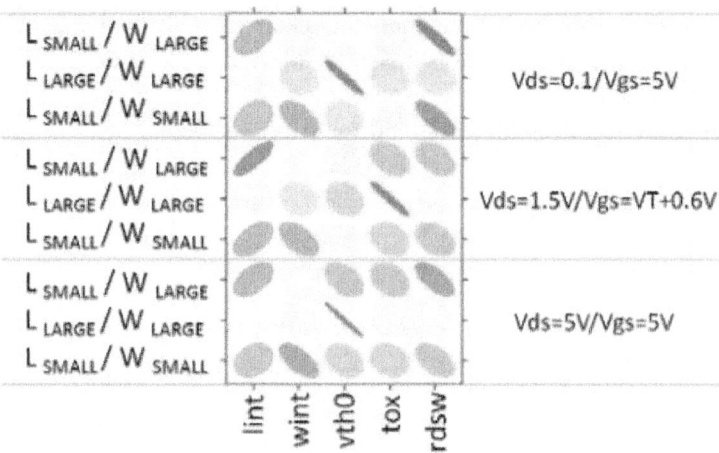

Figure. 4: Some Underlying MOS I_{DS} Sensitivities vs. Device Size and Bias Conditions

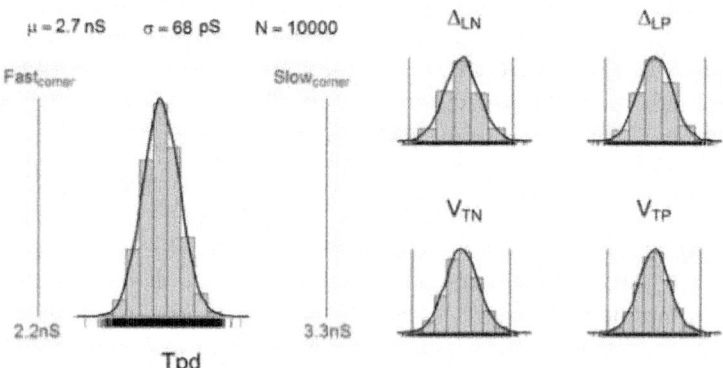

Figure. 5: All Parameters Simultaneously at Worst Case Yields Unrealistic Corners

Complicating the issue of corner selection is the fact that the worst case conditions may be completely different for circuit performance criteria that are sensitive to different process perturbations, such as the propagation delay of a CMOS digital logic circuit versus the gain of an operational amplifier. Even between related circuit performance parameters within the same circuit cell there can be notable differences. Consider the enable and disable propagation delays of a sample CMOS digital logic circuit as present in Table 1. When set to the worst case corners for disable (HZ/LZ) delay, TpZH encompasses less than 25% of the delay window obtained when using worst case enable corners (0.4nS vs. 1.8nS). The difference between the two corners is the placement of Tox. Ordinarily, Tox would be reduced for a Fast corner as it provides higher drive. However, thinner Tox also means higher oxide capacitance. The benefit of higher drive more than compensates for the penalty of higher capacitance in active delays, but the impact of the higher capacitance dominates for disable delays. Statistical models are not tied to a particular fixed choice of conditions as corner models are. They are generally formulated to reflect underlying process interactions by re-expressing the correlated model parameters as functions of an appropriate set of uncorrelated

Table 1: Different Circuit Parameters may have Opposing Corner Conditions

Worst Case Corner Setting	TpHZ (nS)	TpLZ (nS)	TpZH (nS)	TpZL (nS)
Corner 1: Worst Case Disable Times	3.2/4.6	3.3/4.1	2.3/2.7	2.0/2.5
Corner 1: Δ Slow - Fast	1.6	0.8	0.4	0.5
Corner 2: Worst Case Enable Times	3.6/4.2	3.5/3.9	1.6/3.4	1.5/2.9
Corner 2: Δ Slow - Fast	0.6	0.4	1.8	1.4

parameters. When exercising a statistical model, the uncorrelated parameters are perturbed, rather than the model parameters directly. These changes are then propagated through to the model parameters to generate properly correlated model decks. While statistical models do not inherently resolve the issues of circuit dependencies in and of themselves, they do enable the use of exploratory statistical simulation strategies including design of experiments and response surface model (DOE/RSM) techniques that can efficiently evaluate the response of a given circuit over the entire process/design space to determine the particular worst case conditions for a given circuit (Rappitsch et al., 2004; Sengupta et al., 2004; Zhang et al., 2009).

The Issue of Localized Matching Variation

It is imperative for analog/mixed-signal designs, and is becoming increasingly important for digital designs as well, that today's simulation methodologies have the means to evaluate the effects of localized device mismatch on circuit performance. Fixed corner models applied uniformly across all device instances in a circuit do not provide any allowance for mismatch. As seen in Figure 6, the impact of mismatch on analog circuit blocks can easily exceed the variation that would otherwise be expected due to global variation over the entire process range. Simulating under the effects of global process variation only, the current mirror output current, IO, exhibited a standard deviation of ~50nA, traced predominantly to V_T, with some residual sensitivity to L_{EFF}/W_{EFF} and

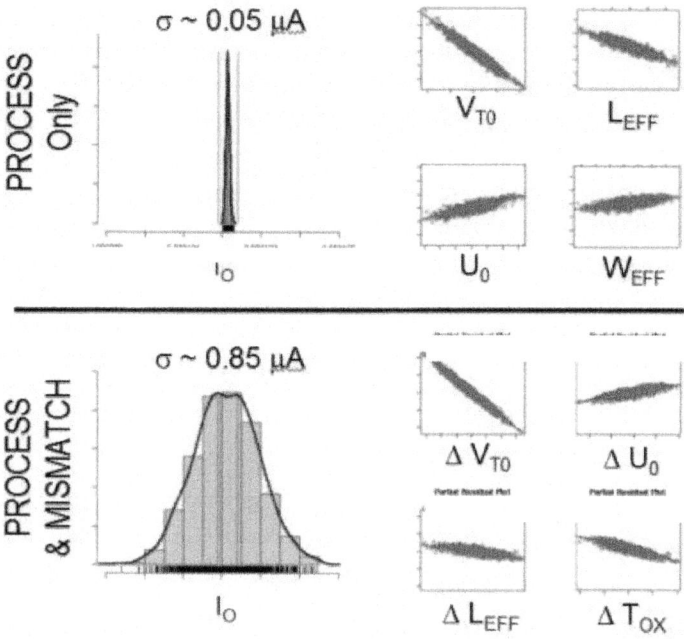

Figure. 6: Statistical Simulation of Basic Current Mirror

mobility. Adding in additional slight perturbations to the values of these parameters as applied each individual device in the circuit, the standard deviation of I_O increased about 17x to 0.85uA, almost entirely attributed to the slight difference in V_T applied between the critically matched MOS devices: Local mismatch variation is observed by comparing two or more identical devices on a die. In the absence of systematic variation, a normally distributed random mismatch variation would induce a normal distribution upon a given parameter, P, such that P would be expected to have a mean of μ_P, the average value of P across that die, and a standard deviation of σ_P:

$$P \sim \eta(\mu_P, \sigma_P^2)$$ (1)

The observed difference in P between any two identical devices would be expected to be distributed with a mean of 0 and standard deviation of $\sqrt{2}\sigma_p$ (variance of $2\sigma_p 2$):

$$\Delta_P \sim \eta(0, 2\sigma_P{}^2) \tag{2}$$

(Lakshmikumar et al., 1986) derived a $1/\sqrt{(LW)}$ scaling dependence for threshold voltage and conductance mismatch. Using Fourier techniques, (Pelgrom et al., 1989) postulated a generalized expression for the variance of Δ_p between two rectangular devices as:

$$\sigma^2(\Delta_P) = \frac{A_P^2}{WL} + S_P^2 D_x^2 \tag{3}$$

where: W and L are the width and length of each rectangle

D_x is the separation distance between the rectangles
A_P, also known as *A factor*, is the area coefficient and
S_P is the spacing coefficient

As indicated that model, the variance of Δ_p would be expected to increase as the device sizes decrease and as the devices are spaced farther apart from one another. The magnitude of the A factor is typically a reflection of the process design itself as opposed to specifically controllable manufacturing components (Tuinhout, 2002). For MOS devices, V_T, gm and ID matching is affected by multiple process architectural components, including S/D and channel doping (Tuinhout et al., 2000 & Dubois et al., 2002) and gate poly/oxide definition (Difrenza et al. 2003; Brown et al., 2007; Cathignol et al., 2008). For analog designs in MOS technologies, threshold voltage mismatch is of particular concern. (Pelgrom et al., 1998) presents a physical representation of AVT, the A factor for MOS threshold voltage mismatch, as:

$$A_{VT} = \frac{q \cdot t_{ox}\sqrt{2Nt_{depl}}}{\varepsilon_0 \varepsilon_{ox}} \tag{4}$$

where: N represents the total number of doping in the depletion region ($N_a + N_d$)

t_{depl} represents the width of the depletion region
t_{ox} represents the gate oxide thickness

A direct relationship between t_{ox} and A_{VT} is clearly evident. A former rule of thumb for technology nodes over 0.1µm gate length suggested A_{VT}, in saturation regions, would run at about 1 mV$_{µm}$ per nm of gate oxide thickness (Pineda de Gyvez & Rodríguez-Montañés, 2003). Within equation (4), the reduction of t_{ox} is somewhat offset by the required increases in doping levels at reduced geometries. Deep sub-100nm processes bring increasing effects from lithography and other gate region uniformity challenges (Brown et al., 2007; Cathignol et al., 2008 & Lewyn et al., 2009). Layout effects and neighbouring topology can all induce additional mismatch deviations beyond those accounted for in AVT (Drennan et al., 2006 & Wils et al., 2010). From a design perspective, it is important to take in account the relationship of circuit bias selections on resulting mismatch performance (Kinget, 2004). For instance, as VGS approaches V_T, the relative mismatch variation in I_D increases, peaking in subthreshold region as shown in Figure 7:

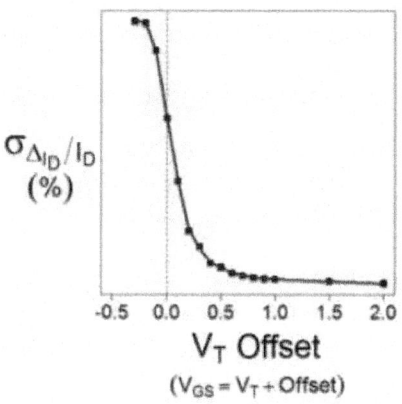

Figure. 7: MOS I_D Relative Mismatch Variation Increases in Subthreshold Region

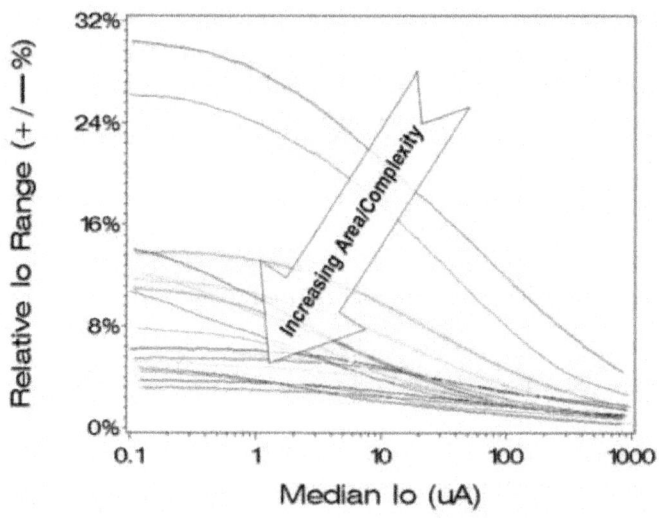

Figure. 8: Comparison of Current Mirror Data

The influence of biasing impacts can be seen in sample current mirror data. Figure 8 shows results, measured over multiple mirror configurations and sizes, for the total observed range of Io (expressed as +/- %) relative to the median operating Io value under various test conditions. Mirrors intended to run at very low currents will be exhibit proportionately greater mismatch sensitivities. Reducing this variation requires larger devices and/ or more complex mirror configurations, either of which can adversely impact manufacturing costs due to a larger die area. Statistical models can offer the designer the opportunity to evaluate and compare the effects of mismatch on circuit performance under different design scenarios. Relative to corner models, statistical models offer improved accuracy, by properly retaining key parameter correlations, improved coverage, by not being tied to some arbitrary set of corners, and improved capability, by incorporating localized mismatch as well as global process variation effects.

IMPLEMENTING STATISTICAL DESIGN

Implementing statistical design requires the development or procurement and integration of 3 key components: a simulation tool capable of exercising statistical models, the statistical models themselves and finally the appropriate methodologies to use them efficiently and cost effectively to validate and improve a circuit's design (Duvall 2000). The goal of statistical circuit modeling is to be able to replicate the observed pattern of global and local variances such that their effects on a particular circuit design can be simulated and, if necessary, design enhancements introduced prior to committing the design to silicon.

Extracting Statistical Models

Statistical models are formulated to retain correlation by re-expressing the correlated model parameters as functions of an appropriate set of uncorrelated parameters. When exercising a statistical model, the uncorrelated parameters are perturbed, rather than the model parameters directly. These changes are then propagated through to the model parameters to generate properly correlated model decks. In its most generic representation, a statistical model would define the value of some parameter P within the j th device on the i th die as:

$$P_{ij} = \mu_{PROCESS} + G_{OFFi} + L_{OFFij} \tag{5}$$

where: $\mu_{PROCESS}$ = overall process mean for that parameter.

G_{OFFi} = global offset associated with the ith die:

$(\mu=0, \sigma^2=\sigma^2_{GLOBAL})$

L_{OFFij} = local offset for the jth device on ith die:

$(\mu=0, \sigma^2=\sigma^2_{LOCAL})$

As indicated in Figure 9, variations in the independent fabrication process variables (eg: implant dose and energy, furnace temperature, ramp time, flow rate, etc.), interact to create statistical distributions of the process characteristics (eg: junction depths, doping profiles, etc.). Different characteristics may exhibit some degree of correlation to one another due to common influences. For example, the annealing temperature/time of a poly implant will have some effect on the ultimate doping profiles of earlier source/drain and well implants/diffusions. The process architecture design and implementation will influence the nature and strength of these correlations. The statistical variations and inter-correlations of process characteristics will drive the statistical variations and inter-correlations of the device characteristics, as influenced by the device architecture design, and so on.

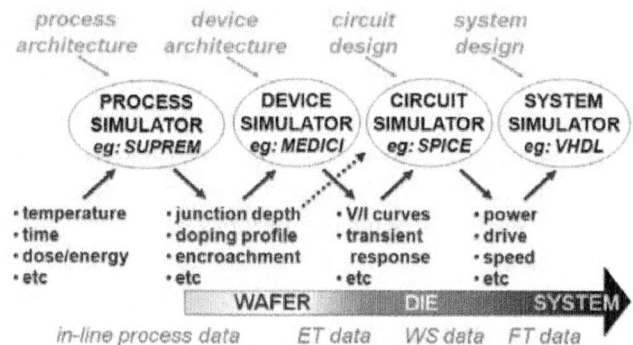

Figure. 9: Progression of Increasingly Complex Parameter Interactions

It is effectively impossible to precisely track the propagation of the variation and their impacts throughout the levels. We can get a general assessment of process variation from inline process data, device variation from wafer electrical test (ET) and circuit performance variation via wafer sort (WS) and final test (FT) data, but we have no way of knowing what specific process conditions any particular die experienced. TCAD simulators can be coupled together to cover the entire process (Hanson et al., 1996), but that requires very well calibrated models as the effects of any errors/omissions would be compounded throughout the system. The inputs to the circuit simulator (referred to hereafter as the model parameters) are a mixture of inter-correlated pseudo-physical as well as non-physical (fitting) parameters. Since they are inter-correlated, it is not statistically (or physically) appropriate to perturb their values independently of each other. Proper correlation between the model parameters can be maintained by expressing the model parameters as functions of other independent parameters which are more suitable for applying direct statistical perturbations. These parameterized model expressions can be thought of as behavioral models, developed to provide suitable proxy for device characteristic/model parameter distributions as inputs to the circuit simulator such that reasonably realistic circuit performance projections can be expected. Establishing appropriate distributions and intercorrelations of the model parameters can be a significant challenge. Wafer electrical test (ET) data is used to characterize a process and extract the circuit model parameters. The cumulative effect of the underlying variations in the process is manifest in the observed distribution of ET parametric data. That is:

$$E = f(\mathbf{p}) \tag{6}$$

where: E = an ET parameter
         $\mathbf{p}$ = a vector of process parameters

ET data is used to extract the circuit model parameters. This is generally done by creating a large database of ET results obtained over a wide array of device geometries, architecture and operating conditions and using a specialized extraction tool, such as ICCAP, to optimize the model parameters via curve fitting. Hence, we have:

$$M = f(E) \tag{7}$$

where: M = a device model parameter
         E = a vector of ET parameters

For statistical modeling, the challenge is to define how to alter the model parameters in a statistically realistic manner. As stated earlier, it is not appropriate to vary the model parameters directly since they are correlated with one another. It is also not feasible to estimate the correlation between the model parameters from the model files themselves as they are usually only directly extracted for a very limited number of ET sites (and even if a suitably large set of model files were generated, there would be concerns over whether the model extraction methodology itself might have influenced the results). TCAD simulation can be used to develop models tied back to independent physical components, but this introduces additional, compounding sensitivities to the inherent accuracy of each modeled stage. Circuit designers and modelers often have less access to and familiarity with those TCAD tools. They are generally quite familiar with ET data, however, and large samples are often readily available from which the necessary statistical information can be determined and utilized for statistical modeling (Chen et al., 1996; Potts & Luk, 1998; Singhal &

Visvanathan, 1999). The variation of several model parameters can be directly mapped to the variation in measured or extracted ET characteristics, including vth0 (to measured threshold voltage), xl/lint and xw/wint (to extracted L_{EFF} and W_{EFF} calculations, respectively), tox (to inverse of gate oxide capacitance) and the sheet resistances of various layers. Others can be proportionally mapped to functions of measured data, including mobility (u0 ~ Gm/Cox*[L/W]) and saturation current (is ~ ln(vbe)). The first step of the extraction process is to validate the ET data, removing any invalid outliers, and transforming each parameter to a standardized normal distribution (keeping track of the transformations so that we know how to reverse transform it back later). Next, we perform principal component analysis (PCA) on the transformed data. PCA is a technique that can be used to re-express a correlated set of variables in terms of uncorrelated components [16]. An orthogonal transformation matrix, B, is found such that:

$$Y=B(E-\bar{E}) \tag{8}$$

$$Z=\Lambda^{-\frac{1}{2}} B(E-\bar{E})=AX \tag{9}$$

$$S=B'\Lambda B \tag{10}$$

where: E = matrix of correlated ET data, with means $\bar{E}$
Y = matrix of principal components
Z = standardized PCA components
S = covariance matrix of $X_1, X_2, ..., X_n$
$B'\Lambda B$ = spectral decomposition of S
Λ = diagonal matrix, diag($\lambda_1, ..., \lambda_n$), with $\lambda_1 > \lambda_2 > ... > \lambda_n$ the eigenvalues of S
$A = \Lambda^{-\frac{1}{2}} B$ and $X = (E-\bar{E})$

Each of the principal components in Y and Z has a mean of 0 and is uncorrelated with all other principal components (that is, each Yi is uncorrelated with all other Yi and each Z_i is uncorrelated with all other Z_j). The variance of each Yi is the value of the corresponding ith eigenvalue, while the standardized PCA components, Z_i, each have a variance of 1. If all Xi are normal, then each of the Z_i is standard normal, which is convenient for formulating the statistical models. For example, to run a monte carlo, the statistical simulation tool would generate vectors of Z, with each Z_i being a random normal value. These random vectors of Z would then be reverse transformed back into corresponding vectors of E, from which we can map random, but properly correlated, perturbations of M! Figure 10 demonstrates this technique. The black data points represent an actual sample of data collected over a 4 month period. The original 6 correlated E_T parameters are decomposed into 4 uncorrelated PCA components. The matrix between them on the lower right graphically depicts that transformation relation. L_{EFFN} and L_{EFFP} are strongly related to PCA parameter A, TOXN and TOXP are strongly related to B, VTP is strongly related to D and VTN is related to C with dependance on A and D as well. While the PCA solution is entirely a mathematical construct, it may offer insights into the underlying physical relationships. Physically, L_{EFFN} and L_{EFFP} would be highly dependent on the gate poly CD, TOXN and TOXP on the gate oxide thickness, V_{TN} would be dependent on multiple parameters, including NA, T_{OX}, xj and, for short/narrow devices, L/W, while V_{TP} would have a strong dependence on V_T adjust implant. A PCA solution that does not appear to bear any resemblance to a logical underlying physical relationship should merit greater scrutiny of the data for a possible invalid readings or a need for normality transformation.

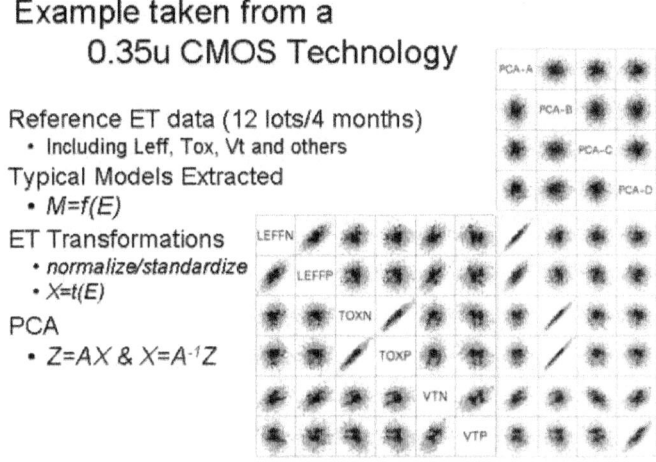

Figure. 10: Example of PCA Transformations: ET > PCA & PCA > ET

For parameters that cannot be directly mapped to physical data, it will be necessary to indirectly estimate appropriate values that will yield appropriate results when used in simulation. This includes all mismatch parameters. The backward propagation of variance (BPV) technique is quite helpful in this process (McAndrew et al., 1997; Telang & Higman, 2001, Drennan & McAndrew 2003; McAndrew et al., 2010). Measured ET data is collected over a wide spectrum of device geometries and bias conditions. Simulations are then set up covering the same set of parameters. For the first pass of simulations, a small arbitrary value of variation is assigned to each of the independent mismatch model parameter (such as 1% of its corresponding global variance). These initial simulations are used to determine the covariance matrix (or squared correlations) between the mismatch models parameters and the resulting simulated mismatch variance. Regression analysis is then performed to fit an appropriate vector of mismatch model parameter variance such that the simulated ET mismatch variances would approximate the actual measured ET mismatch variances:

$$\sigma^2_{ET} = S * \sigma^2_{Model}$$ (11)

where: σ^2_{ET} = vector of observed ET variance in measured data
           S = covariance matrix of simulated ET results vs model parameters
           σ^2_{Model} = vector of (fitted) variance to assign to model parameters

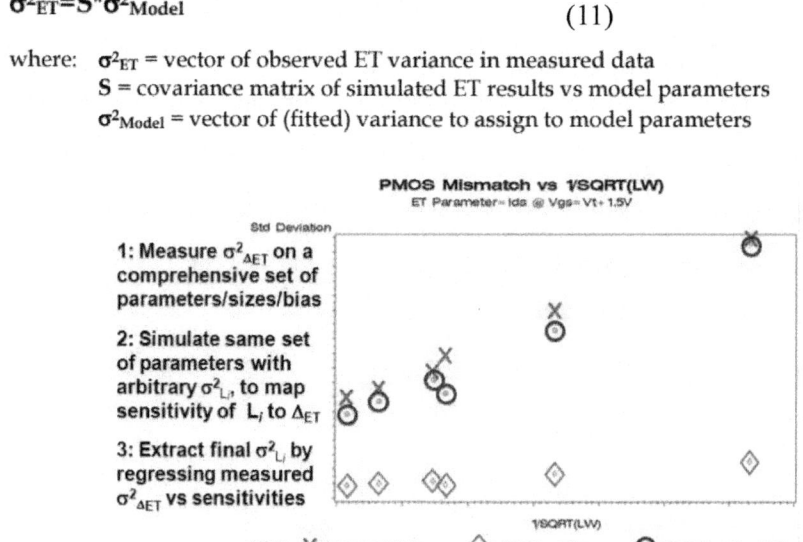

Figure. 11: Example of BPV to Fit Observed ET Data

Implementing Statistical Models

Over the past decade or so, Monte Carlo and other statistical simulation capabilities have been added to commercial SPICE simulators. They enable the use of specially parameterized and formulated expressions to implement the desired statistical model behavior (Lu et al., 2009). Recent compact models are also incorporating new parameters that, when combined with extracted layout information, can better predict important mismatch sensitivities, such as stress and well proximity effects (Watts et al., 2006; Yang et al., 2008). We have implemented our parameterized statistical models within the Cadence Analog Design Environment, utilizing the monte carlo features available within their Statistical Analysis Tool (Potts & Luk, 2005). This tool offers the ability to designate random variables into two groups, process and mismatch, as declared within a statistics block within the model library, prior to the models themselves:

```
statistics {
    process {
            vary G1    dist=gauss   std=1
            ....
            vary Gn    dist=gauss   std=1
    }
    mismatch {
            vary L1    dist=gauss   std=1
            ....
            vary Lm    dist=gauss   std=1
    }
}
```

Within the models, we then encode the ith model parameter, P_i, as a functions of these independent variables by applying the statistical models we have derived for global variation , e.g.: $f_{Gi}(G_1,...,G_n)$, and mismatch, e.g.: $f_{Li}(L_1,...,L_m)$, such that:

$$P_i = P_{TYPICAL} + f_{Gi}(G_1,...,G_n) + f_{Li}(L_1,...,L_m) \tag{5}$$

Since we have formulated our statistical models as functions of independent normal variables, each of our global variables $(G_1 - G_n)$ has been declared as Gaussian distributions with a mean of 0 and a standard deviation of 1. The local variables $(L1 - Ln)$ are declared as Gaussian distributions with a mean of 0 and a standard deviation of δi $(0< \delta_i$ where $<1)\delta_i$ are fitted through a backwards propagation of variance technique. With statistical SPICE models in hand, the simplest and most generic analysis methodology, equally applicable to dc, transient or any other simulation set-up, utilizes Monte Carlo simulations to detect and isolate potential trouble spots in the circuit. With the Cadence 6.x ADE-XL/GXL platform, traceability can be enabled to monitor Monte Carlo values applied to each instance during each trail, providing a means to quickly locate any design weaknesses. The major drawback to Monte Carlo analysis is simulation time. A large number of trails are needed, especially if one needs to accurately evaluate the tails of the distribution. This is less of an issue for small circuits or individual circuit blocks which can be simulated on the order of seconds or less per trial. As such, one strategy for larger circuits would be to break it down into blocks, and fitting behavioural macromodels to express the variation of the output of one block, which could then be applied as the input to the next block. Ignoring correlation, this could simply be done by redefining a fixed voltage or current as a design variable, say V1, set by an additional random variable of desired location and spread, e.g.:

```
parameters V1 = {desired mean value}
statistics {
      process {   ....
                  vary V1    dist=gauss    std={desired standard deviation}
      }
```

A more proper solution, however, would retain correlation by expressing the V1 voltage as a function of the same Monte Carlo variables used in defining the SPICE statistical models themselves. This would be done by running Monte Carlo simulations on the circuit block that generates the V1 signal, applying regression techniques to fit the resulting V1 over the values for the Monte Carlo parameters from each trial, and then using that regression equation to define the V1 input to apply to the next block, e.g.:

parameters $V1 = f_{Gi}(G_1,...,G_n) + f_{Li}(L_1,...,L_m)$

There are alternative methods that do not require Monte Carlo, including sensitivity analysis, design of experiments (DOE) and response surface modelling (RSM) techniques. Typically, a sensitivity analysis is performed to isolate the critical model inputs and then a DOE is run over those variables (which generally requires far fewer trials than a Monte Carlo), and then RSM is employed to analyze/optimize the results. These methodologies are not as readily implemented within standard commercial SPICE simulators, requiring significant additional pre-/post-processors for set-up and analysis. Commercial solutions are available from 3rd party vendors, however, including Circuit Surfer® (PDF Solutions), Variation Designer (Solido Design Automation) and WiCkeD™ (MunEDA GmbH).

DEMONSTRATIONAL ANALYSIS OF A BAND GAP CIRCUIT

In this section, we will demonstrate the use of our statistical CAD tools and methodologies to characterize and optimize a Bi-CMOS band gap circuit consisting of a MOS bias generator, PNP band gap reference and MOS op amp, as shown in Figure 12. The circuit was initially designed and simulated to produce a stable reference voltage, V_{BGOUT}, of about 1.18 +/- 20 mV over corner models.

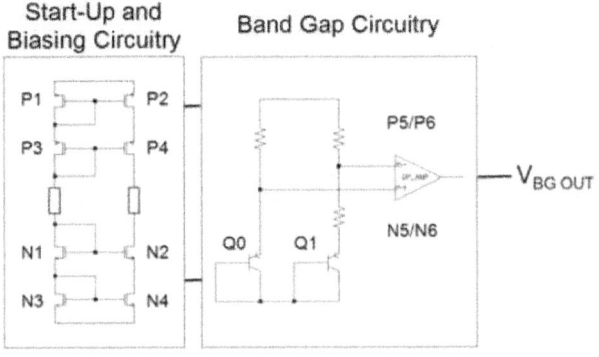

Figure. 12: Band Gap Circuit used in this Example

The baseline process Monte Carlo projected a VBGOUT σ of 9.5mV – virtually all traced to PNP Is variation.

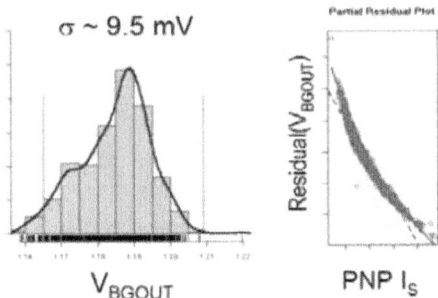

Figure. 13: Process-Only Monte Carlo Results

The combined process and mismatch Monte Carlo generated a much larger variation along with a prominent asymmetric low tail:

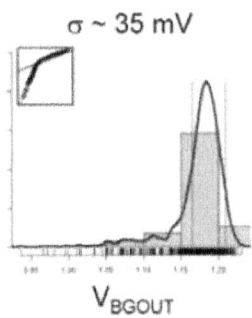

Figure. 14: Combined Process & Mismatch Monte Carlo

Partitioned mismatch Monte Carlos quickly pinpointed the source of the tail to MOS mismatch sensitivities within the start-up & biasing block:

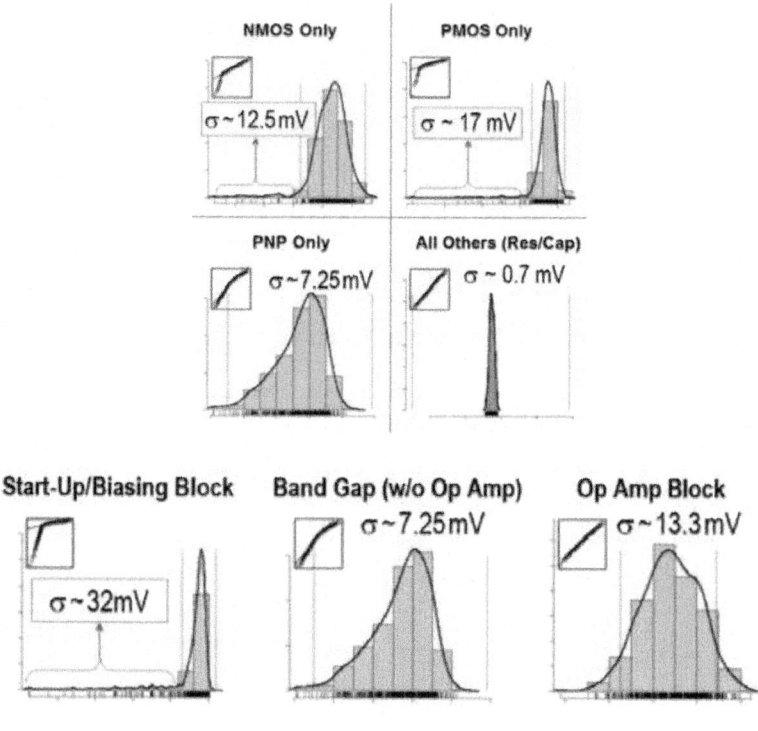

Figure. 15: Partitioned Mismatch Monte Carlo Results

Probing in the biasing block revealed "lurking cliff" ΔVt sensitivities between devices P1 & P2 and N3 & N4 (where P1,P2,... refer to devices as labelled in Figure 12):

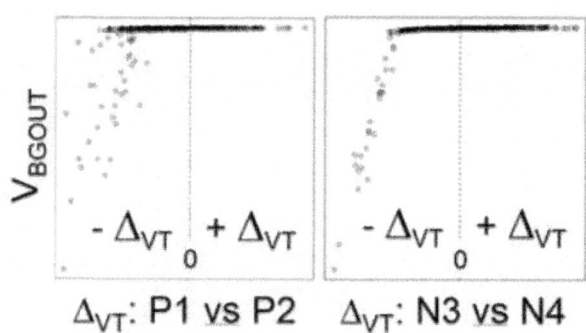

Figure. 16: Tail Traced to Δ_{VT} in Bias Circuit

After removing the outlying values in the tail, the remaining mismatch sensitivities are traced to the differential pair (P5/P6) and mirror (N5/N6) in the op amp and the PNP pair (Q0/Q1) in the band gap:

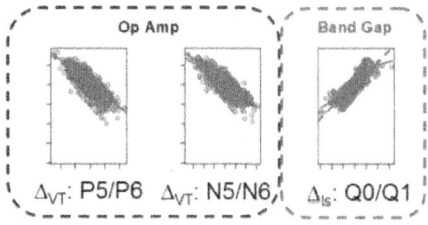

Figure. 17: Non-Tail Sensitivities: Op Amp & Band Gap

Increasing the sizes of these identified critical devices by about 2x to 3x from their original values reduces the Vbgout standard deviation under combined Process & Mismatch Monte Carlo from ~ 35mV to ~ 10mV. At that point, the PNP Is process sensitivity becomes the dominant factor in overall V_{BGOUT} variability and any additional mismatch reduction yields minimal benefit.

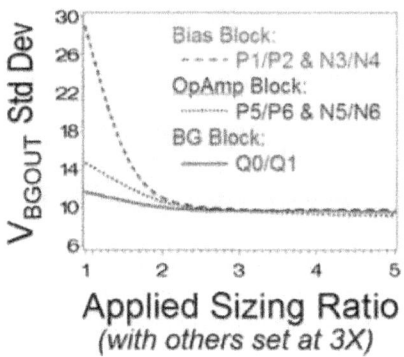

Figure. 18: Overall Variation Optimized @ 2x-3x

CONCLUSION

Statistical design offers considerable improvements over traditional worst case design methodologies. New tools and methodologies are being developed and offered in the EDA market that will enable the designer to use statistical models efficiently. A statistical design simulation framework enables the opportunity to make more intelligent design choices up front that will result in a more robust and manufacturable circuit design.

REFERENCES

1. Brown, A.; Roy, G. & Asenov, A. (2007). Poly-Si-Gate-Related Variability in Decananometer MOSFETs with Conventional Architecture. IEEE Trans. on Electron Devices, Vol. 54, No. 11, (Nov 2007) pp. 3056-3063, ISSN 0018-9383.

2. Cathignol, A.; Cheng, B.; Chanemougame, D; Brown, A; Rochereau, K; Ghibaudo, G. & Asenov, A. (2008). Quantitative Evaluation of Staistical Variability Sources in a 45 nm Technological Node LP N-MOSFET. IEEE Electron Device Letters, Vol. 29, No. 6, (Jun 2008) pp. 609-611, ISSN 0741-3106.

3. Difrenza, R.; Vildeuil, J.; Llinares, P. & Ghibaudo, G. (2003). Impact of Grain Number Fluctuations in the MOS Transistor Gate on Matching Performance. International Conference on Microelectronic Test Structures (ICMTS'03), pp. 244-249, ISBN 0-7803-7653-6, Monterey, CA, USA, Mar 2003.

4. Drennan, P. & McAndrew, C. (2003). Understanding MOSFET Mismatch for Analog Design. IEEE J. of Solid-State Circuits, Vol. 38, No. 3, (Mar 2003) pp. 450-456, ISSN 0018-9200.

5. Drennan, P.; Kniffen, M. & Locascio, D. (2006). Implications of Proximity Effects for Analog Design. 2006 Custom Integrated Circuits Conference (CICC'03), pp. 169-176, ISBN 1-4244-0076-7, San Jose, CA, USA, Sep 2006.

6. Dubois, J.; Knol, J.; Bolt, M.; Tuinhout, H.; Schmidtz, J.; & Stolk, P. (2002). Impact of Source/Drain Implants in Threshold Voltage Matching in Deep Sub-micron CMOS Technologies. 32nd European Solid-State Device Research Conference (ESSDERC 2002),pp. 115-118, ISBN 88-900847-8-2, Bologna, Italy, Sep 2002.

7. Duvall, S. (2000). Statistical Circuit Modeling and Optimization. 5th International Workshop on Statistical Metrology, pp. 56-63, ISBN 0-7803-5896-1, Honolulu, HI, USA,Jun 2000.

8. Hanson, D.; Goosens, R.; Redford, M.; McGinty, J.; Kibarian, J. & Michaels, K. (1996). Analysis of Mixed-Signal Manufacturability with Statistical Technology CAD (TCAD). IEEE Trans. on Semiconductor Manufacturing, Vol. 9, No. 4, (Nov 1996) pp. 478-488, ISSN 0894-6507.

9. Lakshmikumar, K.; Hadaway, R. & Copeland, M. (1986). Characterization and Modeling of Mismatch in MOS Transistors for Precision Analog Design. IEEE Journal of SolidState Circuits, Vol. 21, No. 6, (Dec 1986) pp. 1057-1066, ISSN 0018-9200.

10. Lewyn, L.; Ytterdal, T.; Wulff, C. & Martin, K. (2010). Analog Circuit Design in Nanoscale CMOS Technologies. Proceedings of the IEEE, Vol. 97, No. 10, (Oct 2010) pp. 1687-1714, ISSN 0018-9219.

11. Lu, N, Watts, J. & Springer, S. (2009). Elements of Statistical SPICE Models. NSTI-Nanotech 2009, Vol 3, pp. 616-619, ISBN 978-1-4398-1784-1, Houston, TX, USA, May 2009.

12. Kinget, P. (2005). Device Mismatch and Tradeoffs in the Design of Analog Circuits. IEEE J. of Solid-State Circuits, Vol. 40, No. 6, (Jun 2005) pp. 1212-1224, ISSN 0018-9200.

13. McAndrew, C.; Bates, J.; Ida, R. & Drennan, P. (1997). Efficient Statistical Modeling, Why β is More Than Ic/Ib. Bipolar/BiCMOS Circuits and Technology Meeting (BCTM'97), pp.28-31, ISBN 0-7803-3916-9, Minneapolis, MN, USA, Sep 1997.

14. McAndrew, C.; Stevanović, I.; Li, X.; & Gildenblat, G. (2010). Extensions to Backward Propagation of Variance for Statistical Modeling. IEEE Design & Test of Computers, Vol. 27, No. 2, (Mar/Apr 2010) pp. 36-43, ISSN 0740-7475.

15. Mutlu, A. & Rahman, M. (2005). Statistical Methods for the Estimation of Process Variation Effects on Circuit Operation. IEEE Trans. on Electronics Packaging Manufacturing, Vol. 28, No. 4, (Oct 2005) pp. 364-375, ISSN 1521-334X.

16. Nardi, A.; Neviani, A.; Zanoni, E.; Quarantelli, M. & Guardiani, C. (1999). Impact of Unrealistic Worst Case Modeling on the Performance of VLSI Circuits in Deep Submicron CMOS Technologies. IEEE Transactions on Semiconductor Manufacturing, Vol. 12, No. 4, (Nov 1999) pp. 396-402, ISSN 0894-6507.

17. Pelgrom, M.; Duimaijer, A. & Welbers, A. (1989). Matching Properties of MOS Transistors. IEEE Journal of Solid-State Circuits, Vol. 24, No. 5, (Oct 1989) pp. 1433-1439, ISSN 0018-9200.

18. Pelgrom, M.; Tuinhout, H. & Vertregt, M. (1998). Transistor Matching in Analog CMOS Applications. International Electron Devices Meeting (IEDM'98), pp. 915-918, ISBN 0- 7803-4774-9, San Francisco, CA, USA, Dec 1998.

19. Pineda de Gyvez, J. & Rodríguez-Montañés, R. (2003). Threshold Voltage Mismatch (ΔVT) Fault Modeling. Proceedings of the 21st VLSI Test Symposium (VTS'03), pp. 145-150, ISBN 0-7695-1924-5, Napa Valley, CA, USA, May 2003.

20. Potts, D. & Luk, T. (2005). Extraction and Implementation of Effective Mismatch Models. DesignCon East 2005, Worcester, MA, USA, Sep

2005.

21. Rappitsch, G.; Seebacher, E.; Kocher, M. & Stadlober, E. (2004). SPICE Modeling of Process Variation Using Location Depth Corner Models. IEEE Transactions on Semiconductor Manufacturing, Vol. 17, No. 2, (May 2004) pp. 201-213, ISSN 0894-6507.

22. Sengupta, M.; Saxena, S.; Daldoss, L.; Kramer, G.; Minehane, S. & Cheng, J. (2005). Application Specific Worst Case Corners using Response Surfaces and Statistical Models. IEEE Transactions on Computer-Aided Design of Integrated Circuits and Systems, Vol. 24, No. 9, (Sep 2005) pp. 1373-1380, ISSN 0278-0070.

23. Telang, N & Higman, J. (2001). Statistical Modeling Techniques: FPV vs. BPV. Proceedings of the 2001 International Conference on Microelectronic Test Structures (ICMTS 2001), pp.71-75, ISBN 0-7803-6511-9, Kobe, Japan, Mar 2001.

24. Tuinhout, H. (2002). Impact of Parametric Mismatch and Fluctuations on Performance and Yield of Deep-submicron CMOS Technologies. European Solid-State Device Research Conference (ESSDERC'02), pp. 95-102, ISBN 88-900847-8-2, Bologna, Italy,Sep 2002.

25. Tuinhout, H.; Wils, N. & Andricciola, P. (2010). Parametric Mismatch Characterization for Mixed-Signal Technologies. IEEE Journal of Solid-State Circuits, Vol. 45, No. 9, (Sep 2010) pp. 1687-1696, ISSN 0018-9200.

26. Watts, J.; Su, K. & Basel, M. (2006). Netlisting and Modeling Well-Proximity Effects. IEEE Trans. on Electron Devices, Vol. 53, No. 9, (Sep 2006) pp. 2179-2186, ISSN 0018-9383.

27. Wils, N. ; Tuinhout, H. & Meijer, M. (2010). Influence of Metal Coverage on Transistor Mismatch and Variability in Copper Damascene Based CMOS Technologies, 2010 IEEE International Conference on Microelectronic Test Structures (ICMTS'10), pp. 182-187, ISBN 978-1-4244-6915-4, Hiroshima, Japan, Mar 2010.

28. Yang, L. ; Cui, M. ; Ma, J. ; He, J. ; Wang, W. & Wong, W. (2008). Advanced Spice Modeling for 65nm CMOS Technology, 9th International Conference on Solid-State and IntegratedCircuitTechnology(ICSICT'08), pp. 436-439, ISBN 978-1-4244-2185-5, Beijing, China,Oct 2008.

29. Zhang, H. ; Chen, T. ; Ting, M. & Li, X. (2009). Efficient Design-Specific Worst-Case Corner Extraction for Integrated Circuits, 46th ACM/IEEE Design Automation Conference (DAC '09), pp. 386-389, ISBN 978-1-6055-8497-3, San Francisco, CA, USA, Jul 2009.

Chapter 3

A NEW APPROACH TO BIASING DESIGN OF ANALOG CIRCUITS

Reza Hashemian

Northern Illinois University United States

INTRODUCTION

A new approach for biasing analog circuits is introduced in this chapter. This approach is an attempt to address some of the biasing complexities that exist today in biasing large analog circuits. There are three steps involved in this methodology. First, in circuit analysis, the methodology separates nonlinear components (transistors), particularly drivers, from the rest of the circuit. Second, it uses local biasing introduced in the previous chapter to bias the transistors individually and to the specs provided for the design. Finally, the method presents a new way to change the local biasing into normal (global) circuit biasing with choices of DC supplies at right locations in the circuit. It is the last step that will be our main topic of discussion in this chapter. Here we see how we can remove all sources related to the local biasing and replace them with normal circuit supplies without altering the design specifications. These circuit supplies can be voltage sources, current sources or mirrors. In case the supplies are already specified and in place, this method can still maintain the design specs by re-evaluating some of the power-conducting components in the circuit. Power-conducting components are those circuit components, such as resistors, that conduct DC power (current) from the power supplies to the circuit drivers (transistors), for biasing purposes. Limitations in local Biasing - We fully discussed local biasing, its properties and applications in the previous chapter. Despite all the advantages that local biasing offers one problem still remains unresolved and that is: how to deal with so many DC sources generated due to local biasing, known as distributed supplies? To see the problem, just take a single bipolar transistor: it normally needs four (voltage and current) sources to get locally biased; however, with coupling capacitors used this number reduces to two current sources and two capacitors (taking care of the voltage drops). Similarly, we may need to use four sources to locally bias a MOS transistor. Again, with coupling capacitors this number can get as low as one source-drain current source. The problem, however, is that for the gate, and possibly the substrate, the coupling capacitors need to have charging paths (a resistive path to a DC supply). One way to handle the case and bring the number of DC supplies down to a minimum of one or two is to use source transformation and replacement techniques, such as voltage dividers, Δ-Y transformation, and current sources/mirrors. Nevertheless, the sheer number of such sources in a fairly complex circuit can get so high that unless we find a shortcut to the final solution the validity of local biasing as an effective methodology is undermined.

A new strategy - We are introducing a different strategy for biasing analog circuits in this chapter. The core of this strategy lies on the fact that in an analog circuit design environment we only need to anchor down certain critical biasing specs and not all. By critical biasing specs we mean those operating conditions that are essential in achieving the design criteria, such as gains, undistorted output signals or power consumption. Other design criteria usually adhere to these critical specs and adapt to the situation fairly well enforced by the critical specs. The fact that DC supplies are present in a circuit only to bias the nonlinear components reveals the fact that for each biasing (critical) spec we need to provide a path to a DC power supply, controlled by the spec. With this in mind, the proposed strategy makes a one-to-one correspondence between the circuit biasing requirements (specs) and those DC (voltage or current) supplies needed to support these requirements. Hence, we need at least as many path to DC power supplies as we have biasing specs in a circuit. Consequently, the first task in this strategy is to pair each biasing spec with a biasing supply (voltage, current or a power-conducting component). Second, the method must be capable of replacing "distributed supplies" -- if a local biasing strategy is already in use -- with normal circuit supplies, such as VCC and VDD. The idea here is to keep the main properties of local biasing – translated into the critical biasing specs -- while removing local biasing sources to be replaced with the normal biasing supplies. The main advantages in employing this strategy are: i) to pin down the operating conditions for the critical transistors while replacing the local biasing sources with a much fewer designated DC supplies, ii) to minimize design efforts to fulfill only critical specs, hence speeding up the process, and iii) the possibility to perform biasing entirely linearly. The last point is particularly important and makes biasing almost a one-step process. This chapter introduces two new circuit elements, fixator and norator, that are the center pieces in our biasing design strategy. Fixators and norators come in pairs as effective tools to perform a targeted biasing. It is shown that these pairs are very instrumental in matching biasing critical specs with DC power resources. The method simply associates a designated supply source (or a power-conducting component) with an arbitrary biasing spec. Fixatornorator pairs cause local biasing sources (distributed supplies) to be entirely replaced with normal circuit supplies designated by the designer. It is shown that the pair, when used properly and in combination, will adhere to Kirchhoff laws as well. Important properties of fixator-norator pairs are introduced in this chapter, and the relationships between a fixator-norator pair and other circuit components (such as resistors, voltage sources, and current sources) are discussed. Rules and regulations corresponding to the use of fixator-norator pairs in a circuit are investigated. Being special circuit components, fixators and norators must be used so that KVL and KCL are not violated in a circuit. However, it is important to note that the use of fixator-norator pairs is only temporary in this methodology; i.e., the pairs are removed as soon as the final circuit biasing is established and the DC power is provided for the circuit. This is important in a sense that ideal controlled sources, with very high gains, can be used to mimic fixator-norator pairs without any restrictions. Because fixators can model fixed-biased ports, these devices can also model nonlinear components for specified biasing situations. These nonlinear components can be p-n junction diodes (as single port devices), bipolar transistors (as two port devices), and MOS transistors (as three port devices). An algorithm that explains the biasing design procedure of analog circuits is also introduced in this chapter. This algorithm classifies circuit design procedure into two areas: the performance (AC) design and the biasing design. The performance design (gain, bandwidth, SNR, power, distortion, and so on) is done first. Here is where the circuit topology and the major circuit components are determined to achieve the design goals. In the performance design the circuit is treated entirely linear, where the transistors are replaced with their linear models at specifies operating points. Upon finishing the performance design the circuit biasing design begins by providing a set of critical biasing specs. It is in this stage that the linear models of the transistors (used in the performance design) are replaced with the fixator models. Next, the designer needs to accommodate for the norators that must pair with the fixators. He/she has variety of choices to place the pairing norators in the circuit; having in mind that they are place holders for the power supplies, current sources/mirrors, or power-conducting components. When finished, the circuit is ready for simulation, while still linear. The results from the DC solution contain the voltage and current values for each norator; where, each in turn can be replaced with an appropriate component. This completes the DC design procedure.

Circuit Biasing

Biasing is a major step in designing analog circuits [1 – 3]. In large and complex circuits biasing has always been a great challenge for designers. The challenge is normally in two areas. First, to get the number of iterations minimized and make the convergence possible and fast; second, to move to the right regions of operations for active components (transistors) so that acceptable performance is attained and the output signals are far from being distorted or clipped during the AC operation. Both problems grow in complexity as the number of transistors increases, design requirements become tighter, or more efficient designs are in demand. One difficulty in the traditional approach appears to be the lack of separation between linear and nonlinear components, as well as between the nonlinear components themselves during the biasing process. Typical biasing techniques deal with the entire circuit as a whole with no classification or circuit partitioning; hence, the complexity quickly increases as the circuit grows. In case of analog ICs, where almost all circuit components are nonlinear, distinction between linear and nonlinear components becomes meaningless. Instead we can categorize components into two categories: i) drivers, and ii) supporting components. In conventional methods used for the analysis and simulation of analog circuits all nonlinear components, regardless of their categories and functionalities, are included in a global biasing (DC) analysis. Whereas in more advanced methods we can distinguish between the drivers and those supporting components, such as current sources/sinks, current mirrors, and active loads [4 - 6]. Drivers typically reside along the signal path directly shaping the output waveforms. They are strongly influencing the design specifications, and are more sensitive to signal conditionings. Consequently, drivers must be biased with more care and precision compared to the supporting components in a circuit.

NULLATORS, FIXATORS AND NORATORS

We first need to define terms that are used in this chapter. In addition, all our discussions here apply to DC power unless stated otherwise.

Port Parameter Fixing

The methodology introduced here is based on assigning specific operating points to nonlinear ports (in diodes and transistors) during the biasing design of an analog circuit.

These operating points are considered the critical specs for the design. Once the critical Qpoints are assigned to the ports of the transistors the methodology holds them fixed during the entire design period. Now, the question is how to keep a Q-point fixed while other variables (voltages and currents) in the circuit are changing? As we will see, the answer to this question lies in the use of fixator-norator pairs. A fixator is an expanded version of a nullator. A nullator is a two-terminal element with both its current and voltage equal to zero. A norator is a two-terminal element with unspecified current and voltage [7 – 12]. Consider two networks N_1 and N_2 connected through a port $j(V_j, I_j)$, as shown in Fig.1(a). Nullify port $j(V_j, I_j)$ from both sides by augmenting the port with voltage and current sources that have the same port values, V_j, I_j, as discussed in the previous chapter. As a result a new null port $k(V_k, I_k)$ is created in the process, as shown in Fig.1(b). Now, because port k is a null port ($V_k = 0$ and $I_k = 0$) we can split the two networks from port k and attach each with a nullator, as depicted in Fig.2. Apparently, the operation has not changed any current or voltage inside N_1 or N_2. In addition, it has fixed the port operating point (I_j and V_j) so that any internal changes inside N_2 (or N_1) do not change the port's Q-point. This simply means that we can replace port j by a fixator.

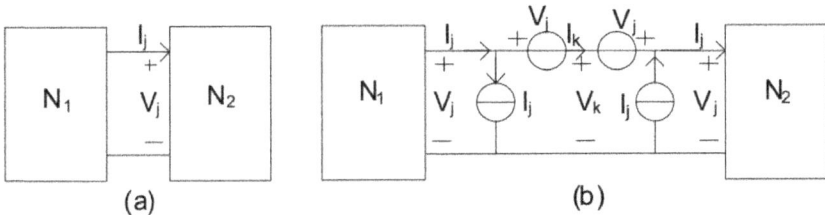

Figure. 1: Port nullification procedure

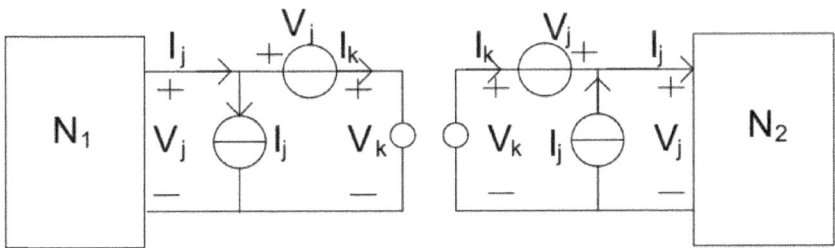

Figure. 2: Two networks N_1 and N_2 disjointed at port $k(v_k, i_k)$ and each terminated by a nullator.

Fixator: A two-terminal component[1] in a circuit is called a fixator if both the voltage across the component and the current through the component represent independent sources [4]. Figures 3(a) and 3(b) represent two types of fixators and Fig. 3(c) is a symbol representing a fixator. Note that a nullator is a special case of a fixator represented by Fx(0, 0), where both the device voltage and current are zero. Also, note the difference between the two fixators $F_x(V_j, I_j)$ and $F_x(I_j, V_j)$; in $F_x(V_j, I_j)$ the voltage source V_j provides (or consumes) power and the current source Ij is inactive2; whereas, in $F_x(I_j, V_j)$ the current source I_j provides (or consumes) power and the voltage source V_j is inactive. Note also the similarity between a fixator and an H-model, discussed in the previous chapter. Both fixator and H-model model a port, representing the existing situation of the port. The major difference, however, is that in a fixator the equivalent impedance Req in the H-model is replaced with a nullator, stamping on the port variables. This is because in an H-model the current going through the Req is also zero making the voltage zero, as well. However, the replacement of Req with a nullator removes the dynamics of the terminal and fixes the port values, I_j and V_j, for the entire operation of the circuit; whereas in the case of Req the H-model behaves normally as the Thevenin or Norton equivalent circuits behave. In fact, we can think of a fixator as a snapshot of a port's behavior, whereas an H-model represents the entire dynamics of the port during the circuit operation. For example, take the case of two networks N_1 and N_2 connected through a port j, as in Fig.1(a); we can replace N_1 by its H-model or alternatively we can replace it with a fixator Fx(Vj, Ij), as shown in Fig. 4. In the later case we are bounded with fixed values of Vj and Ij for the port; hence, the idea of fixing the design specs is born! To further expand the idea, we need to look for a different role for a fixator. Notice that in Fig. 4 we replaced the linear circuit N1 (or its H-model) with a fixator $F_x(V_j, -I_j)$. Now we can do the opposite; a fixator can replace a nonlinear component (or port) N2 in a circuit. This is stated in Property 1. Property 1: A two-terminal component, linear or nonlinear, in a circuit that is biased by a current I and exhibits a terminal voltage V can be replaced with a fixator $F_x(I, V)$ without causing any change in the currents and voltages within the rest of the circuit. One important conclusion from Property 1 is that, fixators are not only helping to fix the design specs for biasing purposes, they also linearize a circuit by replacing all the nonlinear components with fixators that are constructed from linear components. In addition, fixators

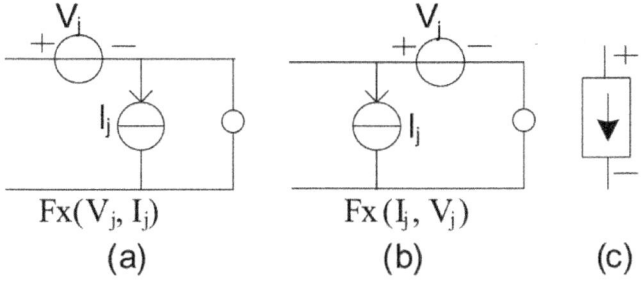

Figure. 3: (a) Voltage Fixator; (b) current Fixator; (c) Symbol representing a Fixator.

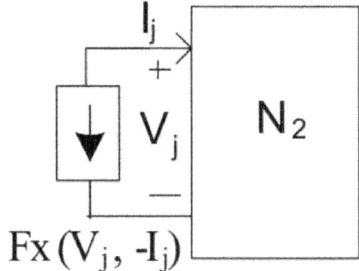

Figure. 4: A Fixator replaced for the biasing circuit N_1.

add to the stability of the design by performing a controlled approach to the design criteria. For example, if for a certain specified biasing situation the circuit behaves unstably, one can simply search for a more stable situation by slightly modifying the Q-points of certain transistors. This can be done by modifying their corresponding fixators without really touching any other parts in the circuit, or leaving the linearity conditions in the circuit. In using fixators for port specification and stability, we realize that for each fixator used we need to have one norator in the circuit to pair it with. As it turns out, fixator-norator pairs provide an effective tool for us to perform the biasing strategy we are looking for in this chapter. Here we show that the pair is the foundation for biasing circuits according to biasing design specifications. The method shows how, through the use of fixator-norator pairs, we can solve the problem of distributed supplies, generated because of local biasing. It actually shows how a pair can be used to couple a biasing spec with a supporting supply source; and in case the supply source is already specified in the design, the match is done with a power-conducting component. Note that a fixator provides a solution and a pairing norator finds, through the analysis, the resource needed for the solution. Hence, when used in combination, the pair will adhere to Kirchhoff's laws. In short, when a biasing criterion requires inclusion in a design, a fixator keeps this criterion fixed while a norator provides, allocated in an arbitrary location, the sourcing needed for the requirement. This is, of course, only possible if the fixator can control the norator and, conversely, the fixator must also be sensitive to the changes in the norator. Again, in case a designated DC supply is already in place for the design, the norator can be placed in a location designated for a powerconducting component, say a resistor, and then find its value through the analysis. There is a different interpretation of fixator-norator pairs that is worth discussing. In general, each circuit component is identified by its two variables, voltage and current. From the two usually only one variable is specified, such as the voltage in a voltage source or the current in a current source; alternatively the two may be related such as ohms law in a resistor. This indicates that from the two variables one must be found through the circuit laws, KVL and KCL. What makes fixators and norators different is that, in a fixator both component variables are specified but in a norator neither is specified. Hence, none of them can live alone in a circuit; whereas, when they pair they complement each other; i.e. overall, the two carry two specified variables and two are left for the circuit to find. This description of fixator-norator pairs suggests that the pair are no longer limited to DC operations and they can be used in any circuit operation including linear and AC circuits. What it means is that, in any type of circuit (linear or nonlinear) with any operation (DC or AC) one can set (fix) some circuit variables in exchange for some component values. To think of it differently, we can argue that fixator-norator pairs change a circuit analysis procedure to a design procedure that guaranties certain design specifications, if obtainable. This is because in circuit analysis we are given all component values and resources needed to analyze a circuit; whereas, in a design procedure there are some component values or resources to be determined in exchange for achieving some design specs. Example 1: To show how the process works, we start with a simple diode circuit depicted in Fig. 5 with an unspecified supply voltage V_1. Suppose the design requirement in this example is to find the value for V_1 so that the diode current reaches 1mA. Figure 6 shows the circuit arrangement for this design using a fixator-norator pair to satisfy the design criteria. As shown, the added fixator -- a current source $I_D = 1$ mA in parallel with a nullator -- forces the assigned current through the diode. Now, because the voltage across the current source is kept zero, the added fixator has no effect on the overall operation of the circuit. In addition, a norator is substituted for the unknown supply voltage V_1. Next, we simulate the circuit and get a voltage of $V_1 = 2.2$ V across the norator with a current $I_1 = 1.2$ mA through it. This suggests that although we have aimed for the voltage source V1 to replace the norator, we have in fact two more choices to make: i) replace the norator with a current source $I_1 = 1.2$ mA, or ii) replace the norator with a resistor $R_1 = -V_1/I_1 = -2.2/1.2 = -1.8$

KΩ. However, the last choice of a negative (active) resistance is not definitely acceptable for this design.

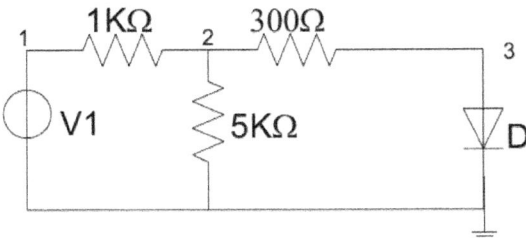

Figure. 5: A diode circuit with an unspecified supply voltage V_1

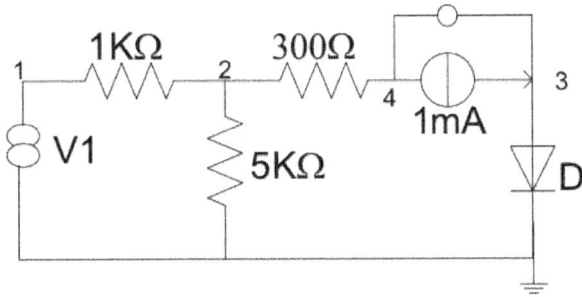

Figure. 6: The diode circuit arrangement using a nullor pair to satisfy the design criteria $I_D = 1$ mA

Note that after the supply $V_1 = 2.2$V (or the current source $I_1 = 1.2$ mA) is replaced with the norator, the fixator-norator pair are removed from the circuit without inflecting any changes to the circuit operation, i.e., still the current through the diode remains $I_D = 1$ mA. Note that in the case of replacing the norator with a current source $I_1 = 1.2$ mA, the circuit operation is not changed but the circuit structure (topology) can get modified. For instance, the 1 KΩ resistor in series with the source becomes redundant and could be removed. Now we are going to examine a third alternative. Let us assume that the voltage supply in the original circuit, Fig.5, is already assigned for $V_1 = 2.5$ V, but it is still necessary to have $I_D = 1$ mA, as a design requirement. This is the case that we need to decide on the value of a "power-conducting" component. To proceed, let us assume the resistor R_2 is the "powerconducting" component that we need to adjust. We replace R_2 with a norator, Fig.7, and simulate the circuit. As usual, we replacing the norator with a very high gain controlled source (VCVS), which is controlled by the fixator. From the simulated results we get a voltage of $V_2 = 1.0$ V across the norator and a current of $I_2 = 0.485$

mA through it. This simply means that the choice is to replace the norator with a resistor $R_2 = V_2/I_2 = 2.09$ KΩ.

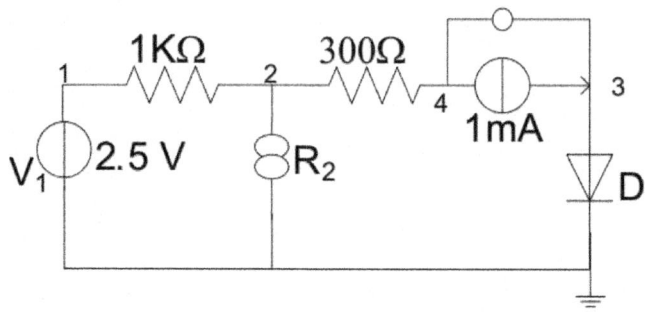

Figure. 7: The diode circuit arrangement using a nullor pair to satisfy the design criteria $I_D = 1$ mA

In general, in a circuit a norator with computed voltage V_1 and current I_1 can be replaced with i) a voltage source of V_1 volts, ii) a current source of I_1 amps, or iii) a component, such as a resistor $R = V_1/I_1$. Before we continue further we must realize that although our main use of fixator-norator pairs here is for biasing purposes their application goes beyond this. The following simple example goes one step further. Example 2: Take the case of the diode circuit discussed in Example 1 (Fig. 5). There are two design criteria to fulfill for this example: i) the power supply is specified with $V_1 = 3.3$ V, and the supply current is also fixed at $I_1 = 1.5$ mA; ii) the diode current still remains fixed at $I_D = 1$ mA. Now, because we have two criteria to meet we must use two fixators, $F_x(0, I_1)$ and $F_x(0, I_D)$, to keep the specified values fixed during the circuit biasing. The two fixators need to match with two norators to make two fixator-norators pairs. Within several choices we have we select two resistors R_2 and R_3 as "power-conducting" resistors to be recalculated. Hence, we replace them with two norators, as depicted in Fig. 8. Now, we need to decide which fixator is pairing which norator, as we have two choices to select; either (I_1 with R_2, I_D with R_3) or (I_1 with R_3, I_D with R_2). As it turns out, both choices work fine, except the choice (I1 with R_2, I_D with R_3) is preferred because it converges faster.

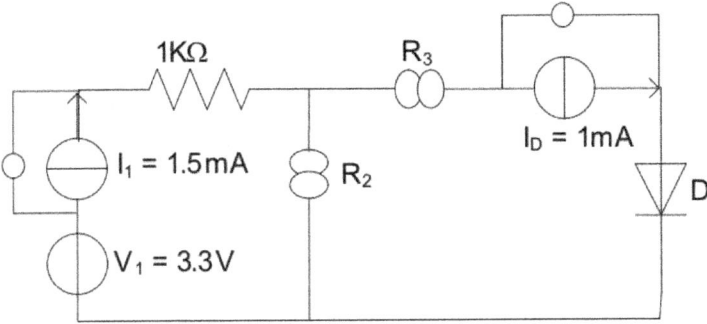

Figure. 8: The diode circuit arrangement using two nullor pairs to satisfy the design criteria of $I_1 = 1.5$ mA and $I_D = 1$ mA.

After simulating the circuit with the fixator-norator pairs we can find all the current and voltages for the circuit components including the two norators. With VR_2 and IR_2 found for the norator R_2, and VR_3 and IR_3 found for the norator R_3 we get the actual resistor values as:

$$R_2 = V_{R2} / I_{R2} = 1.8 / 0.5 = 3.6 \ K\Omega$$

And

$$R_3 = V_{R3} / I_{R3} = 1.08 / 1.0 = 1.08 \ K\Omega$$

Rules Governing Fixators and Norators in a Circuit

Following the introducing of fixators and norators two major issues come up. First, how shall we deal with fixators and norators in a circuit that contains other circuit components so that the KVL and KCL are not violated? Second, for n fixators and n norators in a circuit, how can we pair them for an effective performance? We discuss the first issue as the properties of fixator-norator pairs, and leave the other issue for a later investigation. As we already know fixators must pair with norators in order to have computational stability in a circuit. We should also remember that a fixator represents a current source as well as a voltage source combined; hence, it must adhere to both rules governing voltage sources and current sources. For instance, a current source in series with a fixator may violate the KCL, and a voltage source in parallel with a fixator may violate the KVL. In general, a cutset of fixators with or without current sources may violate the KCL and a loop of fixators with or without voltage sources may also violate the KVL. On the other hand, norators can be considered a current source, a voltage source or a resistive component. As such they can form a cutset with other current sources, and they can make loops with other voltage sources with no restrictions. However, the problem with norators is independency, and it becomes a serious issue when multiple numbers of norators are used in a circuit. For example, two norators in series or in parallel do not violate the Kirchhoff's laws but one loses its independency. In general, a loop of all norators does not violate the KVL but we can always remove (open) one from the loop without changing the circuit results. Similarly, a node or cutset of all norators does not violate the KCL, but we can always short circuit one norator in the group without changing the circuit performance. Other properties of fixatornorator pairs are as follows [13]:

- The power consumed in a fixator $Fx(V, I)$ is $P = V*I$; and the power is delivered by only one of the sources, V (for $F_x(V, I)$) or I (for $F_x(I, V)$)

- A resistance R in series with a fixator $Fx(V, I)$ is absorbed by the fixator and the fixator becomes $F_x(V_1, I)$, where $V_1 = V + R*I$. A resistance R in parallel with a fixator $F_x(V, I)$ is absorbed by the fixator and the fixator becomes $F_x(V, I_1)$; where $I1 = I + V/R$

- A current source IS in parallel with a fixator $Fx(V, I)$ is absorbed by the fixator and the fixator becomes $Fx(V, I1)$, where $I1 = I + IS$

- A voltage source VS in series with a fixator $Fx(V, I)$ is absorbed by the fixator and the fixator becomes $Fx(V1, I)$, where $V1 = V + VS$

- Connecting a fixator $Fx(V, 0)$ across a port with the port voltage V does not affect the operation of the circuit; it only fixes the port voltage

- Connecting a fixator $Fx(0, I)$ in series with any component in a circuit with current I does not affect the operation of the circuit; it only fixes the current going through that component

- In general, any two-terminal element in series with a fixator losses it's current to the fixator; and any two-terminal element in parallel with a fixator losses its voltage to the fixator

- A current source in series with a norator absorbs the norator; and a voltage source in parallel with a norator absorbs the norator. In addition, a current source in parallel with a norator is absorbed by the norator; and a voltage source in series with a norator is absorbed by the norator

- A resistance in series or in parallel with a norator is absorbed by the norator

- A norator in series with a fixator $Fx(V, I)$ becomes a current source I; and a norator in parallel with a fixator $Fx(V, I)$ becomes a voltage source V.

CIRCUIT SOLUTIONS CONTAINING FIXATOR-NORATOR PAIRS

Selective Biasing

Selective biasing is a procedure that fixes part of or the entire operating regions of a nonlinear component (say a transistor) during the circuit operation. To fix a biasing current, I, in a port we can use a fixator Fx(0, I). Similarly, to fix a biasing voltage, V, across a port we can use a fixator Fx(V, 0). However, as we discussed earlier, the use of fixators alone is not permissible in a circuit; we must pair each with a norator. On the other hand, both fixators Fx(0, I) and Fx(V, 0) carry zero power; hence, they alone cannot provide the biasing power to the serving component they are attached to. This simply means that for each fixator that is used to anchor certain biasing value in a circuit we need to provide the supplying power and direct it to the component. Our solution is either i) find a location for the supply power (voltage or current) and have the circuit find its magnitude, or ii) route the required power from an existing power supply through a power-conducting component. As it turns out the norators paring with the fixators can do both, provided that the pair are mutually sensitive, i.e., change in one causes the other to change accordingly.

Sensitivity in Fixator-Norator Pairs

In a circuit, each fixator can only work with a norator in a pair. A norator can be a source of power, a consumer of power or a power-conducting component. This means a norator must share power with a port that is anchored by a fixator. However, to satisfy this property the following condition must hold. A fixator paring with a norator must be "sensitive" to the changes happening in the norator and vice versa. This simply means that between a fixator and its pairing norator there must be a feedback. We can think of a norator as a placeholder for a DC supply or a power conductor in the circuit that must somehow "reach" to the corresponding fixator. In a way, when we replace a transistor port with its fixator model, we are getting a ticket, in exchange, to assign a DC source in the circuit wherever we like. This is true provided that the DC source is "reachable" by the fixator. Apparently, considering this property the choice of a norator pairing a fixator is not unique. In a connected circuit a (voltage or current) change within a component normally causes (voltage or current) changes throughout the circuit, although there are exceptions, particularly in cases of controlled sources without feedback. Therefore, in pairing a fixator with a norator we may have multiple numbers of choices to make; only avoiding those with zero feedback. This brings us to another issue, mentioned earlier, that can be stated as follows: for n fixators and n norators in a circuit how can we pair them for an effective design performance? This is certainly a challenging problem and we do not intend to make a comprehensive study on the subject here. What we would like to address is to find an acceptable relationship between a fixator and a norator in a pair so that it helps to speed up the biasing process in a circuit. The core issue in this relationship is the "sensitivity" issue [14, 15]. Simulating fixator-norator pairs - Before we continue further on the sensitivity issue we need to know how we can analyze or design a circuit that has fixator-norator pairs. Or simply, how can we simulate a circuit that contains nullator-norator pairs? As far as we know the existing circuit simulators, such as SPICE, do not have the means to directly handle the cases [16, 17, 18]. Traditionally, transistors and high gain operational amplifiers have been used for the purpose, and have done the job fairly successfully within acceptable accuracies [7, 9, 12]. However, in our case the situation is different. The fixator-norator pairs are only used symbolically in a circuit in order to establish the design criteria we have adopted. They are acting as catalyst and will be removed after the biasing is established in the circuit. Hence, we can assume the pairs to be ideal in order to provide the component values accurately. Within circuit components acceptable by a circuit simulator such as SPICE, controlled sources with very high gains are the ideal candidates for the job. Now, the question is what type of controlled sources must be used to simulate fixator-norator pairs? Evidently, if a fixator is used to fix a specified current in a circuit component, the source replacing the corresponding norator must be controlled by the voltage across the fixator. Similarly, if a fixator is used to fix a specified voltage in the circuit, the source replacing the corresponding norator must be controlled by the current through the fixator. Finally, the choice of the controlled source itself can be arbitrary. For example, if the job is to find the supply voltage V_{CC} in response to a fixed current I_B in the circuit then the controlled source is a voltage controlled voltage source (V_{CVS}). On the other hand, if in the previous case the supply voltage V_{CC} is already specified but we need to know how much current, I_C, is conducted from V_{CC}, then we can use a voltage controlled current source (V_{CCS}) to manage to find I_C, instead.

Paring Fixators and Norators in a Circuit

As mentioned earlier, one of the conditions to pair a fixator with a norator is to have feedback from the norator to the fixator. The purpose of this feedback is to harness the growth of the voltage or current in the pairing norator. In fact, because we are simulating a fixator-norator pair with a very high gain controlled source, the lack of feedback between them can cause serious instability and cause blow up values; i.e., it can generate a very high (negative or positive) voltage or current at the norator location or elsewhere in the circuit. The only way to control this growth is to establish feedback between the two in the pair. The following two examples show this feedback effects in dealing with fixator-norator pairs. A detailed analysis on the subject is also given in the Appendix. Example 3: - To see the feedback effect between a norator and its pairing fixator, let us consider the biasing circuit of a simple common emitter BJT amplifier with feedback, shown in Fig 9(a). In this example we assume the transistor operates linearly in its active region, so that we can linearize the biasing circuit accordingly, as shown in Fig. 9(b). Table I provides the component values for the linearized amplifier.

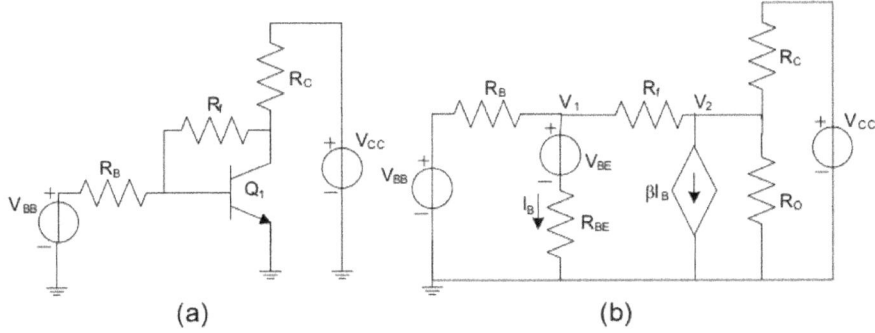

(a)   (b)

Figure. 9: (a) The biasing circuit of a common emitter BJT amplifier with feedback; (b) linearized biasing circuit for the amplifier;

Table I: Component Values for the Linearized Amplifier

V_{CC} V	V_{BB} V	V_{BE} V	R_B KΩ	R_{BE} KΩ	R_O KΩ	β
5	0.83	0.64	16.7	2	50	120

Now, in our first step we assume $R_C = 2$ KΩ and do two experiments with this amplifier. In the first experiment we remove the feedback resistance R_f from the circuit (no feedback), and in the second experiment we assign $R_f = 200$ KΩ. Table II provide the simulation results for the two experiments.

Table II: Simulation Results for the Linearized Amplifier

R_f KΩ	V_1 V	V_2 V	I_B μA
Open	0.66	2.42	10.36
200	0.668	1.526	9.9

In the next step we take the case with feedback ($R_f = 200$ KΩ) and try to find the powerconducting resistor R_C for a fixed $I_B = 9.9$ μA. Figure 10 shows the circuit constructed for this situation. As shown the fixator $F_x(V_{BE}, I_B)$ is paired with the norator R_C. The simulation results for this case provides VRC $= 3.474104$ V, and $I_{RC} = 1.737051$ mA, where V_{RC} and I_{RC} are the voltage across and the current through the norator R_C. This brings us to $R_C = V_{RC} / I_{RC} = 2$ KΩ, as we expected. Now we remove the feedback and repeat the circuit simulation with a fixed IB $= 10.36$ μA, that is slightly different from the previous value. This time the results from the simulation become surprisingly different. We get $V_{RC} = 53.3$ V, and $I_{RC} = 0.2762$ mA, which are obviously not correct and unstable. Again, the reason for this instability and defective result is due to the lack of feedback between the norator R_C and the fixator $F_x(V_{BE}, I_B)$. That is, changes in the current through R_C and the voltage across it is not "sensed" by the controlling fixator $F_x(V_{BE}, I_B)$.

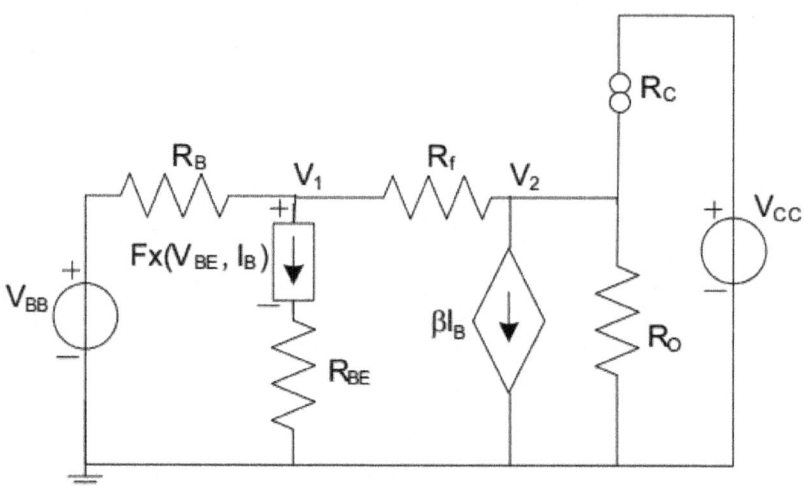

Figure. 10: The common emitter amplifier circuit with fixator-norator pair

Example 4: Consider a two stage BJT amplifier shown in Fig. 11(a). First we run the SPICE simulation on the circuit with the component values as specified. The results, displayed below, show the operating conditions for the two transistors.

$V_{BE1} = 5.790227e-01$
$V_{CE1} = 7.225302e-01$
$V_{BE2} = 6.434079e-01$
$V_{CE2} = 2.382333e+00$
$I_{B1} = 4.405489e-07$
WinSpice 1 ->

Next, we make the following changes in the circuit. i) Keep $I_{B1} = 4.405489e\text{-}07$ fixed, as it resulted from the simulation. This is done by adding a fixator $F_x(0, I_{B1})$ to the base of Q_1. ii) Remove $R_{C2} = 5\ K\Omega$ and replace it with a pairing norator R_{C2}, as depicted in Fig. 11(b). Next, we simulate the new circuit with SPICE, and the following is the simulation results listed.

$$V_{BE1} = 5.790105e\text{-}01$$
$$V_{CE1} = 7.229068e\text{-}01$$
$$V_{BF2} = 6.434051e\text{-}01$$
$$V_{CE2} = 2.547247e\text{+}00$$
$$V_{RC2} = 2.013071e\text{+}00$$
$$I_{C2} = 3.867745e\text{-}04$$
$$R_{C2} = V_{RC2}/\ I_{C2} = 5.204765e\text{+}03$$
WinSpice 2 ->

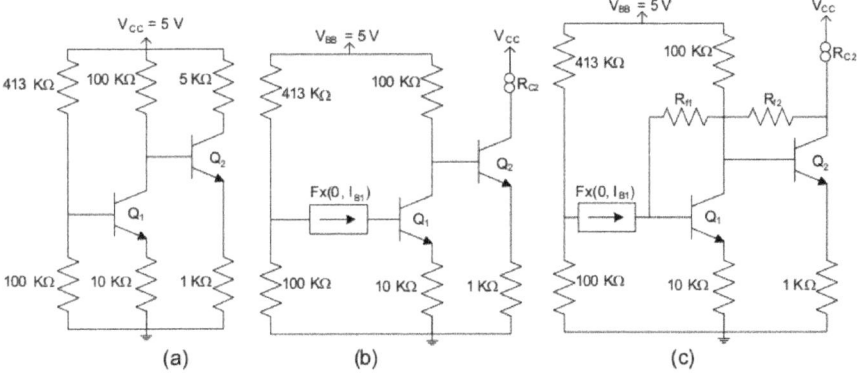

Figure. 11: (a) Two stage BJT amplifier; (b) amplifier circuit with fixator-norator pair; (c) amplifier circuit with feedback.

Note that the results in this case are just slightly different from that of the original circuit (Fig. 11(a)), with difference of about 4%. Now, if we change the base current I_{B1} by a tiny amount of 0.5 PPM (part per million) the responses take unrealistic values, as displayed in the following SPICE responses. For example, the negative resistance R_{C2} cannot be correct. This is of course expected because there is almost no feedback from the norator to the fixator.

$V_{BE1} = 5.789974e\text{-}01$
$V_{CE1} = 7.619999e\text{-}01$
$V_{BE2} = 6.398944e\text{-}01$
$V_{CE2} = 2.206873e\text{+}01$
$I_{B1} = 4.405491e\text{-}07$
$R_{C2} = -3.11725e\text{+}04$
WinSpice 3 ->

In another try we modify the circuit by incorporating feedback into the circuit; one from the output to the second stage and one from the second stage to the first stage, so that changes in the norator R_{C2} reach the fixator $F_x(0, I_{B1})$, as depicted in Fig. 11(c). The following SPICE simulation shows the results after the base current I_{B1} is changed by 100 PPM. The results are shown to be more reasonable, this time. For example, we notice that the power-conducting resistance R_{C2} replacing the norator, is $R_{C2} = 4.73$ KΩ, changed only by about 5%. Again, due to the feedback from the norator to the fixator, the circuit stability is back to normal now.

$$V_{CE2} = 5.802151e\text{-}01$$
$$V_{BE2} = 7.020994e\text{-}01$$
$$V_{CE1} = 6.432040e\text{-}01$$
$$V_{BE1} = 2.509425e\text{+}00$$
$$V_{RC2} = 2.054483e\text{+}00$$
$$I_{C2} = 4.343896e\text{-}04$$
$$R_{C2} = 4.729587e\text{+}03$$
WinSpice 4 ->

COMPONENT MODELING WITH FIXATOR

As stated in Property 1, a fixator can model a two-terminal device for a fixed biasing condition (snapshot). For example, for a diode biased at (I_D, V_D) the fixator that replaces it is $F_x(I_D, V_D)$, where for positive I_D and V_D, the diode consumes power. However, because the device is not locally biased (as discussed in the previous chapter) it must get power from the supplies in the circuit, i.e., global biasing. Property 1 can also be extended to include devices with multiple ports such as bipolar and MOS transistors. Here, for a fix component biasing the original component can be removed from the circuit and be replaced with fixators that mimic the same biasing; hence, imposing no change to the rest of the circuit. In general, there are two types of fixator modeling for nonlinear devices. In the first type, called complete modeling, the component is entirely removed from the circuit and replaced with one or more fixators that represent the component with their intended biasing. In the second method, called partial modeling, the component remains in the circuit but one or more fixators keep its biasing fixed at the specified values. We will discuss each type separately.

Complete Modeling of Devices

As stated in Property 1 a two-terminal device (or network) can be modeled by a single fixator. Likewise, for a multiple port device or network we can model each port separately with a fixator [19]. Hence, an n-port device can be removed from a circuit and replaced by n fixators with the same biasing currents and voltages without inflicting any changes within the rest of the circuit. For example an MOS device can be completely modeled by using three fixators. Figure 12 shows the complete fixator-models for nMOS and pMOS transistors, neglecting the substrate effects. Similarly, Fig. 13 depicts the complete fixator-models for npn and pnp transistors. Again, the models represent the devices with the same voltages

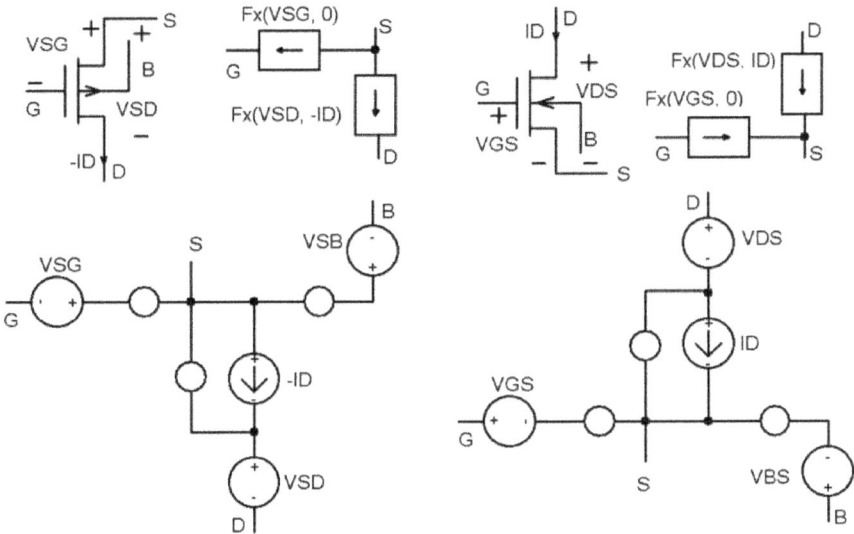

Figure. 12: Fixator models of nMOS and pMOS transistors when globally biased for V_{GS} (V_{SG}), V_{DS} (V_{SD}), I_D, and V_{BS} (V_{SB}). Both symbolic and expanded versions are shown.

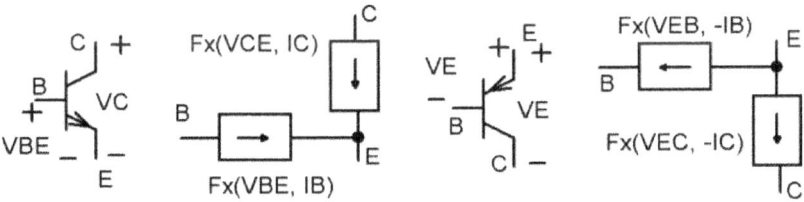

Figure. 13: Fixator models of npn and pnp transistors when globally biased for V_{BE} (V_{EB}), V_{CE} (V_{EC}), and I_C.

and currents that they need to get biased to the specified Q-points. Note that two changes are taking place in the circuit after the modeling is done: i) the resulted circuit becomes linear, and ii) the circuit is DC-freezed at fixed biasing conditions. What it means is that, addition (or removal) of any source or signal to the circuit may change signal conditions within the circuit but no change in inflicted on the modeled transistors. Hence, circuits with fixator-modeled components are not prepared for AC analysis.

Partial Modeling of Devices

In partial modeling the device remains biased in the circuit. In addition one or more fixators are used to freeze one or more device (port) variables at given Q-points. We have already used partial modeling in previous examples; for instance, in Example 4 we have freezed the base current I_{B1} of Q_1 during the entire biasing process. The advantage here is that we can limit the number of fixators to the number of biasing specs provided for the design. Also, a limited number of fixators makes it easier to match the number of fixators with that of norators in the circuit. This helps to speed up the biasing procedure in a large circuit. Another advantage in using partial modeling is that, in partial modeling the fixators are only responsible to provide some critical biasing requirements and the rest are left to the actual device, placed in the circuit, to adjust. For example, in a bipolar transistor only base current IB and the collector-emitter voltage V_{CE} might be considered critical; because with IB given the transistor will decide on the value of V_{BE}. Similarly, with the gain factor β known the collector current I_C is automatically established through the device characteristics. However, the disadvantage here is that the circuit remains nonlinear. In contrast with partial modeling, in complete modeling the transistors are totally absent from the circuit and have been replaced with the fixators. This means the fixators are fully in charge to accurately place the Q-points on the characteristic curves. This produces an extra work for the designer, who, prior to the actual design, needs to run the transistors individually and record the port values for the Q-points he/she has in mind. Then he/she needs to place the port values into the fixators and exchange the fixators with the corresponding transistors for the actual design. The third option is to have a mixture of the two; i.e., some transistors get complete modeling by fixators, while others are partially modeled. However, we are not allowed to have partial modeling on a port of a transistor and apply complete modeling on another port of the same transistor for obvious reasons.

Example 5: The objective in this example is to design a cascade CMOS amplifier, shown in Fig. 14(a). The transistor sizes and the critical specs given for the design are listed in Table III.

Table III: The design Critical specs for the amplifier

Devices	W/L μm	V_{GS} V	V_{DS} V
M_1	150/5	-2.0	-4.4
M_2	50/5	1.4	2.4

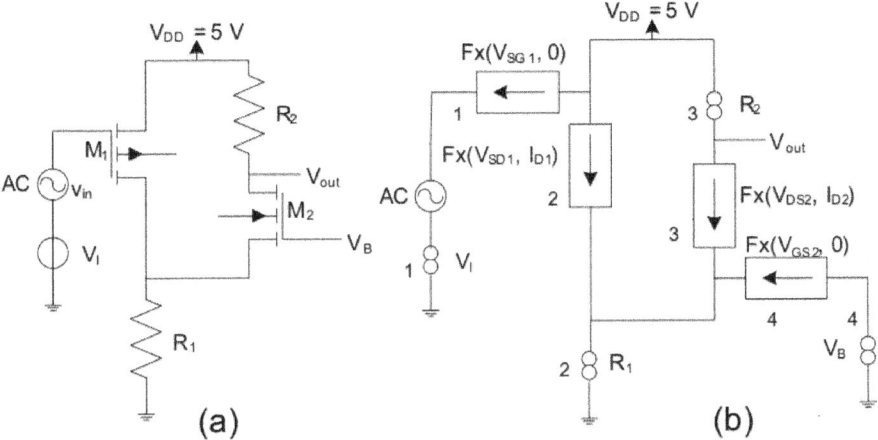

Figure. 14: (a) A cascade CMOS amplifier; (b) the amplifier with complete fixator modeling of the transistors.

To demonstrate different schemes, we are going to design the amplifier once using complete modeling of both devices using fixators, and next we will use mixture of complete and partial modeling. Complete modeling – To perform the design by complete device modeling we first remove the MOS transistors from the circuit and replace them with the fixator models shown in Fig. 12. Note that the fixators carry the critical specs given in Table III. They also include the drain currents I_{D1} = 289 μA and I_{D2} =30 μA that are computed when the transistors are individually simulated using the design specs (refer to "Complete modeling of devices"). Figure 14(b) shows the amplifier after the fixators have replaced the transistors. Note that the circuit is linearized after the transistors are replaced with fixator-norator pairs. Another important observation is the equality of the number of norators -- representing the unspecified component values -- and fixators -- representing the design specs. After pairing the fixators with the norators (identified by the same numbers in the figure) we represent each pair by a high gain controlled source for simulation purposes. Table IV shows the design values resulted from the SPICE simulation.

Table IV: The Amplifier design Values for the Norators

R₁ KΩ	R₂ KΩ	V_GG V	V_B V
1.9	66.3	3.0	2.0

Mixture modeling – In this design procedure we use the mixture of complete and partial modeling devices by fixators. As displayed in Fig. 15(a) the transistor M_1 is partially modeled whereas the transistor M2 is complete modeled. Note that the number of fixatornorator pairs is reduced to three but the circuit remains nonlinear. Similar to the previous case, the fixators carry the critical specs for both transistors plus the drain currents I_{D1} and I_{D2} for both transistors, as given in Table V. After pairing the fixators with the norators and following the same routine as explained in the previous case we get the circuit simulated by SPICE. The results from the simulation provide the component values as listed in Table VI.

Table V: The design specs for the amplifier

Devices	W/L μm	V_GS V	V_DS V	I_D μA
M₁	150/5	-	-4.37	289
M₂	50/5	1.37	2.4	24.7

Table VI: The Amplifier design Values for the Norators

R₁ KΩ	R₂ KΩ	V_B V
2.0	80.0	2.0

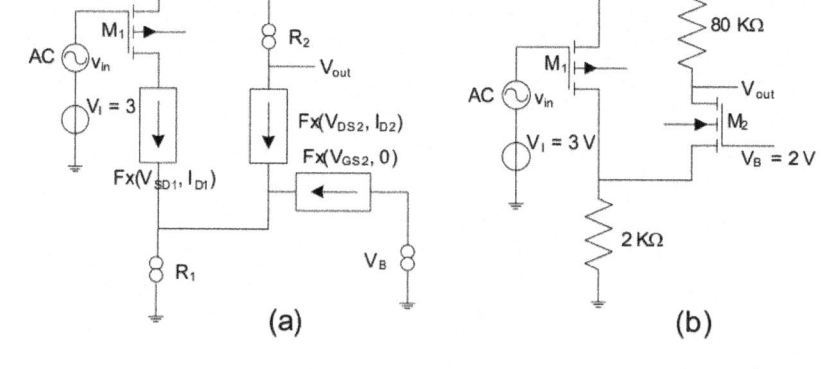

Figure. 15: (a) mixture of complete and partial modeling in the cascade CMOS amplifier; (b) the amplifier with biasing design completed.

Finally, a complete design of the cascade amplifier is depicted in Fig. 15(b). Figure 16 shows the transient response of the amplifier with a full output swing with negligible distortion. Discussion - This study still needs to address two questions. First, what is the solution if the DC supplies (mainly the voltage sources) so obtained are beyond the conventional and standard values – such as 12V, 5V, 3.3V...? In the case of smaller voltage values techniques such as voltage dividers can help to generate the right choices. For larger values, however, the solution may get more complecated. An adjustment in the "power-conducting" resistors is one possible solution. Because of the linearity involved, scaling is another simple tool to adjust the circuit supplies to match the conventional supply values. The second question is:

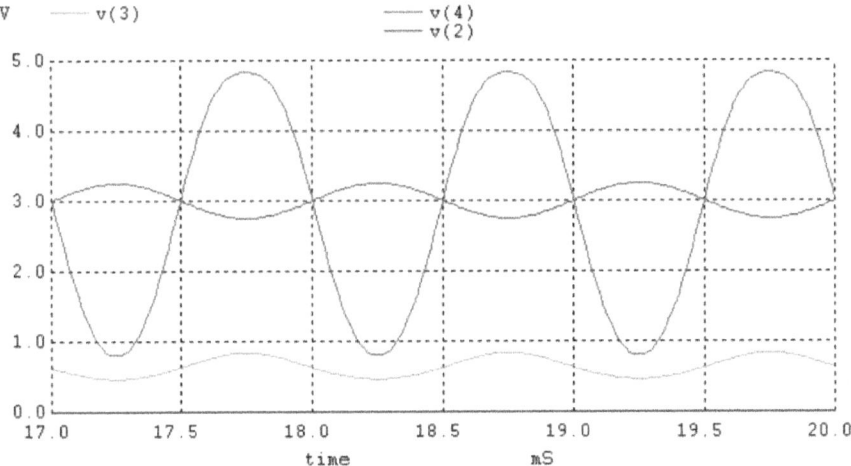

Figure. 16: The transient response of the amplifier for a full output swing that displays negligible distortion.

how to deal with the cases in which the number of fixators and norators are not equal? Typically the number of fixators exceeds the number of norators. For example, in a three stage amplifier with three driving transistors, we might need to have as many as six fixators; whereas one power supply V_{CC} or V_{DD}, can be represented by only one norator. The good news is that there are other components in the circuit that can be represented by norators. In general, norators can represent three types of components, i) voltage sources, ii) current sources/mirrors, or iii) power conducting devices, which are represented by resistors in lumped analog circuits, and in the case of integrated circuits they can also be represented by active loads. A second approach to achieve equality between the number of fixators and norators is to limit the number of fixators to the number of critical biasing specs in a circuit. In this approach we can identify the biasing design specs first; then classify the nonlinear ports as critical and non-critical, where the critical ports carry the design specs. In the second step, fixators are assigned only to those critical ports, which is necessary to keep those design specs protected (fixed) during the biasing procedure. We will be covering this subject in the next section in more detail.

Singularity and Circuit Divergence

Before leaving our discussion on the subject, there are issues that must be dealt with regarding fixator-norator pairs. First, as mentioned earlier, the equality between the number of fixators and norators is necessary to solve the circuit equations but it is not sufficient. The problem is related to the independency of the circuit (KCL and KVL) equations. There is always the possibility of inequality that may occur between the number of independent fixators and nullators, even though they may have originally been set equal. The problem is often caused by violating the rules related to fixators or nullators as discussed in Section 3. Both fixators and norators are relatively new elements in circuit theory; and the rules of engagement in KVL and KCL for these components are different from those of conventional elements, such as resistors, voltage sources, and current sources. The following example explains a similar case. Example 6: Consider a simple nMOS circuit shown in Fig. 17(a). With the circuit values specified the (SPICE) circuit simulator produces the biasing specs that are listed in Table VII. Further test shows that these biasing values well respond to the A_C operation. Next, we keep the voltages V_{GS} and V_{DS} as two critical biasing values and fix them by using two fixators, as depicted in Fig. 17(b). Next we need to assign two independent norators to match the fixators. We first select two resistors RD and RS to be reevaluated for the given design specs (V_{GS} and V_{DS}). To do this, we place the two norators in R_D and RS locations. After simulating the circuit with fixator-norator pairs, we get the resistors calculated as: $R_S = 997.6009\ \Omega$, and $R_D = 9997.974\ \Omega$, which are almost exactly as originally assigned for the circuit.

Table VII: The Biasing specs for the NMOS Circuit

W/L μm	V_{GS} V	V_{DS} V	I_D μA
50/5	1.966961	2.436567	233

Next, we still keep V_{GS} and V_{DS} the same two critical biasing values and represent them by the same two fixators, except, this time, we change the location of one norator switching from R_s to the supply voltage V_{DD}, as shown in Fig. 17(c). We definitely have not violated the K_{CL} by creating a node of two norators but when we run the circuit we get unacceptable responses. The

SPICE simulation results produce: R_D = -11411.8 Ω, and V_{DD} = 10.42594 mV, which both values are invalid! This is again because the two norators are in series and this leave the voltage of the node common between the norators floating.

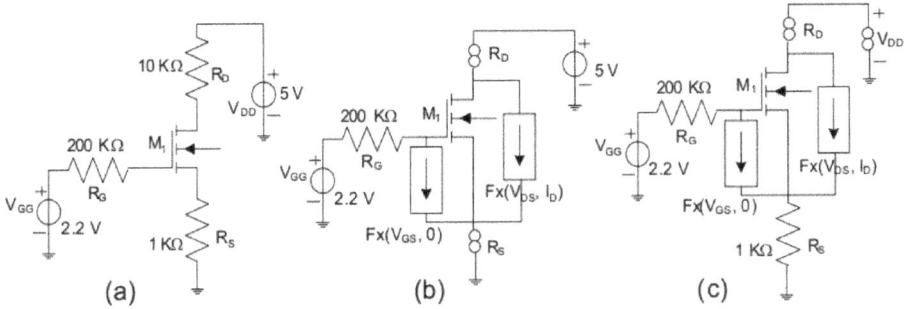

Figure. 17: (a) A simple nMOS circuit; (b) biasing design of the circuit with two fixator-norator pairs; (c) the same as (b), except the norators form an illegal common node.

CIRCUIT DESIGN FOR BIASING

Design of high performance analog circuits can be a complex and often multi stage process – noise, distortion, gain, bandwidth, biasing and so on. One approach to simplify the design and cut loops and feedbacks between the stages is to use as much orthogonality as possible [3]. This orthogonality is practiced in this chapter, between the circuit performances and the biasing of the nonlinear components, or simply between AC and DC circuit designs. The first task is to design for the circuit performances, mainly noise, signal power, and bandwidth [3]. The biasing design typically comes last, except for possible circuit modification that may require us to go back to the performance design, repeatedly. We only deal with the biasing situation in this chapter. A full discussion on the performance design and other related circuit design issues can be found in the literature [3]. Our approach to designing analog circuit biasing starts with a circuit topology (structure) that is suitable for the design. There is, of course, no restriction on this topology and structural modifications are acceptable during the design, as long as the final structure can fulfill the design criteria. In case the circuit structure for the performance design is different from that of the biasing design -- such as those with coupling or bypass capacitors -- we restrict ourselves only to the bias (DC) handling structure. Our next move is to select regions of operations for the transistors that fulfill the design requirements. This step may need some individual testing of the transistors to make sure of their behavior in the circuit. In the third step, and because the operating points for the transistors are specified, the components can be replaced with their small signal linear models; and here is where the performance (AC) design can start and continue until the design criteria are met. Following the performance design we need to bias the components in the circuit so that each one operates at the regions (Q-points) specified by the circuit performances. Algorithm 1 provides a systematic procedure to do the circuit biasing using fixator-norator pairs.

Algorithm 1

Preparation - Given the design specification, we begin with the performance design by selecting a working circuit topology. We then choose the desired operating points for the drivers3 that best meet the design requirements. Then we replace all the transistors with their small signal linear models, to make the circuit entirely linear and ready for the AC design. Note that as long as the linear models, representing locally biased devices, are not altered the circuit topology as well as the component values (including the W/L ratios in MOS transistors) can be changed for an optimal performance of the circuit. Finally, upon the completion of the performance (AC) design, we can start the biasing design as follows:

- Assign one fixator, carrying the biasing spec, to each "critical" transistor port. Also assign one norator to a location in the circuit that is a candidate for i) a DC supply voltage, b) a DC supply current, or iii) a power-conducting component such as a resistor. Note: be sure that the number of fixators and norators match.

- Pair each fixator with a norator in the circuit. This step is rather critical and needs to be handled with care (see Sensitivity in fixator-norator pairs in Section 3). In general, any pair must work (although may not be optimal), except for the cases where a fixator is not sensitive to the changes in the norator.

- Assign one controlled source with high gain to each pair of fixator-norator so that the fixator controls the source at the norator location. It is permissible to assume an ideal controlled source with very high gain; this is because these controlled sources will disappear afterwards, leaving the actual DC supplies or power-conducting components in place. A controlled source can be one of the four types: VCVS, VCCS, CCVS, or CCCS. The choice depends on the individual situation as follows:

- For a fixator keeping a specified current fixed the controlled source is either VCVS, or VCCS

- For a fixator keeping a specified voltage fixed the controlled source is either CCVS, or CCCS

- For a norator holding the place for a voltage supply the best choice is either a VCVS, or CCVS

- For a norator holding the place for a current (mirror) supply the best choice is either a VCCS, or CCCS.

- For a norator holding the place for a power-conducting component any of the four will work.

- Solve the linear circuit equations as prepared. The DC solution (simulation) provides the currents and voltages for the circuit components including those of the norators that are represented by the controlled sources.

- Remove all the controlled sources from the circuit and replace each with an appropriate voltage supply, V_j, a current supply, I_j, or a resistor $R_j = V_j/I_j$; where V_j and I_j are the voltage and current found for that controlled source (norator). This concludes the biasing design algorithm.

DESIGN EXAMPLES

The following examples provide a systematic procedure for biasing design of analog circuits using the new approach, given in Algorithm 1. Example 7: This example presents a negative feedback BJT amplifier; fully explained in reference [3]. Figure 18 shows a simplified AC schematic of the amplifier after it has gone through the performance design in three areas: noise reduction, clipping/distortion reduction, and high loop-gain-poles-product4. To perform the biasing of the circuit we need to first specify the values of the DC supplies and their locations in the circuit. Next, we need to select the operating points for the transistors so that they can fulfill the design specs. For the actual power supplies, we choose two DC sources of 4V and - 4V, as assigned in the reference [3]. Next we need to select DC power-conducting components that provide biasing power to the drivers. However, there are certain performance design criteria that must be given priority in this selection so that the biasing is smoothly aligned with the rest of the design. These major performance design criteria are as follows:

- The emitters of Q_1 and Q_2 must be driven by a high impedance current source, Ie.

- The base of Q_2 must be driven by a low impedance voltage source, V_{b2}.

- The collector of Q_1 can be driven directly by V_{CC}.

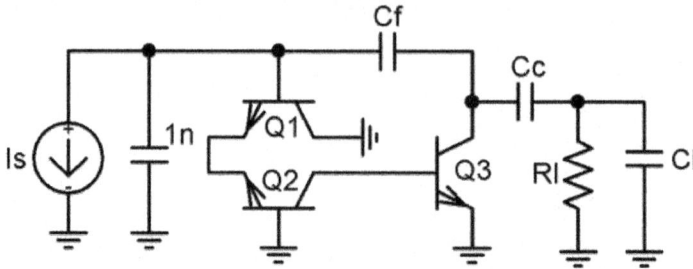

Figure. 18: A three stage amplifier topology after going through the performance, AC, design [3].

- The collector of both Q_2 and Q_3 must be driven by high impedance current sources I_{S2} and I_{S3}, to maximize the gain
- The base current of Q_1 can be provided through a feedback resistor R_f^5.

For this particular design we choose the collector-emitter voltages of two transistors Q_2 and Q_3 (vce$_2$ and vce$_3$) as the "critical" design values. The collector-emitter voltage of Q_1 (Vce$_1$) is considered "non-critical" because it is directly connected to V_{CC}. Also all three collector currents ic$_1$, ic$_2$, and ic$_3$ are considered "critical" for this design. Table VIII, columns 1 and 2, provides all five critical values for the selected operating points; also all five fixators that keep these critical values fixed during the design are listed. Column 3 shows the matching norators that are later replaced with computed components: a voltage source, three current sources and one feedback resistor (DC power-conducting component). Figure 19 is extracted from Fig. 18 after the fixator-norator pairs, specified in Table VIII, are added to the circuit.

Table VIII: Bias design specs and fixator-norators

Critical specs	Fixator representations	Norator representations
$I_{C1} = 0.1$ mA	Fx(0, 0.1mA)	R_F
$V_{CE2} = 0.67$ V	Fx(0.67V, 0)	V_{B2}
$I_{C2} = 0.5$ mA	Fx(0, 0.5mA)	I_E
$V_{CE3} = 2.2$ V	Fx(2.2V, 0)	I_{S3}
$I_{C3} = 3.6$ mA	Fx(0, 3.6mA)	I_{S2}

Below is a piece of the WinSPICE program code simulating the DC biasing of the amplifier. Note that each fixator-norator pair is simulated by a very high gain controlled source (namely VCVS, CCVS, VCCS, CCCS, and VCCS in sequence).

```
ic1    2          a          DC          1.0e-04
e1     4          51         a           2        1000MEG
vce2   c          7          DC          0.67
hb2    Vb2        0          vce2        1000MEG
ic2    3          c          DC          0.5m
ge     7          11         3           c        1000MEG
vce3   e          0          DC          2.2
fc3    21         4          vce3        1000MEG
ic3    4          e          DC          3.6m
gc2    12         3          4           e        1000MEG
```

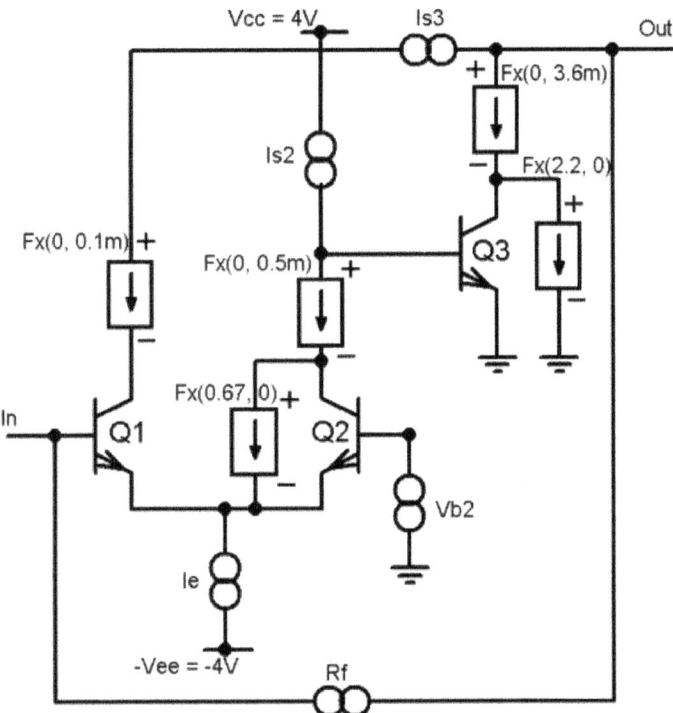

Figure. 19: The three stage amplifier with fixator-norator pairs indicating the biasing design specs.

The results from the WinSPICE simulation are shown below and listed in Table IX.

```
TEMP=27 deg C
DC analysis ... 100%
(v(4)-v(5))/vf#branch = 1.528640e+06
vb2 = 6.770538e-01
ve#branch = 6.068945e-04
vs3#branch = 3.601024e-03
vs2#branch = 5.229127e-04
WinSpice 6 ->
```

Table IX: Component Values for the Specified Biasing

R_F = 1.53 MEGΩ
V_{B2} = 0.677 V
I_E = 0.607 mA
I_{S3} = 3.601 mA
I_{S2} = 0.523 mA

Finally, we remove the controlled sources (representing the fixator-norator pairs) from the circuit and replace each with the computed voltage source, current sources, and one feedback resistance. The final amplifier so designed is depicted in Fig. 20[6]. As expected, the resulted DC sourcing matches with those in [3].

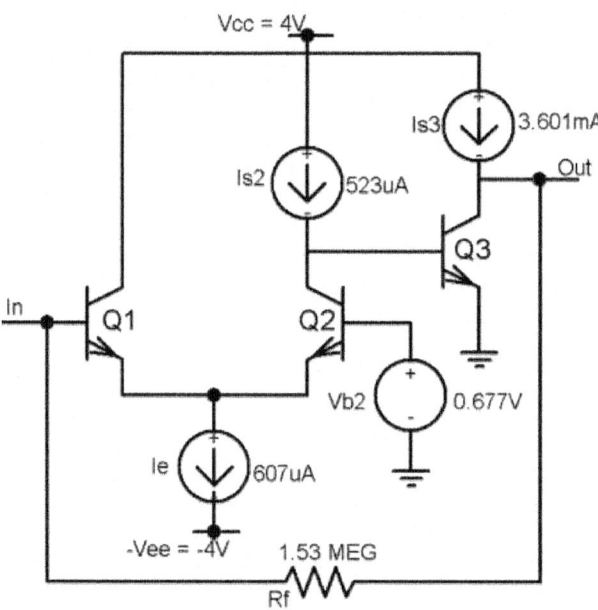

Figure. 20: The three stage amplifier with complete biasing.

Example 8: The purpose here is to complete the biasing design of a two stage CMOS differential amplifier shown in Fig. 21. The design criteria set for this amplifier requires that both the input offset voltage V_G and the output offset voltage V_O, remain stable at 0.5V. Hence, we have two design criteria to fulfill and need two fixators to fix V_{IN} = 0.5 V and V_{OUT} = 0.5 V. The circuit with fixator (or rather nullator)-norator pairs is shown in Fig. 22. Next, because the supply voltage V_{DD} is already specified for the design at V_{DD} = 1$_v$, we need to focus on finding the two current sources (mirrors), as power-conducting components. So we can replace the current sources with two norators and simulate the circuit (Fig. 22). The SPICE simulation finds the currents flowing through the norators as $I_1 = 1.26\ \mu A$ and $I_2 = 21\ \mu A$. This means we can replace the norators with two current sources at the designated locations, as they were before.

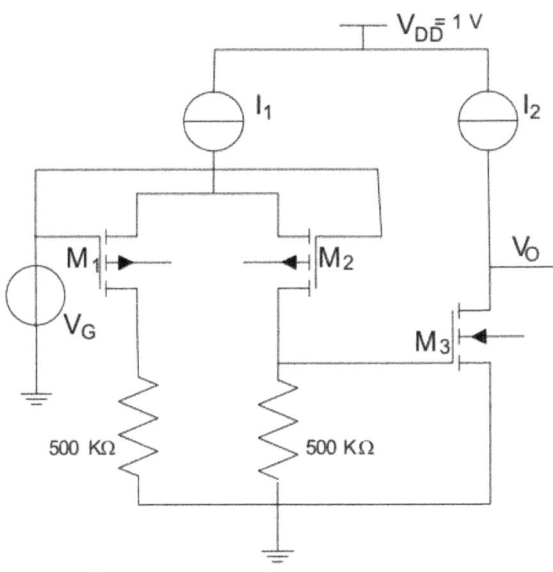

Figure. 21: A two stage CMOS differential amplifier.

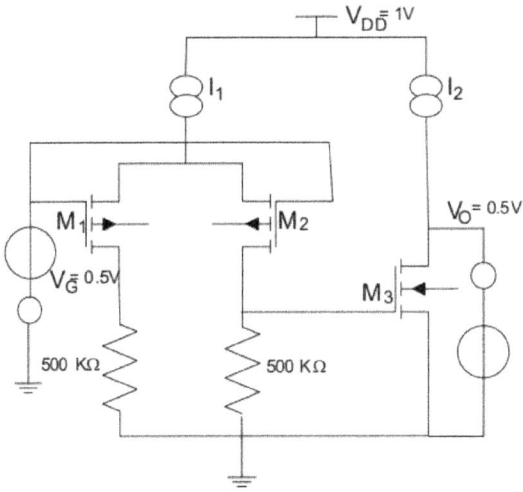

Figure. 22: Design stages of a CMOS differential amplifier

Note, in this example, that the choice of two current sources replacing the norators is only one option. Here the source resistance for each current source happens to be infinity, but this is not a requirement. In fact, any component, or combination of components as a twoterminal circuit, is permissible to replace the norator, say norator I_1, provided that the D_C current through the two-terminal amounts to the current I_1, and the voltage across the twoterminal is the same as that obtained for the norator I_1, in the circuit simulation. For instance, let us take the following case: let us assume that in doing the A_C performance design of the amplifier, we have come up with a resistance of R_{I1} necessary to place it at the location of the current source I_1. Now, to make this resistance also available for the D_C biasing, all we need to do is to add R_{I1} in parallel with the current source I_1. The only correction we need to make is to reduce the current in the source from I_1 to $I'1$; where $I'1 = I_1 - I_{RI1}$, and I_{RI1} is the D_C current that is conducted through the resistance R_{I1}. In short, the overall branch current must stay fixed at I_1. The significance of this issue is in providing link between design of D_C and A_C in analog circuits. It simply opens a new procedure in the design where both D_C and A_C design are pursued in combinations, but they may differ in some component values. This is more apparent in design of integrated circuits, where the roles of active loads and current mirrors are different from D_C biasing to A_C signal loading. However, this is a topic of further investigation. Let us get back into our design. Now that we have substituted for the norators, the design is complete, after removing the fixator-norator pairs from the circuit. Next, to perform the transient operation, we apply 0.5V D_C supply to the gate of M_2 and run the amplifier with an input signal Vi = 500 + 5*sinωt mV applied to the gate in M1. As shown in Fig. 23, the generated output voltage Vout,pp = 0.8 V still remains undistorted. Note that the output offset voltage stays at 0.5V, as expected.

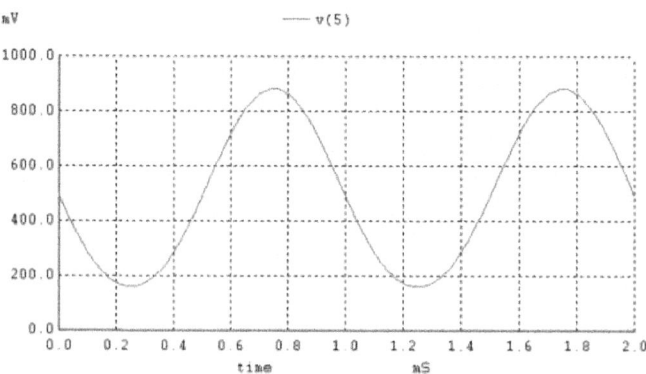

Figure. 23: The undistorted output waveform for the CMOS differential amplifier

Example 9: The purpose of this example is to complete the design of a C_{MOS} differential amplifier with a buffer stage. Figure 24(a) depicts the circuit configuration. As shown, the performance design of the amplifier is completed giving the transistor sizes listed in Table X.

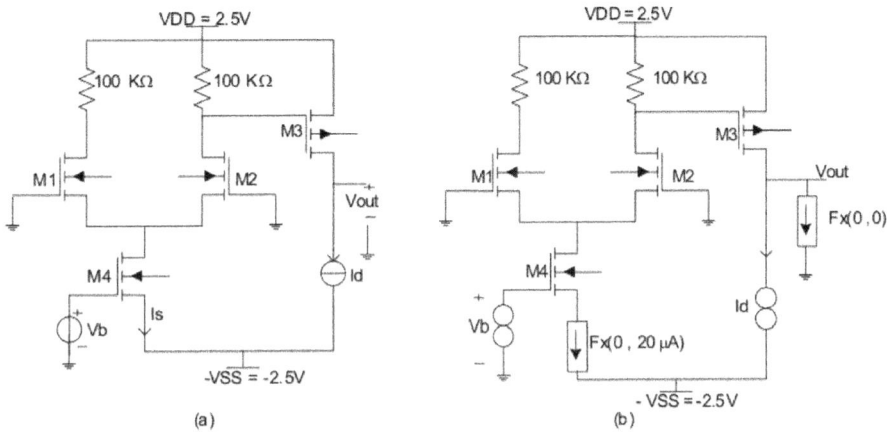

Figure. 24: (a) A CMOS differential amplifier with buffer stage; (b) biasing design procedure for the amplifier

To complete the biasing design we need to do the following: i) specify the biasing voltage V_b so that we can get a current sink of $I_S = 20$ μA, and ii) specify the current mirror I_D in the buffer stage so that the output offset voltage $V_{out} = 0$. Figure 24(b) shows the biasing design procedure, where two fixator-norator pairs are used for I_S and V_{out}, and V_b and I_d. Again, because of the two fixator-norator pairs used in this example the problem is to find the best pairing situation among the four so that it provides the fastest and most accurate solution. Within the two existing choices it turns out that the fixator $F_x(0, 20mA)$ and the norator V_b make a good match; likewise, $F_x(0, 0)$ and the norator Id also produce good results. Again, the fixator-norator pairs are replaced with two high gain controlled sources, prepared for circuit simulation. Following the SPICE simulation of the circuit the two unknown values are computed as: $V_b = -1.56V$, and Id $= 48$ μA. Next, the amplifier circuit is completed by making $V_b = -1.56V$, and Id $= 48$ μA in Fig. 24(a). Because the two voltage supplies $V_{DD} = 2.5V$ and $-V_{SS} = -2.5_V$ are available in this design we can simply generate $V_b = -1.56_V$ through a voltage referencing (divider) circuit; and for Id $= 48$ μA a current mirror circuit can be put in place. This completes the biasing design of the amplifier.

Table X: The CMOS Transistor Sizes

M1 W/L - μm	M2 W/L - μm	M3 W/L - μm	M4 W/L - μm
20/2	20/2	200/2	40/2

Some Challenges and Potential Impacts of the Proposed Methodology7

We believe the proposed methodology can have a profound impact on the research and development of techniques for designing analog circuits. It provides circuit designers a collection of choices and short cuts to create better designs in shorter time periods. The design tools and procedures introduced in this and a previous chapter are new and expandable. The proposed tools can be interpreted as the beginning of a new methodology in analog circuit designs. Through this methodology, one can see the challenges that exist for more direct, faster and cost effective designs of otherwise complex analog circuits. What it brings to a designer is simplicity, time and management. It brings simplicity because no matter how complex the circuit might be, it can be partitioned and linearized. The designer can save time because by linearization he/she has entirely removed the nonlinear iterations from the analysis. The designer is in full control of the management of the design because he/she is not faced with a complex network of mixed linear and nonlinear components, but individual transistors to assume the right operating points for. By a mixture of global and local biasing (see the previous chapter) a skilled designer can maneuver around and find a selective path for gradually applying DC supplies in the circuit, aiming at a smooth and fast converging biasing. Finally, because of the exact and selective environment that is provided by this methodology, the designer is capable of accurately calculating for possible distortions, noise, bandwidth, power and other design attributes. Last but not least, this study introduces new missions and roles for some virtual components: nullator, fixator and norators, that have not been practiced in the past. Here are some of the evidences for the challenges discussed:

- No matter how complex, the nonlinearity is entirely removed and replaced with the linearized equivalent circuits for biasing

- If selected, each transistor (nonlinear component) is individually biased to the selective and desirable operating points without affecting the rest of the circuit

- Local biasing minimizes the DC power consumption in the circuit. In general, the methodology can be used to monitor the DC power consumption in a circuit and direct it so that one can reduce the power effectively

- Through the use of fixator-norator pairs a circuit designer can specify and fix the design criteria (pertinent to the biasing) all throughout the design. The pair also serves to locate and find values for voltage/current supplies or components that conduct the DC power

- Although fixator-norator pairs, as non realistic circuit components, are used in the biasing design, they only act as a catalyst and removed after the proper components are substituted.

A mixture of the traditional and the new method is also possible for the design; which is in fact recommended for circuit modification and debugging.

APPENDIX

Feedback effect in fixator-norator pairs: - In pairing fixators with norators in a circuit, one of the essential conditions is to have mutual feedback between the two. In one direction, it is the fixator that generates the current and voltage values of the pairing norator; but in the other direction it is the feedback from the norator to the fixator that controls the event and puts harness into the growth of the voltage or the current in the pairing norator. The following analysis is an attempt to show this effect through an example by using feedback theory. Analysis - To see the feedback effect between a norator and its pairing fixator, let us consider the biasing circuit of a simple common emitter BJT amplifier with feedback, shown in Fig A1(a). With the assumption that the transistor operates close to its linear regions on the characteristic curves we can linearize the biasing circuit according to Fig. A1(b). Next, we can even simplify the circuit more as represented in Fig. A1(c); where we can easily find the circuit values as

$$I_1 = \frac{V_{BB}}{R_B} + \frac{V_{BE}}{R_{BE}},$$

$$V_1 = R_{BE}I_B + V_{BE},$$

$$R_{in} = \frac{R_B R_{BE}}{R_B + R_{BE}},$$

$$G_m = \frac{\beta}{R_{BE}},$$

$$I_2 = \frac{V_{CC}}{R_C},$$

$$I_{CE} = G_m V_{BE}, \text{ and}$$

$$R_{out} = \frac{R_C R_O}{R_C + R_O} \tag{1}$$

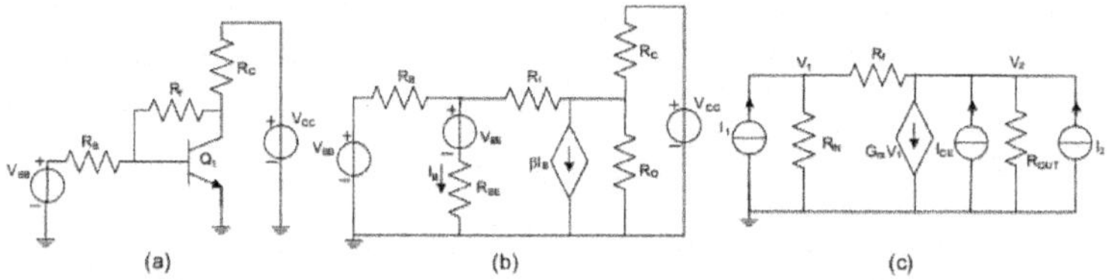

(a)   (b)   (c)

Figure. A1: (a) The biasing circuit of a common emitter BJT amplifier with feedback; (b) linearized biasing circuit for the amplifier; (c) reduced equivalent circuit.

Now, we can start writing the node equations for the circuit (Fig. A1(c)), and after solving the equations we get

$$I_2 = (G_{out}G_{in} / G_f + G_{out} + G_{in} + G_m) V_1 - (G_{out} / G_f + 1) I_1 - I_{CE}$$

$$G_i = 1 / R_i \text{ for all i.} \tag{2}$$

We substitute from Eqs. (1) into Eq. (2), and after proper simplification we get

$$V_B = RG_C V_{CC} - (1 - RG_{BE}(G_{out}G_{in} / G_f + \beta + 1))V_{BE} + RG_B(G_{out} / G_f + 1) V_{BB} \tag{3}$$

Where

$$V_B = R_{BE}I_B, \ R = 1 / G,$$

and

$$G = G_{out}G_{in} / G_f + G_{out} + G_{in} + G_m. \tag{4}$$

The assumption is that the supply voltage V_{BB} is already given and stays constant; also V_{BE} stays constant. Suppose the design requires having I_B stay fixed at its specified value. Then according to Eq. (3) the amount of feedback voltage that V_{CC} can contribute to the base voltage of the transistor is.

$$V_B' = RG_C V_{CC} \qquad (5)$$

Equation (5) provides the feedback effect from V_{CC} (the norator) to the transistor base where the fixator is located. Now, to complete the loop we need to get the feed forward effect, i.e., how the fixator in the transistor base generates V_{CC}. As mentioned earlier, for simulation purposes we can use a very high gain controlled source (V_{CVS}, in this case) to handle the case. Hence, for a gain of Av we can write the relationship as

$$V_{CC} = A_v V_B \qquad (6)$$

This is how the norator voltage (V_{CC}) is generated due to the variation across the fixator IB, i.e. V_B. Now, to get the feedback part strait we first substitute for R from Eq. (4) into Eq. (5). Next, we simplify Eq. (5), for very high feedback resistance R_f, to get

$$V_B' = \frac{R_{OUT}R_{IN}}{R_f R_C} V_{CC} = F V_{CC} \qquad (7)$$

The variable F is the feedback coefficient. From the feedback control systems we know that, for high gain A_v, where $F*A_v \gg 1$, the closed loop gain A_C can be approximated as

$$A_C = \frac{1}{F} = \frac{R_f R_C}{R_{OUT} R_{IN}} \qquad (8)$$

As Eq. (8) indicates, A_C will be limited for limited values of the feedback resistance R_f. On the other hand, if Rf grows high the system become more unstable; eventually with broken feedback a fixator fails to generate the required DC supply (V_{CC}) as a substitution for the pairing norator.

ACKNOWLEDGMENT

The author would like to thank Ms. Golnaz Hashemian for her valuable suggestions and editing the chapter.

REFERENCES

1.   A.S. Sedra, and K.C. Smith, Microelectronic Circuit 6th ed. Oxford University Press, 2010.

2. R.C.Jaeger, and T.N. Blalock, Microelectronic Circuit Design 4th ed. Mc Graw-Hill Higher Education, 2010.

3. C. J. Verhoeven, Arie van Staveren, G. L. E. Monna, M. H. L. Kouwenhoven, E. Yildiz, Structured Electronic Design: Negative-Feedback Amplifiers, Kluwer Academic Publishers, 2003.

4. R. Hashemian, "Local Biasing and the Use of Nullator-Norator Pairs in Analog Circuits Designs," VLSI Design, vol. 2010, Article ID 297083, 12 pages, 2010. doi:10.1155/2010/297083. http://www.hindawi.com/journals/vlsi/2010/297083.html

5. ___, "Analog Circuit Design with Linearized DC Biasing ", Proceedings of the 2006 IEEE Intern. Conf. on Electro/Information Technology, Michigan State University; Lancing, MI, May 7– 10, 2006.

6. ___, " Designing Analog Circuits with Reduced Biasing Powe", Proceedings of the 13th IEEE International Conference on Electronics, Circuits and Systems, Nice, France Dec. 10– 13, 2006

7. R. Kumar, and R. Senani, "Bibliography on Nullor and Their Applications in Circuit Analysis, Synthesis and Design", Analog Integrated Circuit and Signal Processing, Kluwer Academic Pub, 2002.

8. H. Schmid, "Approximating the universal active element", IEEE Trans. on Cir. and Sys. II, Volume 47, Issue 11, Nov 2000, pp 1160 – 1169.

9. E. Tlelo-Cuautle, M.A. Duarte-Villasenor, C.A. Reyes-Garcia, M. Fakhfakh, M. Loulou, C. Sanchez-Lopez, and G. Reyes-Salgado, "Designing VFs by applying genetic algorithms from nullator-based descriptions", ECCTD 2007, 18th European Conference on Circuit Theory and Design, Volume, Issue , 27-30 Aug. 2007, pp 555 – 558.

10. E. Teleo-Cuautle, L.A. Sarmiento-Reyes, "Biasing analog circuits using the nullor concept", Southwest Symp. on Mixed-Signal Design, 2000.

11. D.G. Haigh, and P.M. Radmore, "Admittance Matrix Models for the Nullor Using Limit Variables and Their Application to Circuit Design", IEEE Transactions on Circuits and Systems I: Regular Papers, Volume 53, Issue 10, Oct. 2006, pp 2214 – 2223.

12. Claudio Beccari "Transmission zeros", Departimento di Electronica, Turin Institute of Technology, Turino, Italy; December 6, 2001.

13. D.G. Haigh, T.J.W. Clarke, and P.M. Radmore, "Symbolic Framework for Linear Active Circuits Based on Port Equivalence Using Limit Variables", IEEE Transactions on Circuits and Systems I: Regular Papers, Volume 53, Issue 9, Sept. 2006, pp 2011 – 2024. www.intechopen.com 46 Advances in Analog Circuits

14. T.L. Pillage, R.A. Rohrer, C. Visweswariah, Electronic Circuit & System Simulation Methods, McGraw-Hill, Inc., 1995.

15. J. Vlach and K. Singhal, computer methods for circuit analysis and design, Van Nostrand Reinhold Electrical/Computer Science and Engineering Series, 1983.

16. L.W. Nagel, "SPICE2, A computer program to simulate semiconductor circuits," Univ. of California, Berkeley, CA, Memorandum no. ERL-M520, 1975.

17. Mike Smith, "WinSpice3 User's Manual, v1.05.08", http://www. ousetech.co.uk/winspice2/, May 2006.

18. R. Jacob. Baker, CMOS, Circuit Design, Layout, and Simulation, 2nd ed. IEEE Press, Wiley Interscience, 2008, pp. 613 – 823.

19. R. Hashemian, "Source Allocation Based on Design Criteria in Analog Circuits", Proceedings of the 2010 IEEE International Midwest Symposium On Circuits And Systems, Seattle, WA, August 1 - 4, 2010.

Chapter 4

ADVANCED STATISTICAL METHODOLOGIES FOR TOLERANCE ANALYSIS IN ANALOG CIRCUIT DESIGN

Bruno Apolloni[1] , Simone Bassis[2] , Angelo Ciccazzo[3] , Angelo Marotta[4] , Salvatore Rinaudo[5] and Orazio Muscato[6]

[1,2]Department of Computer Science, University of Milan, Via Comelico 39/41, 20135 Milano

[3,4,5]STMicroelectronics, Stradale Primo Sole 50, 95121 Catania

[6]Department of Mathematics and Informatics, University of Catania, Viale Andrea Doria 6, 95125 Catania Italy

INTRODUCTION

The influence of process variations is becoming extremely critical for nano technology nodes (90nm and below), due to geometric tolerances and manufacturing non-idealities (such as edge or surface roughness, or the fluctuation in the number of doping atoms). The most worrying of all is the statistical variability introduced by discreteness of charge and granularity matter in the transistors approaching molecular and atomic scale dimensions. The main sources of statistical variability are the random distributions of discrete dopants and charged defects, the line edge roughness of the photo resist and the granularity of the materials (Bernstein et al., 2006; Boning & Nassif, 1999). As a result, production yields and circuit figures of merit (such as performance, power, and reliability) have became extremely sensitive to incontrollable statistical process variations (PV). The main sources of variations are: environmental factors, whose transient arises during the operation of a circuit (e.g. power supply or temperature variations), and physical factors due to the manufacturing process, which result in a (permanent or aging) variation of the device structure and interconnections. The latter reflect into random (possibly spatial) drifts of the design parameter. Although already considered in the past, the increasing impact of these drawbacks constitutes a completely new challenge. While process engineers have traditionally coped with die-to-die fluctuations, the today within-die variations are more subtle since they imply that different areas of the same die exhibit different values of the various parameters. With a further shrinking of process technology, the on-chip variation is getting worse for each technology node, thus having a direct impact on the design flows. By contrast, the latter conventionally rely on deterministic models. At a front end, parameter variability has a significant impact both on the power dissipation and performance of a circuit, with a consequent yield decrease and remarkable cost implications. Indeed, to maintain production efficiency we must raise up control costs and cycle time, a drawback which dramatically increases with the process complexity. To contrast it, the following two joint tasks become essential:

- to characterize statistically integrated circuits (IC) manufacturing process fluctuations;
- to predict reliably circuit performance spreads at the design stage

Failure in the former can result in a low parametric yield, since ICs do not meet design specifications. On the one hand, a successful statistical characterization promotes a robust manufacturability reflecting in a high fabrication yield (i.e. a high proportion of produced circuits which function properly). On the other hand, it requires managing complex design flows in the design-verification-production life-cycle of ICs. Summing up, random and systematic defects as well as parametric process variations have a big influence on the design/production cycle, causing frequent re-spinning of the whole development and manufacturing chain. This leads to high costs of multiple manufacturing runs and entails extremely high risks of missing a given market window. One way to overcome these drawbacks is to implement the DFM/ DFY paradigm (Bühler et al., 2006) where Design for Manufacturability (DFM) mates Design for Yield (DFY) to form a synergistic manufacturing chain to be dealt with in terms of: i) relationships between the statistical circuit parameters matching the production constraints, and ii) performance indicators ensuring correctly functioning dies. This chapter introduces a pair of procedures aimed at identifying these parameters exactly with the goal of maximizing performance indicators defined as a function of the parameters' confidence region. The material is organized as follows. In Section 2 we discuss the statistical aspect of IC design and introduce the lead formalism. In Section 3 we focus on the statistical modeling task with special regard to two advanced solution methods. Hence we introduce benchmarks in Section 4 to both provide a comparison between the performances of the above methods and show their behaviors w.r.t. state-of-the-art procedures introduced by researchers in the last years. Concluding remarks are drawn in the final section.

STATISTICS IN IC DESIGN

Electronic devices are replicated multiple times on a wafer and different wafers are produced, but each device cannot be produced in the same way in terms of electrical performance. Main factors that make the fabrication result uncertain are: the imperfections characterizing the masks and tolerances in their positionings, various changing effects of ion plant temperature during production, tolerances in size, etc. Generally fluctuations' processes produce fluctuations in electrical performance. Consequently, an essential tool for electronic circuit designing is represented by the statistical model which formally relates the former to the latter. A circuit is classified as acceptable in performances if all specifications on its electrical behavior are met. In the context of the microelectronics industry, the term yield phrases the ratio between the number of acceptable chips and total number of produced chips:

$$\text{yield} = \frac{\text{\# accetable chips}}{\text{\# manufactured chips}} \tag{1}$$

The acceptability of each chip is decreed by checking that the questioned electrical parameters individually fall into tolerance intervals. In addition, each wafer contains several sites with special test structures that enable further performance measurements in order to verify the manufacturing process. All the measurements are collected in a database which statistically characterizes the electrical behavior of the devices. As for the final product we may classify the integrated circuits into:

- acceptable chips, which satisfy all performance requests,
- functional failures, when malfunctions affect chips
- parametric failures, when chips fail to reach performances.

Coming to their manufacturing, we are used to distinguish three categories of failures that we synthesize through:

Random Yield (Sometimes Called Statistical Yield)

Concerning the random effects occurring during the manufacturing process, such as catastrophic faults in the form of open or short circuits. These faults may be a consequence of small particles in the atmosphere landing on the chip surface, no matter how clean is the wafer manufacturing environment. An example of a random component is that of threshold voltage variability due to random dopant fluctuations (Stolk et al., 1988);

Systematic Yield (Including Printability Issues)

Related to systematic manufacturability issues deriving from combinations and interactions of events that can be identified and addressed in a systematic way. An example of these events is the variation in wire thickness with layout density due to Chemical Mechanical Polishing/Planarization (CMP) (Chang et al., 1995). The distinction from the previous yield is important because the impact of systematic variability can be removed by adapting the design appropriately, while random variability will inevitably impact design margins in a negative manner;

Parametric Yield (Including Variability Issues)

Dealing with the performance drifts induced by changes in the parameter setting – for instance, lower drive capabilities, increased leakage current and greater power consumption, increased resistance and capacitance (RC) time constants, and slower chips deriving from corruptions of the transistor channels.

From a complementary perspective, the unacceptable performance causes for a circuit may be split into two categories of disturbances:

- local, caused by disruption of the crystalline structure of silicon, which typically determines the malfunctioning of a single chip in a silicon wafer;

- global, caused by inaccuracies during the production processes such as misalignment of masks, changes in temperature, changes in doses of implant. Unlike the local disturbance, the global one involves all chips in a wafer at different degrees and in different regions. The effect of this disturbance is usually the failure in the achievement of requested performances, in terms of working frequency decrease, increased power consumption, etc.

Both induce troubles on physical phenomena, such as electromagnetic coupling between elements, dissipation, dispersion, and the like. The obvious goal of the microelectronics factory is to maximize the yield as defined in (1). This translates, from an operational perspective, into a design target of properly sizing the circuit parameters, and a production target of controlling their realization. Actually both targets are very demanding since the involved parameters π are of two kinds:

- controllable, when they allow changes in the manufacturing phase, such as the oxidation times

- non controllable, in case they depend on physical parameters which cannot be changed during the design procedure, like the oxide growth coefficient.

Moreover, in any case the relationships between π and the parameters φ characterizing the circuit performances are very complex and difficult to invert. This induces researchers to model both classes of parameters as vectors of random variables, respectively Π and Φ [1]. The corresponding problem of yield maximization reverts into a functional dependency among the problem variables. Namely, let $\Phi = (\Phi_1, \Phi_2,..., \Phi_t)$ be the vector of the performances determined by the parameter vector $\Pi = (\Pi_1, \Pi_2,..., \Pi_n)$, and denote with D_Φ the acceptability region of a given chip. For instance, in the common case where each performance is checked singularly in a given range, i.e.:

$$\phi_k^l \leq \Phi_k \leq \phi_k^u \quad k = 1, \ldots, t \tag{2}$$

D_Φ reads:

$$D_\Phi = \left\{ \Phi | \phi_k^l \leq \Phi_k \leq \phi_k^u \quad k = 1, \ldots, t \right\} \tag{3}$$

The yield goal is the maximization of the probability P that a manufactured circuit has an acceptable performance, i.e.

$$\mathscr{P} = P\left[\Phi \in D_\Phi\right] = \int_{D_\Phi} f_\Phi(\phi) d\phi \tag{4}$$

where fΦ is the joint probability density of the performance Φ. To solve this problem we need to know fΦ and manage its dependence on Π. Namely, methodologies for maximizing the yield must incorporate tools that determine the region of acceptability, manipulate joint probabilities, evaluate multidimensional integrals, solve optimization problems. Those instruments that use explicit information about the joint probability and calculate the yield multidimensional integral (4) during the maximization process are called direct methods. The term indirect is therefore reserved for those methods that do not use this information directly. In the next section we will introduce two of these methods which look to be very promising when applied to real world benchmarks.

STATISTICAL MODELING

As mentioned in the introduction, a main way for maximizing yield passes through mating Design for Manufacturability with Design for Yield (DFM/DFY paradigm) along the entire manufacturing chain. Here we focus on model parameters at an intermediate location in this chain, representing a target of the production process and the root of the circuit performance. Their identification in correspondence to a performances' sample measured on produced circuits allows the designer to get a clear picture of how the latter react to the model parameters in the actual production process and, consequently, to grasp a guess on their variation impact. Typical model and performance parameters are described in Table 1 in Section 4. In a greater detail, the first requirement for planning circuits is the availability of a model relating input/output vectors of the function implemented by the circuit. As aforementioned, its achievement is usually split into two phases directed towards the search of a couple of analytic relations: the former between model parameters and circuit performances, and the latter, tied to the process engigneers' experience, linking both design and phisical circuit parameters as they could be obtained during production. Given a wafer, different repeated measurements are effected on dies in a same circuit family. As usual, the final aim is the model identification, in terms of designating the input (respectively output) parameter values of the aforementioned analytical relation. In some way, their identification hints at synthesizing the overall aspects of the manufacturing process not only to use them satisfactory during development yet to improve oncoming planning and design phases, rather than directly weigh on the production. For this purpose there are three different perspectives: synthesize simulated data, optimize a simulator, and statistically identify its optimal parameters. All three perspectives share the following common goals: ensure adequate manufacturing yield, reduce the production cost, predict design fails and product defects, and meet zero defects specification. We formalize the modeling problem in terms of a mapping g from a random vector $X = (X_1, ..., X_n)$, describing what is commonly denoted as model parameters 2 , to a random vector $Y = (Y_1, ..., Y_t)$, representing a meaningful subset of the performances Φ. The statistical features of X, such as mean, variance, correlation, etc., constitute its parameter vector θ_X, henceforth considered to be the statistical parameter of the input variable X. Namely, $Y = g(X) = (g_1(X),..., g_t(X))$, and we look for a vector θY that characterizes a performance population where $P(Y \in DY) = \alpha$, having denoted with DY the α-tolerance region, i.e. the domain spanned by the measured performances, and with α a satisfactory probability value. In turn, DY is the statistic we draw from a sample sy of the performances we actually measured on correctly working dies. Its simplest computation leads to a rectangular shape, as in (3), where we independently fix ranges on the singular performances. A more sophisticated instance is represented by the convex hull of the jointly observed performances in the overall Y space (Liu et al., 1999). At a preliminary stage, we often appreciate the suitability of θY by comparing first and second order moments of a performances' population generated through the currently identified parameters with those computed on s_y.

As a first requisite, we need a comfortable function relating the Y distribution to θ_X. The most common tool for modeling an analog circuit is represented by the Spice simulator (Kundert, 1998). It consists of a program which, having in input a textual description of the circuit elements (transistors, resistors, capacitors, etc.) and their connections, translates this description into nonlinear differential equations to be solved using implicit integration methods, Newton's method and sparse matrix techniques. A general drawback of Spice – and circuit simulators in general – is the complexity of the transfer function it implements to relate physical parameters to performances which hampers intensive exploration of the performance landscape in search of optimal parameters. The methods we propose in this section are mainly aimed at overtaking the difficulty of inverting this kind of functions, hence achieving a feasible solution to the problem: find a θ_X corresponding to the wanted θ_Y.

Monte Carlo Based Statistical Modeling

The lead idea of the former method we present is that the model parameters are the output of an optimization process aimed at satisfying some performance requirements. The optimization is carried out by wisely exploring the research space through a Monte Carlo (MC) method (Rubinstein & Kroese, 2007). As stated before, the proposed method uses the experimental statistics both as a target to be satisfied and, above all, as a selectivity factor for device model. In particular, a device model will be accepted only if it is characterized by parameters' values that allow to obtain, through electrical simulations, some performances which are included in the tolerance region.

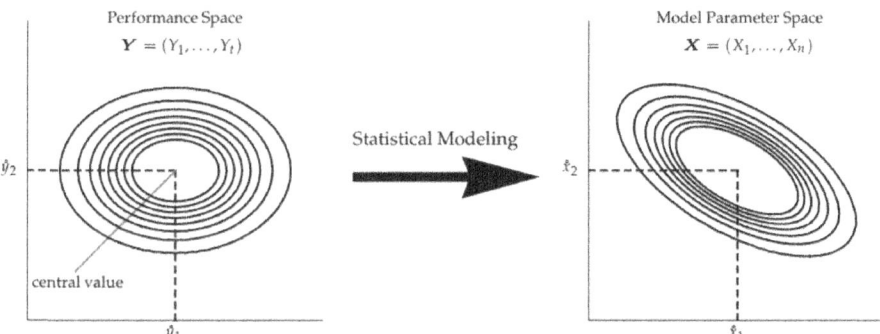

Figure. 1: Proposed flow: from the experimental statistics we determine a statistical Spice model for the device.

The aim of the proposed flow is the following: on the basis of the information which constitutes the experimental statistics, we want to map the space Y of the performances (such as gain and bandwidth) to the space X of circuit parameters (such as Spice parameters or circuit components values), as outlined in Fig. 1. Variations in the fabrication process cause random fluctuations in Y space, which in turn cause X to fluctuate (Koskinen & Cheung, 1993). In other words, we want to extract a Spice model whose parameters are random variables, each one characterized by a given probability distribution function. For instance, in agreement with the Central Limit Theorem (Rohatgi, 1976), we may work under usual Gaussianity assumptions. In this case, for the model parameters which have to be statistically described, it is necessary and sufficient to identify the mean values, standard deviations and correlation coefficients. In general, the flow of statistical modeling is based on several MC simulation steps (strictly related to bootstrap analysis (Efron & Tibshirani, 1993)), in order to estimate unknown features for each statistical model parameter. The method will proceed by executing iteratively the following steps, in the same way as in a multiobjective optimization algorithm, where the targets to be identified are the optimal parameters θX of the model. In the following procedure, general steps (described in roman font) will be specialized to the specific scenario (in italics) used to perform simulations in Section 4.

Step 1. Assume a typical (nominal) device model m0 is available, whose model parameters' means are described by the vector $°v_X$ (central values). Let $\hat{D}_Y$ be the corresponding typical tolerance region estimated on Y observations s_y. Choose an initial guess of X joint distribution function on the basis of moments estimated on given X observations sx. Let M denote the companion device statistical model, and set k = 0. In the specific case of hyper-rectangle tolerance regions defined as in (3), let $v°Y_j \pm 3\sigma°Y_j$, j = 1, . . . , t denote the two extremes delimiting each admissable performance interval. Moreover, since model parameters X of M follows a multivariate Gaussian distribution, assume (in the first iteration) a null cross-correlation between $\{X_1, ..., X_n\}$, hence $\theta_{Xi} = \{v_{Xi}, \sigma_{Xi}\}$, i = 1, . . . , n, where by default $v_{Xi} = v°X_i$, i.e. the same mean as the nominal model is chosen as initial value, and σ_{Xi} is assigned a relatively high value, for instance set equal to the double of the mean value.

Step 2. At the generic iteration k, an m-sized [3] sample $s_{Mk} = \{x_r\}$, r = 1..., m will be generated according to the actual X distribution.

In particular, when X_i are nomore independent, the discrete Karhunen-Loeve expansion (Johnson, 1994) is adopted for sampling, starting from the actual covariance matrix.

Step 3. For each model parameter x_r in s_{Mk}, the target performances y_r will be calculated through Spice circuit simulations.

Step 4. Only those model parameters in s_{Mk} reproducing performances lying within the chosen tolerance region $\hat{D}_Y$ will be accepted. On the basis of this criterion a subsample s_M of s_{Mk} having size m' ≤ m will be selected.

In particular, by keeping a fraction $1 - \delta$, say 0.99, of those models having all performance values included in $\hat{D}_Y$, we are guaranteeing a confidence region of level δ under i.i.d. Gaussianity assumptions.

Step 5. On the basis of the subsample $\tilde{s}_{\tilde{m}_{k'}}$ a new model $\mathcal{M}'_k$ will be computed through standard statistical techniques.

For each model parameter X_i , i = 1, . . . , n, the n standard deviations could be computed on the sample $\tilde{s}_{\tilde{m}}$ through Maximum Likelihood Estimators (MLE) (Mood et al., 1974), Spearman Rank-Order correlation coefficient (Lehmann, 2006; Press et al., 1993) may be used to estimate cross-correlation, while, according to circuit designers' report, the n means will be kept equal to the nominal $v°X_i$, i = 1, . . . , n.

Step 6. If the number $\tilde{m}$ of selected model parameters which have generated M' is sufficiently high (for instance they constitute a fraction 1 − δ, let's say 0.99, of the m instances, then the algorithm stops returning the statistical model M' . Otherwise, set k = k + 1 and goto Step 2.

The iterative procedure described above is based on Attractive Fixed Point method (Allgower & Georg, 1990), where the optimal value of those features to be estimated represents the fixed point of the algorithm. When the number of the components significantly increases, the convergence of the algorithm may become weak. To manage this issue, a two-step procedure is introduced where the former phase is aimed at computing moments involving single features X_i while maintaining constant their cross-correlation; the latter is directed toward the estimation of the cross-correlation between them. The overall procedure is analogous to the previous one, with the exception that cross-correlation terms will be kept fixed until Step 5 has been executed. Subsequently, a further optimization process will be performed to determine the cross-correlation coefficients, for instance using the Direct method as described in Jones et al. (1993). The stop criterion in Step 6 is further strengthen, prolonging the running of the procedure until the difference between cross-correlation vectors obtained at two subsequent iterations will drop below a given threshold.

Reverse Spice Based Statistical Modeling

A second way we propose to bypass the complexity handicap of Spice functions passes through a principled philosophy of considering the region DX where we expect to set the model parameters as an aggregate of fuzzy sets in various respects (Apolloni et al., 2008). First of all we locally interpolate the Spice function g through a polynomial, hence a mixture of monomials that we associate to the single fuzzy sets. Many studies show this interpolation to be feasible, even in the restricted form of using posynomials, i.e. linear combination of monomials through only positive coefficients (Eeckelaert et al., 2004). The granular construct we formalize is the following

Given a Spice function g mapping from x to y (the generic component of the performance vector y), we assume the domain $DX \subseteq R^n$ into which x ranges to be the support of c fuzzy sets $\{A_1, ..., A_c\}$, each pivoting around a monomial mk . We consider this monomial to be a local interpolator that fits g well in a surrounding of the A_k centroid. In synthesis, we have $g(x) \simeq \sum_{k=1}^{c} \mu_k(x) m_k(x)$, where $\mu_k(x)$ is the membership degree of x to A_k , whose value is in turn computed as a function of the quadratic shift $(g(x) - m_k(x))^2$.

On the one hand we have one fuzzy partition of D_x for each component of y. On the other hand, we implement the construct with many simplifications, in order to meet specific goals. Namely:

- since we look for a polynomial interpolation of g, we move from point membership functions to sets, to a monomial membership function to g, so that $g(x) \simeq \sum_{k=1}^{c} \mu_k m_k(x)$. In turn, μ_k is a sui generis membership degree, since it may assume also negative values;

- since for interpolation purposes we do not need $\mu_k(x)$, we identify the centroids directly with a hard clustering method based on the same quadratic shift.

Denoting $m_k(x) = \beta_k \prod_{j=1}^{n} x_j^{a_{kj}}$, if we work in logarithmic scales, the shifts we consider for the single (say the i-th) component of y are the distances between $z_r = (\log x_r, \log y_r)$ and the hyperplane $h_k(z) = w_k \cdot z + b_k = 0$, with $w_k = \{\alpha_{k1}, ..., \alpha_{kn}\}$ and $b_k = \log \beta_k$, constituting the centroid of Ak in an adaptive metric. Indeed, both wk and bk are learnt by the clustering algorithm aimed at minimizing the sum of the distances of the zrs from the hyperplanes associated to the clusters they are assigned to.

With the clustering procedure we essentially learn the exponents α_{kj} through which the x components intervene in the various monomials, whereas the β_k
s remain ancillary parameters. Indeed, to get the polynomial approximation of g(x) we compute the mentioned sui generis memberships through a simple quadratic fitting, i.e. by solving w.r.t. the vector $\mu = \{\mu_1, ..., \mu_c\}$ the quadratic optimization problem: $\mu = \arg\min_{\tilde{\mu}} \sum_{r=1}^{m} (g(x_r) - y_r)^2$, where xrj denotes the j-th component of the r-th element of the training set sx, yrj its approximation, with

$$y_j = \sum_{k=1}^{c} m_{jk}(x) = \sum_{k=1}^{c} \mu_{jk} \prod_{i=1}^{n} x_i^{\alpha_{jki}}$$

(5)

where the index r has been hidden for notational simplicity, and $\mu_{k\,s}$ override $\beta_{k\,s}$.

A Suited Interpretation of the Moment Method

An early solution of the inverse problem: Which statistical features of X ensure a good coverage (in terms of α-tolerance regions) of the Y domain spanned by the performances measured on a sample of produced dies? relies on the first and second moments of the target distribution, which are estimated on the basis of a sample sy of sole Y collected from the production lines as representatives of properly functioning circuits. Our goal is to identify the statistical parameters $\tilde{\theta}_X$ of X that produce through (5) a Y population best approximating the above first and second order moments. X is assumed to be a multidimensional Gaussian variable, so that we identify it completely through the mean vector vX and the covariance matrix ΣX which we do not constrain in principle to be diagonal (Eshbaugh, 1992). The analogous vY and ΣY are a function of the former through (5). Although they could not identify the Y distribution in full, we are conventionally satisfied when these functions get numerically close to the estimates of the parameters they compute (directly obtained from the observed performance sample). Denoting with $v_{X_j}, \sigma_{X_j}, \sigma_{X_{j,k}}$ and $\rho_{X_{j,k'}}$, respectively, the mean and standard deviation of X_j and the covariance/correlation between X_j and X_k, the master equations of our method are the following:

$$v_{Y_i} = \sum_{k=1}^{c} a_{ikj} v_{M_{ik}}$$

(6)

where M_{ik} on the right is a short notation of $m_{ik}(X)$, and vM_{ik} denotes its mean

Thanks to the approximations

$$v_\Xi \simeq \log v_X, \quad \sigma_\Xi \simeq \sigma_X / v_X, \quad \rho_{\Xi_{i,j}} \simeq \rho_{X_{i,j}}$$

(7)

with $\Xi = \log X$, coming from the Taylor expansion of respectively $\Xi, (\Xi - v_\Xi)^2$ and $(\Xi_i - v_{\Xi_i})(\Xi_j - v_{\Xi_j})$ around (v_{X_i}, v_{X_j}) disregarding others than the second terms, the rewriting of Σ_Y reads

$$\sigma_{Y_i}^2 = \sum_{k=1}^{c} \sigma_{M_{ik}}^2 + 2 \sum_{\substack{k,r=1 \\ k<r}}^{c} \sigma_{M_{ik,ir}}$$

(8)

$$\sigma_{Y_{i,j}} = \sum_{k,r=1}^{c} \sigma_{M_{ik,ir}}$$

(9)

with

$$\sigma_{M_{ik}}^2 \simeq v_{M_{ik}}^2 \left(\sum_{j=1}^{n} a_{ikj}^2 \frac{\sigma_{X_j}^2}{v_{X_j}^2} + 2 \sum_{\substack{j,r=1 \\ j<r}}^{n} \rho_{X_{j,r}} a_{ikj} a_{ikr} \frac{\sigma_{X_j}}{v_{X_j}} \frac{\sigma_{X_r}}{\mu_{X_r}} \right)$$

(11)

$$\sigma_{M_{ik,ir}} \simeq v_{M_{ik}} v_{M_{ir}} \sum_{j,w=1}^{n} a_{ikj} a_{irw} \rho_{X_{j,w}} \frac{\sigma_{X_j}}{v_{X_j}} \frac{\sigma_{X_w}}{v_{X_w}}$$

We numerically solve (6) and (8-9) in v_X and Σ_X when the left members coincide with the target values of v_Y and Σ_Y, respectively, and $v_{M_{ik}}$ is approximated with its sample estimate computed on samples artificially generated with the current values of the parameters. Solving equations means minimizing the differences between left and right members, so that the crucial point is the optimization method employed. The building blocks are the following. The steepest descent strategy. Using the Taylor series expansion limited to second order (Mood et al., 1974), we obtain an approximate expression of the gradient components of v_Y w.r.t. v_X through

$$\frac{\partial v_{Y_i}}{\partial v_{X_j}} \simeq \sum_{k=1}^{c} \alpha_{ikj} \left(\frac{1}{v_{X_j}} + \frac{\sigma_{X_j}^2}{v_{X_j}^3} \right) v_{M_{ik}}$$

(12)

Thus we may easily look for the incremental descent on the quadratic error surface accounting for the difference between computed and observed means. Expression (12) confirms the scarce sensitivity of the unbiased mean v_X, and its gradient as well, to the second moments, so that we may expect to obtain an early approximation of the mean vector to be subsequently refined. While analogous to the previous task, the identification of X variances and correlations owns one additional benefit and one additional drawback. The former derives from the fact that we may start with a, possibly well accurate, estimate of the means. The latter descends from the high interrelations among the target parameters which render the exploration of the quadratic error landscape troublesome and very lengthy. Identification of second order moments. An alternative strategy for X second moment identification is represented by the evolutionary computation. Given the mentioned computational length of the gradient descent procedures, algorithms of this family become competitive on our target. Namely, we used Differential Evolution (Price et al., 2005), with specific bounds on the correlation values to avoid degenerate solutions. A brute force numerical variant. We may move to a still more rudimentary strategy to get rid of the loose approximations introduced in (6) to (12). Thus we: i) avoid computing approximate analytical derivatives, by substituting them with direct numerical computations (Duch & Kordos, 2003), and ii) adopt the strategy of exploring one component at a time of the questioned parameter vector, rather than a combination of them all, until the error descent stops. Spanning numerically one direction at a time allows us to ask the software to directly identify the minimum along this direction. The further benefit of this task is that the function we want to minimize is analytic, so that the search for the minimum along one single direction is a very easy task for typical optimizers, such as the naive Nelder-Mead simplex method (Nelder & Mean, 1965) implemented in Mathematica (Wolfram Research Inc., 2008). We structured the method in a cyclic way, plus stopping criterion based on the amount of parameter variation. Each cycle is composed of: i) an iterative algorithm which circularly visits each component direction minimizing the error in the means' identification, until no improvement may be achieved over a given threshold, and ii) a fitting polynomial refresh on the basis of a Spice sample in the neighborhood of the current mean vector. We conclude the routine with a last assessment of the parameters that we pursue by running jointly on all them a local descent method such as Quasi-Newton procedure in one of its many variants (Nocedal & Wright, 1999).

Fine Tuning Via Reverse Mapping

Once a good fitting has been realized in the questioned part of the Spice mapping, we may solve the identification problem in a more direct way by first inverting the polynomial mapping to obtain the X sample at the root of the observed Y sample, and then estimating θ_X directly from the sample defined in the D_X domain. The inversion is almost immediate if it is univocal, i.e., apart from controllable pathologies, when X and Y have the same number of components. Otherwise the problem is either overconstrained (number n of X components less than t, dimensionality of Y components) or underconstrained (opposite relation between component numbers). The first case is avoided by simply discarding exceeding Y components, possibly retaining the ones that improve the final accuracy and avoid numeric instability. The latter calls for a reduction in the number of questioned X components. Since X follows a multivariate Gaussian distribution law, by assumption, we may substitute some components with their conditional values, given the others.

NUMERICAL EXPERIMENTS

The procedures we propose derive from a wise implementation of the Monte Carlo methods, as for the former, and a skillful implementation of granular computing ideas (Apolloni et al.,

Table 1: Model parameters and performances of the identification problems

device	model parameter		performance parameter	
	label	description	label	description
pMOS	U_0	Mobility at nominal temperature		
	A_0	Bulk charge effect coefficient	GM	conductance
	VTH_0	Threshold voltage at $V_{BS} = 0$ for large L	ID_{SAT}	source drain current
	K_1	First order body effect coefficient	VTH_{25-25}	saturation voltage
	B_{01}	Bulk charge effect coefficient for channel lenght	VTH_{25-08}	saturation voltage
	B_{11}	Bulk charge effect coefficient for channel width		
nMOS	U_0	Mobility at nominal temperature	GM	conductance
	V_{SAT}	Saturation voltage	ID_{SAT}	source drain current
	VTH_0	Threshold voltage at $V_{BS} = 0$ for large L	VTH_{25-25}	saturation voltage
	K_1	First order body effect coefficient	VTH_{25-08}	saturation voltage
NPN-DIB12	Bf	Ideal maximum foward Beta	HFE	Current Gain
	Re	Emitter Resistance	VA	Early Voltage
	Is	Transport Saturation Current	I_c	Collector Current
	Vaf	Forward Early Voltage		

2008), as for the latter, however without theoretical proof of efficiency. While no worse from this perspective than the general literature in the field per se (McConaghy & Gielen, 2005), it needs numerical proof of suitability. To this aim we basically work with three real world benchmarks collected by manufacturers to stress the peculiarities of the methods. Namely, the benchmarks refer to:

- A unipolar pMOS device realized in Hcmos4TZ technology

- A unipolar nMOS device differentiating from the former for the sign (negative here, positive there) of the charge of the majority mobile charge carriers. Spice model and technology are the same, and performance parameters as well. However, the domain spanned by the model parameters is quite different, as will be discussed shortly

- A bipolar NPN circuit realized in DIB12 technology. DIB technology achieves the full dielectric isolation of devices using SOI substrates by the integration of the dielectric trench that comes into contact with the buried oxide layer.

The related model parameter took into consideration and measured performances are reported in Table 1. We have different kinds of samples for the various benchmarks as for both the sample size which ranges from 14, 000 (pMOS and nMOS) to 300 (NPN-DIB12) and the measures they report: joint measures of 4 performance parameters in the former two cases, partially independent measures of 3 performance parameters in the latter, where only HFE and VA are jointly measured. Taking into account the model parameters, and recalling the meaning of t and n in terms of number of performance and model parameters, respectively, the sensitivity of the former parameters to the latter and the different difficulties of the identification tasks lead us to face in principle one balanced problem with n = t = 4 (nMOS), and two unbalanced ones with n = 6 and t = 4 (pMOS) and n = 4 and t = 3 (NPN-DIB12). In addition, only 4 of the 6 second order moments are observed with the third benchmark.

Reverting the Spice Model on the Three Benchmarks

With reference to Table 2, in column $\tilde{\theta}_X$ we report the parameters of the input multivariate Gaussian distribution we identify in the aim of reproducing the θY of the Y population observed through s_y. Of the latter parameter, in the subsequent column $\tilde{\theta}_Y / \hat{\theta}_Y$ we compare

Table 2: Benchmarks used for testing the proposed procedure and analysis of the identification solution Rows: benchmarks. Columns: inferred model distribution parameters (indexed by X) and reconstructed performance parameters (indexed by Y), plus comparative levels of the tolerance regions (as a function of δ).

benchmark			solution						
dataset	(n,t)	m	θ_X			$\theta_Y / \hat\theta_Y$			$1 - \delta /$
benchmark			μ_X	σ_X	ρ_X	μ_Y	σ_Y	ρ_Y	$1 - \delta$
pMOS	(6,4)	14,000	$\begin{pmatrix} 233.424 \\ 0.28798 \\ 0.99185 \\ 0.45255 \\ 4.0662\text{e} \times 10^{-5} \\ 4.67824 \times 10^{-5} \end{pmatrix}$	$\begin{pmatrix} 3.63673 \\ 0.01806 \\ 0.01083 \\ 0.03275 \\ 4.8810\text{e} \times 10^{-6} \\ 9.9003\text{e} \times 10^{-6} \end{pmatrix}$	$\begin{pmatrix} -0.16982 \\ -0.46312 \\ -0.41451 \\ -0.49665 \\ -0.35008 \\ -0.12573 \\ -0.47067 \\ -0.07086 \\ -0.39530 \\ 0.09484 \\ -0.16367 \\ 0.21068 \\ 0.49711 \\ 0.22781 \\ 0.48312 \end{pmatrix}$	$\begin{pmatrix} -0.835824 \\ -0.838496 \\ -0.971835 \\ -0.969196 \\ 0.009973318 \\ 0.00097472 \\ 0.00448103 \\ 0.00447346 \end{pmatrix}$	$\begin{pmatrix} 0.0118109 \\ 0.0187507 \\ 0.0122665 \\ 0.0164674 \\ 0.000029378 \\ 0.000029348 \\ 0.0001866\text{2b} \\ 0.000013048\text{b} \end{pmatrix}$	$\begin{pmatrix} 0.933746 \\ 0.451486 \\ -0.287658 \\ -0.282512 \\ -0.389979 \\ -0.387441 \\ -0.254446 \\ -0.0727698 \\ -0.367477 \\ -0.174543 \\ 0.900391 \\ 0.983658 \end{pmatrix}$	$\begin{matrix} 0.946713 \\ 0.9 \\ 0.900393 \\ 0.8 \end{matrix}$
nMOS	(4,4)	14,000	$\begin{pmatrix} 752.395 \\ 152858.0 \\ 0.68184 \\ 0.521661 \end{pmatrix}$	$\begin{pmatrix} 134.099 \\ 9667.22 \\ 0.0186654 \\ 0.131933 \end{pmatrix}$	$\begin{pmatrix} -0.765278 \\ -0.462972 \\ 0.756786 \\ 0.306389 \\ -0.786377 \\ -0.468842 \end{pmatrix}$	$\begin{pmatrix} 0.552391 \\ 0.550715 \\ 0.66383 \\ 0.664162 \\ 0.00221691 \\ 0.00222077 \\ 0.0100527 \\ 0.0100711 \end{pmatrix}$	$\begin{pmatrix} 0.028568 \\ 0.0276768 \\ 0.0176982 \\ 0.0173677 \\ 0.000030626 \\ 0.0000619134 \\ 0.000355129 \\ 0.000280373 \end{pmatrix}$	$\begin{pmatrix} 0.445091 \\ 0.395429 \\ -0.499279 \\ -0.432434 \\ -0.637963 \\ -0.640323 \\ -0.299401 \\ -0.271932 \\ -0.375841 \\ -0.354887 \\ 0.92015 \\ 0.950419 \end{pmatrix}$	$\begin{matrix} 0.9008 \\ 0.9 \\ 0.8304 \\ 0.8 \end{matrix}$
NPN-DIB12	(4,3)	322	$\begin{pmatrix} 138.302 \\ 0.67258 \\ 5.28102 \times 10^{-18} \\ 136.319 \end{pmatrix}$	$\begin{pmatrix} 8.3859 \\ 0.263238 \\ 4.14306 \times 10^{-19} \\ 13.6538 \end{pmatrix}$	$\begin{pmatrix} -0.192107 \\ 0.00139749 \\ -0.477207 \\ -0.980327 \\ 0.167527 \\ -0.0444712 \end{pmatrix}$	$\begin{pmatrix} 113.244 \\ 113.242 \\ 0.0000654246 \\ 0.0000653275 \\ 110.164 \\ 110.238 \end{pmatrix}$	$\begin{pmatrix} 6.82099 \\ 6.95918 \\ 4.96031 \times 10^{-6} \\ 4.81021 \times 10^{-6} \\ 11.1459 \\ 11.2166 \end{pmatrix}$	$\begin{pmatrix} -0.490798 \\ -0.566678 \end{pmatrix}$	$\begin{matrix} 0.9654 \\ 0.9 \\ 0.8136 \\ 0.8 \end{matrix}$

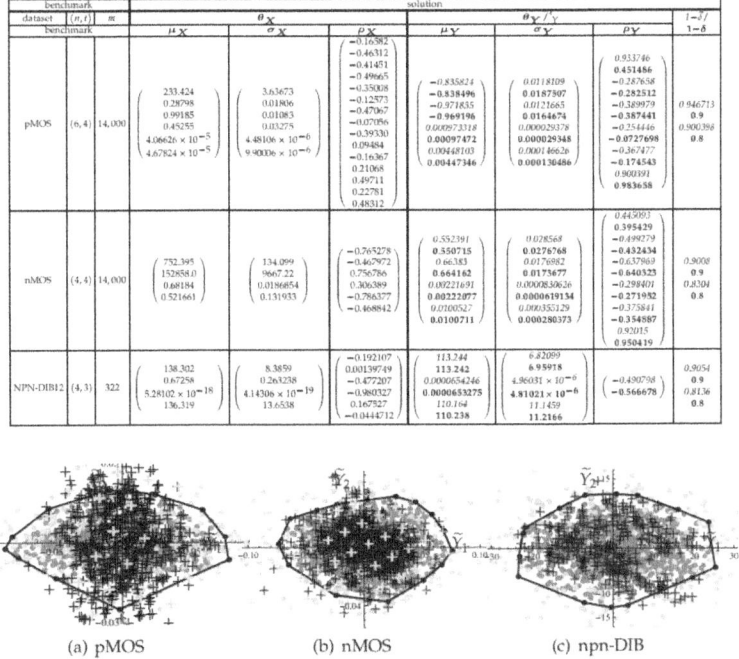

(a) pMOS　　　　　(b) nMOS　　　　　(c) npn-DIB

Figure. 2: Comparison between output data and reconstruction provided by Reverse Spice based procedure for the devices listed in Table 2 when projected on the two principal components of the target. Points: reconstructed population lying within (dark gray) and outside (light gray) 0.90 tolerance region (black curves) identified by black points. Gray crosses: original target output; black crosses: target output uniformly spread with noise terms.

The values computed on the basis of $\tilde{\theta}x$ (referring to a reconstructed distribution – in italics) with those computed through the maximum likelihood estimate from s_y (referring to the original distribution – in bold). As a further accuracy indicator, we will consider tolerance regions obtained through convex hull peeling depth (Barnett, 1976) containing a given percentage $1 - \delta$ of the performance population. In the last column of Table 2, headed by $(1 - \tilde{\delta})/(1 - \delta)$, , we appreciate the difference between planned tolerance rate (in bold), as a function of the identified Y distribution, and ratio of sampled measures found in these regions (in italics). We consider single values in the table cells since the results are substantially insensitive to the random components affecting the procedure, such as algorithm initialization. Rather, especially with difficult benchmarks, they may depend on the user options during the run of the algorithm. Thus, what we report are the best results we obtain, reckoning the overall trial time in the computational complexity consideration we will do later on in this section.

For a graphical counterpart, in Fig. 2 we report the scatterplot of the original Y sample and an analogous one generated through the reconstructed distribution, both projected on the plane identified by the two principal components (Jolliffe, 1986) of the original distribution. We also draw the intercept of this plane with a tolerance region containing 90% of the reconstructed points (hence $\delta = 0.1$). An overview of these data looks very satisfactory, registering a relative shift between sample and identified parameters that is always less than 0.17% as for the mean values, 45% for the standard deviations and 25% for the correlation. The analogous shift between planned and actual percentages of points inside the tolerance region is always less than 2%. We distinguish between difficult and easy benchmarks, where the pMOS sample falls in the first category. Indeed the same percentages referring to the remaining benchmarks decreases to 0.13%, 10% and 9%. Given the high computational costs of the Spice models, their approximation through cheaper functions is the first step in many numerical procedures on microelectronic circuits. Within the vast set of methods proposed by researchers on the matter (Ampazis & Perantonis, 2002a;b; Daems et al., 2003; Friedman, 1991; Hatami et al., 2004; Hershenson et al., 2001; McConaghy et al., 2009; Taher et al., 2005; Vancorenland et al., 2001) in Table 3 we report a numerical comparison between two well reputed fitting methods and our proposed Reverse Spice based algorithm (for short RS). The methods are Multivariate Adaptive Regression Splines (MARS) (Friedman, 1991), i.e. piecewise polynomials, and Polynomial Neural Networks

Table 3: Performance comparison between fitting algorithms Rows: algorithms; main columns: benchmark parameterization; subcolumns: experimental environments (training set, test set)

| | $\tilde{\theta}_x$ | | $\tilde{\theta}'_x$ | |
	train	test	train	test
RS	0.0000125623	0.0000242739	0.000228931	0.000369871
	$\begin{pmatrix} 0.0000350975 \\ 0.0000151476 \\ 3.06034 \times 10^{-10} \\ 3.59774 \times 10^{-9} \end{pmatrix}$	$\begin{pmatrix} 0.0000759397 \\ 0.0000211444 \\ 6.62265 \times 10^{-10} \\ 1.10138 \times 10^{-8} \end{pmatrix}$	$\begin{pmatrix} 0.000751481 \\ 0.000164105 \\ 1.54286 \times 10^{-8} \\ 1.24052 \times 10^{-7} \end{pmatrix}$	$\begin{pmatrix} 0.00131925 \\ 0.000159924 \\ 2.33858 \times 10^{-8} \\ 2.92353 \times 10^{-7} \end{pmatrix}$
MARS	$8.68173 * 10^{-6}$	0.0000168024	0.000124012	0.0002805
	$\begin{pmatrix} 0.0000246876 \\ 0.0000100344 \\ 2.80773 \times 10^{-10} \\ 4.66935 \times 10^{-9} \end{pmatrix}$	$\begin{pmatrix} 0.0000528055 \\ 0.0000143915 \\ 5.92204 \times 10^{-10} \\ 1.19291 \times 10^{-8} \end{pmatrix}$	$\begin{pmatrix} 0.000401349 \\ 0.0000946271 \\ 5.3722 \times 10^{-9} \\ 6.47147 \times 10^{-8} \end{pmatrix}$	$\begin{pmatrix} 0.00100927 \\ 0.000112503 \\ 6.07291 \times 10^{-9} \\ 2.22601 \times 10^{-7} \end{pmatrix}$
PNN	0.0000602061	0.0000769737	0.000125976	0.000280898
	$\begin{pmatrix} 0.000230822 \\ 0.0000100003 \\ 2.7761 \times 10^{-10} \\ 2.38434 \times 10^{-9} \end{pmatrix}$	$\begin{pmatrix} 0.000293665 \\ 0.0000142199 \\ 5.70282 \times 10^{-10} \\ 9.12621 \times 10^{-9} \end{pmatrix}$	$\begin{pmatrix} 0.000409046 \\ 0.0000948249 \\ 4.14671 \times 10^{-9} \\ 2.84136 \times 10^{-8} \end{pmatrix}$	$\begin{pmatrix} 0.00101197 \\ 0.000111354 \\ 7.14833 \times 10^{-9} \\ 2.62591 \times 10^{-7} \end{pmatrix}$

(PNN) (Elder IV & Brown, 2000). Namely, we consider the $\tilde{\theta}_x$ reported in Table 2 as the result of the nMOS circuit identification. On the basis of these parameters and through Spice functions, we draw a sample of 250 pairs (xr, yr) that we used to feed both competitor algorithms and our own. In detail we used VariReg software (Jekabsons, 2010a;b) to implement both MARS and PNN. To ensure a fair comparison among the differente methods, we: i) set equal to 6 the number of monomials in our algorithm and the maximum number of basis functions in MARS, where we used a cubic interpolation, and ii) employ the default configuration in PNN by setting the degree of single neurons polynomial equal to 2. Moreover, in order to understand how the various algorithms scale with the fitting domain, we repeat the procedure with a second set $\tilde{\theta}'_x$ of parameters, where the original standard deviations have been uniformly doubled. In the table we report the mean squared errors measured on a test set of size 1000, whose values are both split on the four components of the performance vector and resumed by their average. The comparison denotes similar accuracies with the most concentrated sample – the actual operational domain of our polynomials – and a small deterioration of our accuracy in the most dispersed sample, as a necessary price we have to pay for the simplicity of our fitting function. As for the whole procedure, we reckon overall running times of around half an hour. Though not easily contrastable with computational costs of analogous tasks, this order of magnitude results adequate for an intensive use of the procedure in a circuit design framework.

Stochastically Optimizing the Third Benchmark Model

The same NPN-DIB12 benchmark discussed in Section 4.1 was also used to run the two-step MC procedure depicted in Section 3.1. In particular, estimation of the sole standard deviations σ_{X_i} s in the former phase alternates with cross-correlation coefficients' in the latter, while the means remain fixed to their nominal values $v_{X_i} = v°_{X_i}$ Namely, at each iteration a sample $sM = \{x_r\}, r = 1..., m = 5000$ was generated, and the whole procedure was repeated 7 times, until over 99% of sample instances were included in the tolerance region. Fig. 3 shows the number $\tilde{m}$ of selected instances for each iteration of the algorithm.

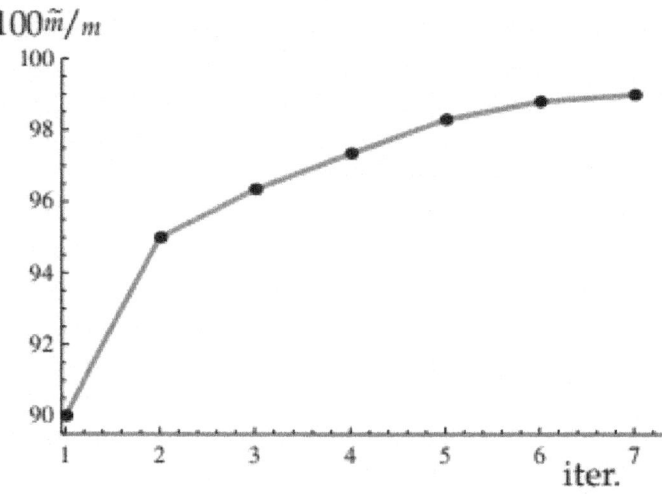

Figure. 3: Percentage of selected instances at each iteration of the two-step MC algorithm.

Comparing the Proposed Methods

In order to grasp insights on the comparative performances of the proposed methods, we list their main features on the common NPN-DIB12 benchmark. Namely, in the first row of Table 4 we report the reference value of the means and standard deviations of both X and Y distributions. As for the first variable, we rely on the nominal values of the parameters for the

Table 4: Comparison between both model and performance moments re reference and reconstructed frameworks

	$\tilde{\theta}_X$		$\tilde{\theta}_Y$	
	μ_X	σ_X	μ_Y	σ_Y
Reference	$\begin{pmatrix} 135 \\ 0.8 \\ 5.12 \times 10^{-18} \\ 138 \end{pmatrix}$		$\begin{pmatrix} 113.242 \\ 6.5328 \times 10^{-5} \\ 110.238 \end{pmatrix}$	$\begin{pmatrix} 6.9592 \\ 4.8102 \times 10^{-6} \\ 11.2166 \end{pmatrix}$
MC	$\begin{pmatrix} 135 \\ 0.8 \\ 5.12 \times 10^{-18} \\ 138 \end{pmatrix}$	$\begin{pmatrix} 8.2375 \\ 7.9064 \times 10^{-2} \\ 3.9744 \times 10^{-19} \\ 9.4 \end{pmatrix}$	$\begin{pmatrix} 110.5854 \\ 6.346 \times 10^{-5} \\ 110.039 \end{pmatrix}$	$\begin{pmatrix} 6.6418 \\ 4.691 \times 10^{-6} \\ 7.507 \end{pmatrix}$
RS	$\begin{pmatrix} 138.302 \\ 0.6726 \\ 5.281 \times 10^{-18} \\ 136.319 \end{pmatrix}$	$\begin{pmatrix} 8.3859 \\ 0.2632 \\ 4.1431 \times 10^{-19} \\ 13.6538 \end{pmatrix}$	$\begin{pmatrix} 113.244 \\ 6.5425 \times 10^{-5} \\ 110.164 \end{pmatrix}$	$\begin{pmatrix} 6.821 \\ 4.9603 \times 10^{-6} \\ 11.1459 \end{pmatrix}$

means, leaving empty the cell concerning the standard deviations. As for the performances, we just use the moment MLE estimate computed on the sample s_y. In the remaining rows we report the analogous values computed from a huge sample of the above variables artificially generated through the statistical models we identify. Both tables denote a slight comparative benefit of using the reverse modeling (row RS), in terms of both a greater variance of the model parameters and a better similarity of the reconstructed performance parameters with the estimated ones w.r.t. the analogous parameters obtained with Monte Carlo method (row MC). The former feature reflects into less severe constraints in the production process. The latter denotes some improvement in the reconstruction of the performances' distribution law, possibly deriving from both freeing the v_X from their nominal values and a massive use of the Spice function analytical forms.

CONCLUSIONS

A major challenge posed by new deep-submicron technologies is to design and verify integrated circuits to obtain a high fabrication yield, i.e. a high proportion of produced circuits that function properly. The classical approach implemented in commercial tools for parameter extraction (IC-Cap by Agilent Technology (2010), and UTMOST by Silvaco Engineered (2010)) requires a dedicated electrical characterization for a large number of devices, in turn demanding for a very long time in terms both of experimental characterization and parameter extraction. Thus, a relevant goal with these procedures is to reduce the computational time to have a statistical description of the device model. We fill it by using two non conventional methods so as to get a speed-up factor greater than 10 w.r.t. standard procedures in literature. The first method we propose is based on a Monte Carlo technique to estimate the (second order) moments for several statistical model parameters, on the basis of characterizated data, collected during the manufacturing process. The second method exploits a granular construct. In spite of the methodology broadness the attribute granular may evoke, we obtain a very accurate solution taking advantage from strict exploitation of state-of-the-art theoretical results. Starting from the basic idea of considering the Spice function as a mixture of fuzzy sets, we enriched its implementation with a series of sophisticated methodologies for: i) identifying clusters based on proper metrics on functional spaces, ii) descending, direction by direction, along the ravines of the cost functions of the related optimization problems, iii) inverting the (X,Y) mapping in case of unbalanced problems through the bootstrapping of conditional Gaussian distributions, and iv) computing tolerance regions through convex hull based peeling techniques. In this way we supply a very accurate and fast algorithm to identify statistically the circuit model. Of course, both procedures are susceptible of further improvements deriving from a more and more deep statistics' exploitation. In addition, nobody may guarantee that they will resist to a further reduction of the technology scales. However the underlying methods we propose could remain at the root of new solution algorithms of the yield maximization problem.

REFERENCES

1.   Agilent Technology (2010). IC-CAP Device Modeling Software – Measurement Control and Parameter Extraction, Santa Clara, CA.URL: http://www.home.agilent.com/agilent/home.jspx

2.   Allgower, E. L. & Georg, K. (1990). Computational solution of nonlinear systems of equations, American Mathematical Society, Providence, RI.

3.   Ampazis, N. & Perantonis, S. J. (2002a). OLMAM Neural Network toolbox for Matlab. URL: http://iit.demokritos.gr/ abazis/toolbox/ Ampazis, N. & Perantonis, S. J. (2002b). Two highly efficient second order algorithms for training feedforward networks, IEEE Transactions on Neural Networks 13(5): 1064–1074.

4.   Apolloni, B., Bassis, S., Malchiodi, D. & Witold, P. (2008). The Puzzle of Granular Computing,Vol. 138 of Studies in Computational Intelligence, Springer Verlag. Barnett, V. (1976). The ordering of multivariate data, Journal of Royal Statistical Society Series A 139: 319–354.

5.   Bernstein, K., Frank, D. J., Gattiker, A. E., Haensch, W., Ji, B. L.,

Nassif, S. R., Nowak, E. J., Pearson, D. J. & Rohrer, N. J. (2006). High-performance CMOS variability in the 65-nm regime and beyond, IBM Journal of Research Development 50(4/5): 433–449.

6. Boning, D. S. & Nassif, S. (1999). Models of process variations in device and interconnect, in A. Chandrakasan (ed.), Design of High Performance Microprocessor Circuits, chapter 6, IEEE Press.

7. Bühler, M., Koehl, J., Bickford, J., Hibbeler, J., Schlichtmann, U., Sommer, R., Pronath, M. & Ripp, A. (2006). DFM/DFY design for manufacturability and yield – influence of process variations in digital, analog and mixed-signal circuit design, DATE'06,pp. 387–392.

8. Chang, E., Stine, B., Maung, T., Divecha, R., Boning, D., Chung, J., Chang, K., Ray, G., Bradbury, D., Nakagawa, O. S., Oh, S. & Bartelink, D. (1995). Using a statistical metrology framework to identify systematic and random sources of dieand wafer-level ILD thickness variation in CMP processes, in CMP processes, IEDM Technology Digest, pp. 499–502.

9. Daems, S., Gielen, G. & Sansen, W. (2003). Simulation-based generation of posynomial performance models for the sizing of analog integrated circuits, IEEE Transactions on Computer-Aided Design of Integrated Circuits and Systems 22(5): 517–534.

10. Duch, W. & Kordos, M. (2003). Multilayer perceptron trained with numerical gradient, Proceedings of the International Conference on Artificial Neural Networks (ICANN) and International Conference on Neural Information Processing (ICONIP), Istanbul, pp. 106–109.

11. Eeckelaert, T., Daems, W., Gielen, G. & Sansen, W. (2004). Generalized simulation-based posynomial model generation for analog integrated circuits, Analog Integrated Circuits Signal Processing 40(3): 193–203.

12. Efron, B. & Tibshirani, R. J. (1993). An Introduction to the Bootstrap, Chapman & Hall, New York.

13. Elder IV, J. F. & Brown, D. E. (2000). Induction and polynomial networks. network models for control and processing, in M. Fraser (ed.), Intellect, Portland, OR, pp. 143–198.

14. Eshbaugh, K. S. (1992). Generation of correlated parameters for statistical circuit simulation, IEEE Transactions on CAD of Integrated Circuits and Systems 11(10): 1198–1206.

15. Friedman, J. H. (1991). Multivariate Adaptive Regression Splines, Annals of Statistics 19: 1–141. Hatami, S., Azizi, M. Y., Bahrami, H. R., Motavalizadeh, D. & Afzali-Kusha, A. (2004).

16. Accurate and efficient modeling of SOI MOSFET with technology independentneural networks, IEEE Transactions on Computer-Aided Design of Integrated Circuits and Systems 23(11): 1580–1587.

17. Hershenson, M., Boyd, S. & Lee, T. (2001). Optimal design of a CMOS OP-AMP via geometric programming, IEEE Trans. on Computer-Aided Design of Integrated Circuits and Systems 20(1): 1–21.

18. Jekabsons, G. (2010a). Adaptive basis function construction: an approach for adaptive building of sparse polynomial regression models, Machine Learning, In-Tech p. 28. In Press.Jekabsons, G. (2010b). VariReg software.URL: http://www.cs.rtu.lv/jekabsons/

19. Johnson, G. E. (1994). Constructions of particular random processes, Proceedings of the IEEE 82(2): 270–285.

20. Jolliffe, I. T. (1986). Principal Component Analysis, Springer Verlag.

21. Jones, D. R., Perttunen, C. D. & Stuckman, B. E. (1993). Lipschitzian optimization without the Lipschitz constant, Journal of Optimization Theory and Applications 79(1): 157–181.

22. Koskinen, T. & Cheung, P. (1993). Statistical and behavioural modelling of analogue integrated circuits, Circuits, Devices and Systems, IEE Proceedings G 140(3): 171–176.

23. Kundert, K. S. (1998). The DesignerâA˘Zs Guide to SPICE and SPECTRE ´, Kluwer Academic Publishers, Boston.

24. Lehmann, E. (2006). Nonparametrics, Statistical Methods Based on Ranks, Vol. XVI, Prentice-Hall. 1st edition in 1975, revised edition in 2006.

25. Liu, R. Y., Parelius, J. M. & Singh, K. (1999). Multivariate analysis by data depth: Descriptive statistics, graphics and inference, The Annals of Statistics 27: 783–858.

26. McConaghy, T. & Gielen, G. (2005). Analysis of simulation-driven numerical performancemodeling techniques for application to analog circuit optimization, Proceedings of IEEE International Symposium on Circuits and Systems.

27. McConaghy, T., Palmers, P., Gao, P., Steyaert, M. & Gielen, G. G. E. (2009). Variation-Aware Analog Structural Synthesis: A Computational Intelligence Approach, Springer.

28. Mood, A. M., Graybill, F. A. & Boes, D. C. (1974). Introduction to the Theory of Statistics, McGraw-Hill, New York.

29. Nelder, J. A. & Mean, R. (1965). A simplex method for function minimization, Computer Journal 7: 308–313.

30. Nocedal, J. & Wright, S. J. (1999). Numerical Optimization, Series: Springer series in operations research, Springer, New York.

31. Press, W. H., Teukolsky, S. A., Vetterling, W. T. & Flannery, B. P. (1993). Numerical Recipes in Fortran; the Art of Scientific Computing, Cambridge University Press, New York, NY,USA.

32. Price, K. V., Storn, R. M. & Lampinen, J. A. (2005). Differential Evolution, A Practical Approach to Global Optimization, Vol. 538 of Natural Computing Series, Springer.

33. Rohatgi, V. K. (1976). An Introduction to Probablity Theory and Mathematical Statistics, Wiley Series in Probability and Mathematical Statistics, John Wiley & Sons, New York.

34. Rubinstein, R. Y. & Kroese, D. P. (2007). Simulation and the Monte Carlo Methods, Probability and Statistics, 2nd edn, John Wiley and Sons Inc.

35. Silvaco Engineered (2010). UTMOST III – SPICE Modeling Software, Santa Clara, CA.

36. Stolk, P. A., Widdershoven, F. P. & Klaassen, D. B. M. (1988). Modeling statistical dopant fluctuations in MOS transistors, IEEE Transactions on Electron Devices 45(9): 1960 –1971.

37. Taher, H., Schreurs, D. & Nauwelaers, B. (2005). Extraction of small signal equivalent circuit model parameters for statistical modeling of HBT using artificial neural networks, Gallium Arsenide Applications Symposium (GAAS 2005) 3-7 ottobre 2005.

38. Vancorenland, P., Van der Plas, G., Steyaert, M., Gielen, G. & Sansen, W. (2001). A layout-aware synthesis methodology for RF circuits, ICCAD'01: Proceedings of the 2001 IEEE/ACM International Conference on Computer-Aided Design, IEEE Press, Piscataway, NJ, USA, pp. 358–362.

39. Wolfram Research Inc. (2008). Mathematica 7.URL: http://www. wolfram.com/products/ mathematica/index.html.

Chapter 5

DESIGN AND DSP IMPLEMENTATION OF FIXED-POINT SYSTEMS

Martin Coors, Holger Keding, Olaf Luthje, Heinrich Meyr

Institute for Integrated Signal Processing Systems, Aachen University of Technology, 52056 Aachen, Germany

ABSTRACT

This article is an introduction to the FRIDGE design environment which supports the design and DSP implementation of fixedpoint digital signal processing systems. We present the tool-supported transformation of signal processing algorithms coded in floating-point ANSI C to a fixed-point representation in SystemC. We introduce the novel approach to control and data flow analysis, which is necessary for the transformation. The design environment enables fast bit-true simulation by mapping the fixed-point algorithm to integral data types of the host machine. A speedup by a factor of 20 to 400 can be achieved compared to C++-library-based bit-true simulation. FRIDGE also provides a direct link to DSP implementation by processor specific C code generation and advanced code optimization.

INTRODUCTION

Digital system design is characterized by ever-increasing complexity that has to be implemented within reduced time, resulting in minimum costs and short time-to-market. This requires a seamless design flow that allows the execution of the design steps at the highest suitable level of abstraction. For most digital systems, the design has to result in a fixed-point implementation, either in HW or SW. This is due to the fact that these systems are sensitive to power consumption, chip size, throughput, and price-per-device. Fixed-point realizations outperform floating-point realizations by far with regard to these criteria. A typical fixed-point design flow is depicted in Figure 1. Algorithm design starts from a floating-point description that is analyzed by means of simulation without taking the quantization effects into account. This abstraction from all implementation effects allows an exploration of the algorithm space, for example, the evaluation of different digital receiver structures. This exploration is well supported by a variety of commercial block-diagram oriented system level design tools [1, 2, 3]. The modeling efficiency on the floatingpoint level is high and the floating-point models offer a maximum degree of reusability. In a next step towards system implementation, a transformation to a bit-true representation of the system is necessary, that is, assigning a fixed word length and a fixed exponent to every operand. This process is quite tedious and error-prone if done manually: often more than 50% of the implementation time is spent on the algorithmic transformation [4] to the fixed-point level for complex designs once the floatingpoint model has been specified. The major reasons for this bottleneck are as follows: (1) There is no unique transformation from floatingpoint to fixed-point. (a) Different HW and SW targets put different constraints on the fixed-point specification. (b) Optimization for different design criteria, like throughput, chip size, memory size, or accuracy are in general mutually exclusive goals and result in a complex design

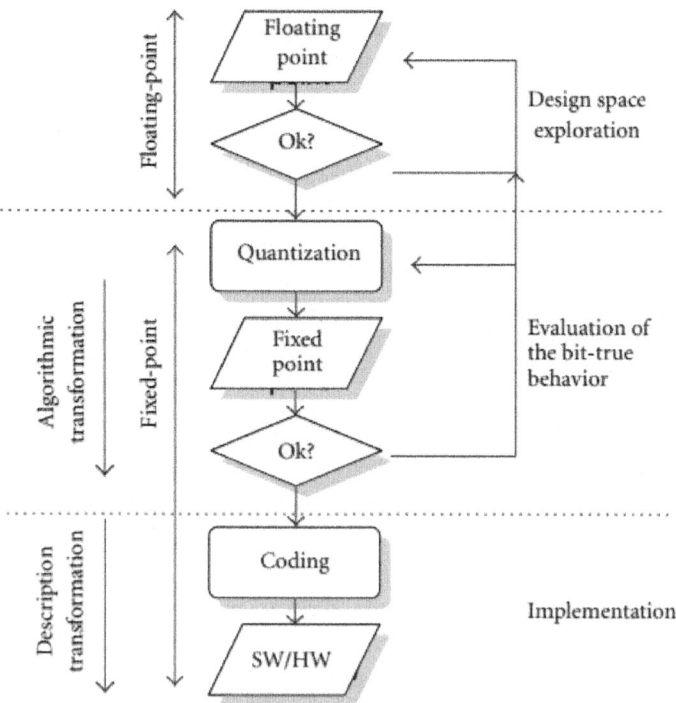

Figure 1: Fixed-point design process.

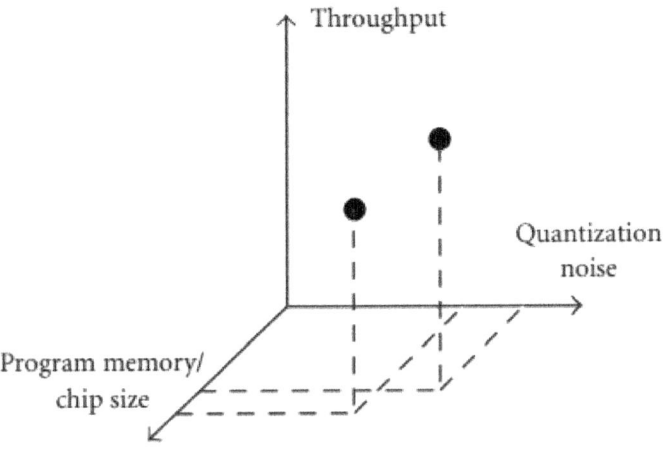

Figure 2: Fixed-point design space.

space as sketched in Figure 2. Furthermore, targets with a given datapath, for example, DSPs put different constraints on the quantization than ASICs where the datapaths are flexible. (c) The quantization is generally highly dependent on the application, that is, on the applied stimuli. (2) Quantization is a nonlinear process. Analytical models based on signal theory are only applicable for systems with a low complexity [5]. An exploration of the fixed-point design space with respect to quantization noise, performance, and operand word lengths cannot be done without extensive system simulation. (3) Some algorithms are difficult to implement in fixedpoint due to high signal dynamics or sensitivity to quantization noise. Thus algorithmic alternatives need to be employed. Finally, the quantized system is implemented, either in hardware or in software on a programmable DSP. The implementation needs to be optimized with respect to chip area, memory consumption, throughput, and power consumption. Here the bit-true system-level model serves as a "golden" reference for the target implementation which yields bit-by-bit the same results. To increase the designer's efficiency, software tool support for fixed-point design is necessary. Ideally the design environment would have the following features: (1) A modeling language supporting generic fixed-point data types to model the fixed-point behavior of the system. It will also provide a means of data monitoring of variables and operands during simulation, for example, range, mean, and variance. (2) A semiautomatic transformation from floating-point to a bit-true representation. The designer can bring in his knowledge about the system and he has full control over the transformation. The tool will accept a set of constraints specified by the designer to model the characteristics of the target hardware. (3) The ability to perform bit-true simulation with a simulation speed close to floating-point simulation. (4) A seamless design flow down to system implementation, generating optimized input for DSP compilers. These requirements have been the motivation for the Fixed-point pRogrammIng and Design Environment (FRIDGE) [6, 7, 8], an interactive design environment for the specification, simulation, and implementation of fixed-point systems. In this article we describe the principles and elements of FRIDGE and outline the seamless design flow as it becomes possible with this design environment. FRIDGE relies on five main concepts which are briefly introduced in the following.

Fixed-Point Modeling Language

DSP system design is frequently done on a PC or a workstation utilizing a C/C++-based system-level design environment. For efficient modeling of finite word length effects, language extensions implementing generic fixed-point data types are necessary. ANSI C does not offer such data types and hence fixed-point modeling using pure ANSI C becomes a very tedious and error-prone task. Fixed-point language extensions implemented as libraries in C++ [9, 10, 11] offer a high modeling efficiency. They supply generic fixed-point data types and various casting modes for overflow and quantization handling. The simulation speed of these libraries on the other hand is rather poor. Some of these libraries also offer data monitoring capabilities during simulation time. In the FRIDGE design environment, the SystemC fixedpoint data types are used for fixed-point modeling and simulation. A more detailed description of the SystemC fixedpoint data types is given in Section 3.

Interpolative Transformation

A central component of the FRIDGE design environment is the interpolative transformation from a hybrid description into a fully bit-true representation. The interpolative transformation, which is presented in detail in Section 4 uses analytical range propagation to determine operand word lengths.

Data Flow Analysis

During the development of the FRIDGE design environment, we have identified a need for accurate data flow analysis. The published approaches for static and dynamic program analysis did not match the requirements of the design environment, thus we have developed a novel approach for control and data flow analysis, which is presented in Section 5

Fast Bit-True Simulation

Existing C++-based simulation libraries model the fixedpoint operands as objects and make extensive use of operator overloading and container data types. Also, for ease of use, many decisions are made during run time. These mechanisms increase the execution time of fixed-point simulations by one to two orders of magnitude compared to floatingpoint arithmetic. This makes the simulation run time a major bottleneck during the fixed-point design process. In Section 7 various approaches for fixed-point simulation are presented and a methodology for fast bit-true simulation by mapping fixed-point algorithms in SystemC to an integer based ANSI C algorithm is introduced.

DSP Target Mapping

The final step in a float-to-fixed design flow is the implementation of the DSP system, either in hardware or in software. As a case study for targeting a high performance DSP, we have developed a FRIDGE back end which addresses the Texas Instruments TMS 320C62x fixed-point DSP processor and its C compiler. The back end generates target specific integer C code which exploits the features of the processor and the compiler to achieve a high efficiency of the compiled code. In Section 9 the FRIDGE C62x back end and the optimization strategies are presented.

THE FRIDGE DESIGN FLOW

The FRIDGE design flow starts from a floating-point algorithm in ANSI C. As illustrated in Figure 3, the designer then annotates single operands with fixed-point attributes. Inserting these local annotations results in a hybrid description of the algorithm, that is, some of the operands are specified bit-true, while the rest remain floating-point. A comparative simulation of the floating-point and the hybrid code within the same simulation environment shows whether the local annotations are appropriate, or if some annotations have to be modified. The integer word length of the local annotations can be derived from operand range monitoring during simulation runs. Typically, the designer manually annotates function parameters and key variables, for example, accumulator variables, which account for approximately 5% of all operands.

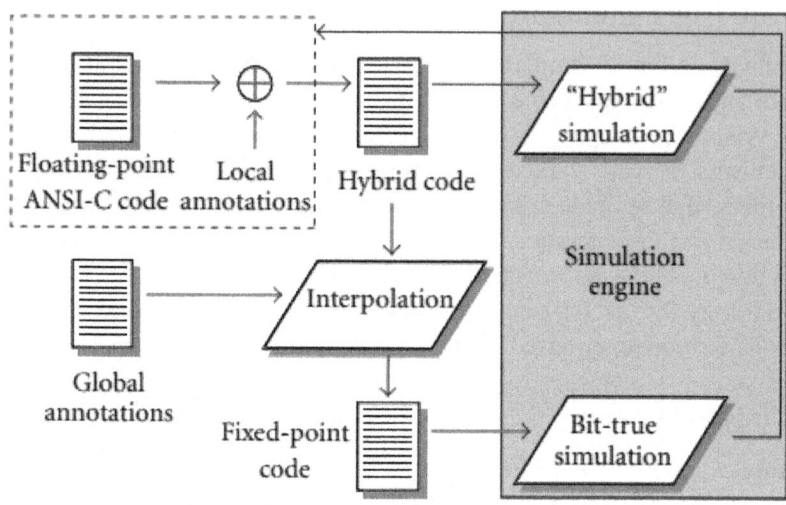

Figure 3: Quantization methodology with FRIDGE

Once the hybrid program matches the design criteria, the remaining floating-point operands are automatically transferred to fixed-point operands by interpolation. Interpolation denotes the process of computing the fixed-point parameters of the nonannotated operands from the information that is inherent to the annotated operands and the operations performed on them. Additionally, the interpolator has to observe a set of global annotations, that is, default restrictions for the calculation of fixed-point parameters. This can be, for example, a default maximum word length that corresponds to the register length of the target processor. The interpolation results in a fully annotated program, where each operand and operation is specified bit-true way. Cosimulating this algorithm with the original floating-point code will give an accuracy evaluation—and for changes now only the set of local and/or global annotations have/has to be modified, while the rest is determined and kept consistent by the interpolator. Described above are the algorithmic level transformations as illustrated in Figure 1, that change the behavior or accuracy of an algorithm. The resulting completely bit-true algorithm in SystemC is not directly suited for implementation, thus it needs to be mapped to a target, such as, a processor's architecture or to an ASIC. This is an implementation level transformation, where the bit-true behavior normally remains unchanged. Within the FRIDGE environment, different back ends map the internal bit-true specification to different formats/targets, according to the purpose or goal of the quantization process.

FIXED-POINT DATA TYPES AND LOCAL ANNOTATIONS

Since ANSI C offers no efficient support for fixed-point data types [12, 13], we initially developed the fixed-point language fixed-C [14] that is a superset of the ANSI C language. It comprises different generic fixed-point data types, cast operators, and interpolator directives. The fixed-C language was licensed to Synopsys, Inc., and Synopsys contributed it as a set of additional fixed-point data types to the Open SystemC

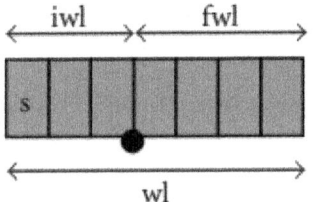

wl : word length
iwl : integer word length
fwl : fractional word length
s : sign encoding/sign bit

Figure 4: Fixed-point attributes of a bit-true description.

Initiative (OSCI) [11]. Together with additional fixed-point language elements from the A|RT Library by Frontier Design Inc., [10] fixed-C has been the base for the development of the SystemC fixed-point data types that are now used in the FRIDGE project as well. The SystemC fixed-point data types are utilized for different purposes in the FRIDGE design flow: • Since ANSI C is a subset of SystemC, the additional fixed-point constructs can be used as bit-true annotations to dedicated operands of the original floating-point ANSI C file, resulting in a hybrid specification. This partially fixed-point code can be used for simulation or as input to the interpolator. • The bit-true output of the interpolator is represented in SystemC as well. This allows a maximum transparency of the results to the designer, since the changes to the code are reduced to a minimum and the effects of the designer's directives, such as local annotations in the hybrid code, become directly visible. The additional fixed-point types and functions are part of a C++ class library that can be used in any design and simulation environment that are based on or can integrate C or C++ code (see, e.g., [1, 2, 3].) For a bit-true and implementation independent specifi- cation of a fixed-point operand, a three-tuple is necessary: the word length wl, the integer word length iwl, and the sign s, as illustrated in Figure 4. For every fixed-point format, two of the three parameters wl, iwl, and fwl (fractional word length) are independent; the third parameter can always be calculated from the other two, wl = iwl + fwl. With a given sign encoding s, we can also compute the minimum and maximum value that the fixed-point format can hold. For example, for a two's complement (tc) signed representation the minimum and maximum compute to

$$\max_{\langle wl,iwl,tc\rangle} = 2^{iwl-1} - 2^{fwl},$$

$$\min_{\langle wl,iwl,tc\rangle} = -2^{iwl-1}.$$

$$(1)$$

For an unsigned representation (us), on the other hand, the minimum and maximum are

$$\max_{\langle wl,iwl,us\rangle} = 2^{iwl} - 2^{fwl},$$

$$\min_{\langle wl,iwl,us\rangle} = 0.$$

$$(2)$$

Note that an integral data type is merely a special case of a fixed-point data type with an iwl that always equals wl— hence an integral data type can be described by two parameters only, the word length wl and the sign encoding s. In the following sections, we provide a short overview of the most frequently used fixed-point data types and functions in SystemC. A more detailed description can be found in the SystemC users manual [11].

The Data Types SC Fixed and SC Ufixed

The two's complement data type sc fixed and the unsigned data type sc ufixed receive their format when they are declared, that is, the fixed-point attributes must be known at compile time (static arguments),

```
sc_fixed<wl,iwl>   d,*e,g[8];
sc_ufixed<wl,iwl>  c;
```

Thus they behave according to these fixed-point parameters throughout their lifetime. This concept is called declaration time instantiation (DTI). Similar concepts exist in other fixed-point languages as well [9, 10, 15]. Pointers and arrays, as frequently used in ANSI C, are supported as well. For every assignment to a DTI variable, a data type check is performed. If the left-hand data type does not match the right-hand data type as illustrated in the code example below, an implicit cast to the left-hand data type becomes necessary,

```
sc_fixed<6,3>      a,b;
sc_ufixed<12,12>   c;
a = b; /* correct, both types match */
c = b;
/* type mismatch -> implicit cast necessary */
```

The data types sc fixed and sc ufixed are the data types of choice, for example, for interfaces to other functionalities or for lookup tables, since they behave like a memory location of a specific length and a known embedding/ scaling.

The Data Type Sc Fxval

Additionally to the DTI data type concept, SystemC provides the assignment time instantiation (ATI) data type sc fxval. This type may hold fixed-point numbers of arbitrary format and is especially tailored for the float-to-fixed transformation process. A declaration of a variable of type sc fxval does not specify any fixed-point attributes and if subsequently in the code a fixed-point value is assigned to a sc fxval variable, the variable is (re-)instantiated with all fixed-point attributes of the assigned value

The Data Types Sc Fix and Sc Ufix

Along with the static attribute types sc fixed and sc ufixed, SystemC also provides the fixed-point types sc fix and sc ufix that may also take nonstatic fixed-point attributes such as variables. The function in the code example below has the word length wl and the integer word length iwl as formal parameters, that is, wl and iwl are not known at compile time.

```
sc_fxval cast_func(int wl, int iwl, sc_fxval in)
{
return sc_fix(in,wl,iwl);
}
```

As shown in this example, the constructor for the types sc fix and sc ufix are often used to cast a value to a different fixed-point format.

Cast Modes

For a cast operation to a fixed-point format , it is also important to specify the overflow and precision reduction in case the target data type cannot hold the original value:

```
a = sc_fix(input,wl,iwl,q_mode,o_mode);
```

The variable a holds a two's complement fixed-point format and the value of input is cast to this fixed-point data type according to the quantization mode q mode1 and the overflow mode o mode. [2] The most important casting modes are listed below. SystemC also specifies many additional cast modes to model target specific behavior.

Quantization Modes

Truncation (SC TRN). The bits below the specified LSB are cut off. This quantization mode is the default for SystemC fixedpoint types and will be used if no other value is specified. Rounding (SC RND). Adds LSB/2 first, before cutting off the bits below the LSB.

Overflow Modes

Wrap-around (SC WRAP). In case of an overflow the MSB carry bit is ignored. This overflow mode is the default for SystemC fixed-point types and will be used if no other value is specified. Saturation (SC SAT). In case the minimum or maximum value is exceeded the result is set to the minimum or maximum value, respectively. With the sc fxval type, every assignment to a variable overwrites all prior instantiations, that is, one sc fxval variable may have different context-specific bit-true attributes in the same scope. This concept of ATI is motivated by the specific design flow: transformation starts from a floating-point program, where the designer abstracts from the fixed-point problems and does not think of a variable as finite length register. The concept of local annotations and ATI is also an effective way to assign context specific information without changing structures or variables when exploring the fixedpoint design space.

INTERPOLATION

The interpolator with its control and data flow analyzer is the core of the FRIDGE design environment. As depicted in Figure 3 it determines the fixed-point formats for all operands of an algorithm, taking as input a user annotated hybrid description of the algorithm and a set of global default rules, the global annotation file. Hence interpolation describes the computation of the fixed-point parameters of the nonannotated operands from the information that is inherent to the annotated operands. The interpolative concept is based on three key ideas: (1) Attribute propagation. The method of using the attributes of the bit-true specified operands in the code to calculate bit-true attributes for the remaining operands and operations in the code. (2) Global annotations. The description of default rules and restrictions for attribute propagation. (3) Designer support. The interpolator supplies feedback and reports to assist the designer to debug or improve the interpolation result. For a better understanding the first two points are explained more detailed in the following. (1) Attribute propagation. Given the information of the fixed-point attributes of some operands, the type and the fixed-point format of other operands can be extracted from this information. For example, if for the inputs to an operation both the range and the relevant fractional word length are specified, the same attributes can be determined for the result.[3]

Consider the following line of code:

```
c = a + b;   d = 1.5;   e = c * d;
```

The corresponding data flow graph is depicted in Figure 5. We assume that the ranges and the precision of the variables a and b are known, for example, by user annotations:

$$a \in [-0.25, 0.75] \implies R_a = [-0.25, 0.75]; \quad \text{fwl}(a) = 2,$$
$$b \in [-1.25, 0.5] \implies R_b = [-1.25, 0.5]; \quad \text{fwl}(b) = 2. \tag{3}$$

To receive the range Rc for the variable c that contains the sum of the variables a and b we add the ranges R_a and R_b (a detailed description of the range arithmetic used here can be found in [14]),

$$R_c = R_a + R_b = \left[\min_a + \min_b, \max_a + \max_b \right] = [-1.5, 1.25]. \tag{4}$$

The precision P_c (fwl) for the sum c computes to the maximum of the precisions P_a and P_b,

$$P_c = \max (P_a, P_b) = 2. \tag{5}$$

The information on the range and on the precision of the variable c is sufficient to calculate the required word length

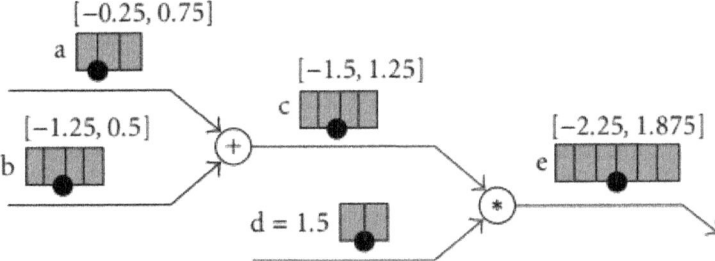

Figure 5: Example for interpolation of ranges/word lengths.

or integer word length for c. The correlation between fwl, range, and iwl yields the iwl of c:

$$iwl_c = \left\lceil \max \left(\log_2 \left| \min_c \right|, \log_2 \left(\left| \max_c \right| + 2^{-\text{fwl}_c} \right) \right) + 1 \right\rceil$$
$$= \left\lceil \max(0.58, 0.58) + 1 \right\rceil = 2. \tag{6}$$

Thus the resulting format for c is , where tc indicates the two's complement representation of c. The next step for the interpolator is to compute the fixedpoint format of the constant d. Since the range of d is $R_d = [1.5, 1.5]$ and the precision is $P_d = \text{fwl}_d = 1$ the iwl of d can be calculated as

$$iwl_d = \lceil \log_2 (\max_d + 2^{-fwl}) \rceil = \lceil \log_2(1.5 + 0.5) \rceil = 1.$$
(7)

After all fixed-point parameters of the input operands to the multiplication e=d*c are known to the interpolator, it continues with the calculation of the bit-true format and parameters for the variable e:

$$R_e = R_c * R_d = [-1.5, 1.25] * 1.5 = [-2.25, 1.875],$$
$$P_e = P_c + P_d = 2 + 1 = 3 \implies iwl_e$$
$$= \lceil \max (\log_2 | \min_e |, \log_2 (| \max_e | + 2^{-fwl_e})) + 1 \rceil$$
$$= \lceil \max(1.17, 1) + 1 \rceil = 3.$$
(8)

Hence we receive a fixed-point format of for the variable e. Note that this is a rather conservative way of interpolation, bits that may contain any information are never discarded. For the MSB side this is called a worst case interpolation, since with the iwl calculated by the interpolator an overflow is impossible, while on the other hand it may lead to iwls much larger than actually needed. In this case the designer may add additional local annotations to cut back the iwl to a more suited value. For the LSB side this is called interpolation, that is, by default every LSB of the operands is kept, maintaining the highest possible accuracy. LSBs are only discarded if the word length exceeds the maximum word length specified in the global annotation file. This can lead to a large increase in the (fwl), but with additional local annotations the designer can also keep the fwl shorter. In [6] we also describe a method to have the interpolator calculate a less conservative value for the fwl.

Global annotations. While local annotations express fixed-point information for single operands, the global annotations describe default restrictions to the complete design. For different targets, different global restrictions apply. For SW, the functional units to perform specific operations are already defined by the architecture of the processor. Consider a 16×16 bit multiplier writing to a 32-bit register. A global annotation can supply the information to the interpolator that the word length of a multiplication operand must not exceed 16 bits, while the result may have a word length of up to 32 bits.

Implementational Issues

In a first step the FRIDGE front end parses in the hybrid description into a C++-based intermediate representation (IR). Then range propagation is performed to determine the bittrue format for all the operands. During this process, control and data flow analysis is also carried out. The information gained is stored in the IR. The advanced algorithms used for the analysis will be described in Section 5. After this process the IR holds a bit-true description of the algorithm with additional control and data flow information. These data structures form the basis for additional transformation steps performed in the FRIDGE back ends that target different languages and platforms.

ADVANCED DATA FLOW ANALYSIS

During the development of the FRIDGE design environment, we have identified a need for accurate data flow analysis to cater the needs of the interpolation, the fast simulation code generation and the target specific code optimization. The published methods were not capable of matching the requirements, thus we have developed a novel approach for data flow analysis that can provide the necessary data for the FRIDGE back ends. Researchers have worked on program analysis techniques since the 1960s and there is, by now, an extensive literature [16]. There are two major approaches to program analysis: (a) There are static analysis techniques that analyze the program code at compile time. Usually, sets of equations are set up according to the program semantics and solved by finding their fixpoint. One of the best known static approaches is Data Flow Analysis. It is treated in depth in standard compiler books [17, 18]. Other techniques such as constraint-based analysis and abstract interpretation are also described in [19]. PAG [20] is a tool for generating interprocedural data flow analyzers that implement these techniques. (b) On the other hand, there are techniques for dynamic analysis that are used for examining the behavior of program code during execution. Typically, these techniques are employed by profiling tools. Profiling information can for example be used by programmers to find critical pieces of code or as input to profile-driven optimizers. Dynamic program analysis techniques have been implemented in tools like Pixie [21] or QPT [22]. By principle, dynamic program analysis relies on input vectors to be processed during execution. Thus the results are of no general nature. Analysis techniques of neither category are suited for the needs of the FRIDGE design environment. Static analysis puts tight constraints onto the code to be analyzed. The use of pointers is usually not supported or yields too conservative results. Implementations of digital signal processing systems usually make extensive use of pointers, even, for example, for iterating over data arrays. Furthermore, static analysis is blind for program properties that result from run time effects. However, especially these properties have to be taken into account by FRIDGE in order to obtain precise results. Dynamic analysis is to some extend capable of detecting these properties. Nevertheless, it is not applicable for the FRIDGE design environment for two reasons. First, the results are of statistical, numerical nature. There is no way to gain information about data flow or control flow properties. Second, the results are not generally valid, that is, they only reflect the behavior of the program running on the given input vectors. FRIDGE requires analysis results that are valid for all possible executions of the program though. The requirements for the analysis employed by FRIDGE are different from those of standard tools like, for example, a general purpose compiler. FRIDGE is focused on digital processing systems. These systems are typically data flow dominated, that is, their execution is to a great extent independent from the data to be processed. Besides, the accuracy and quality of the results are more important than speed (of analysis). This allows for a more comprehensive code analysis than, for example, a general purpose compiler can apply. In order to gain precise results including also run time properties and being able to handle pointer operations, the code is interpreted. Since there is no concrete data to be processed, we process abstract data instead. In the following this methodology is referred to as abstract execution.

The data flow analysis unit in the FRIDGE design environment is based on three main components:

- The concept of data abstraction
- The state controlled memory model
- The concept of coupled iterators.

Data Abstraction

While in concrete execution numeric values are written to and read from memory, we use operations for abstract execution. An operation is a collection of information about possible values. The two most important elements are

- the range, that is, the minimum value and the maximum value, and
- a reference to the expression in the code that corresponds to the operation.4

Furthermore, operations may be ambiguous. Consider the code example below.

```
01 int func(int x, int y, int z){
02    int a, b, c, d;
03
04    switch(y){
05      case 1:
06        a = 8; break;
07      case 2:
08        a = 16; break;
09      case 3:
10        a = 32;}
11
12    if(z>0)
13      b = 0;
14    else
15      b = 1;
16
17    if(x>0){
18      c = 5;
19      d = a;}
20    else {
21      c = b;
22      d = 7;}
23
24    return c + d;
25 }
```

The only information available about parameters x, y, and z is that they are integers. Hence it cannot be decided which branches of the switch- and if-statements in lines 04, 12, and 17 are executed. This results in an ambiguous content, for example, of variable b, namely, values 05 and 1, referring to the expressions in lines 13 and 15, respectively. We combine both operations to an ambiguous operation. In addition, ambiguous operations are associated with conditions, under which the alternatives are chosen. In the example, alternative 0 is chosen if $(z > 0)$ is true, alternative 1 if it is false. In general, there may be more than two alternatives and conditions may be combined by a logical AND. Operations are arranged in graphs similar to binary decision diagrams introduced by Akers [23], where the nodes embody the ambiguous operations and the leafs the unambiguous operations. In general, operations are described by the following rules:

- an operation is either an unambiguous operation or an ambiguous operation;

- an unambiguous operation represents a possible content in memory during concrete execution of a program;

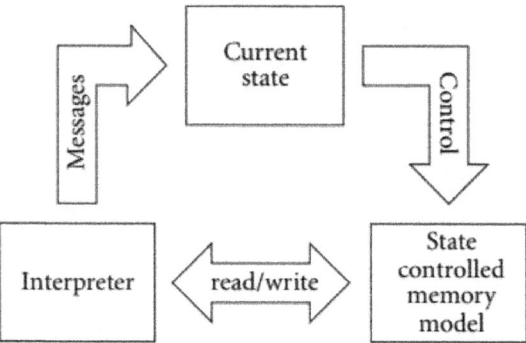

Figure 6: Abstract execution

- an ambiguous operation is associated with a control flow ambiguity in the code (dashed line in Figure 7) and matches each possible branch to an operation.

Thus these trees do not only contain the alternatives, but also the conditions under which the alternatives are taken. The conditions are determined by all the ambiguities along the path from the root to the alternative. Each ambiguity contributes to the condition in this way, that the condition for the execution of the control flow branch must be fulfilled, that is associated with the link to the next operation on the path. A logical AND is applied to the contributions of each ambiguity. For example, the tree in Figure 7 with A_3 as its root shows the ambiguity tree corresponding to variable d in line 24. The path to value 32 (bold line) goes through ambiguities A_3 and A_4. A_3 is associated with the if-statement and the path follows the link that is associated with the true-branch. That yields the condition (x > 0) == true. Further on, the path passes through A_4 and follows the link to 32. A_4 is associated with the switch-statement and the link to 32 with case 3. That yields the condition y == 3. Thus the resulting condition for A3 taking on the value 32 is[6]

```
(x > 0) == true && y == 3
```

The State Controlled Memory Model

As illustrated in Figure 6, the state controlled memory Model serves as a regular memory that can be read and written to. Besides, it is responsible for building the ambiguity trees described in Section 5.1. As long as the current state is in initial state, the behavior of the state controlled memory model does not differ from a regular memory. Once the current state contains a condition, all changes done to memory contents only occur under that condition and result in appropriate ambiguity trees. The state is defined by a set of assumptions about the result of particular expressions in the code. A logical AND is performed on these assumptions. The initial state makes no assumptions at all. Other valid states could for example be "$(x > 0)$ == true" or "$(x > 0)$ == true && y == 3." During abstract execution, the state can be changed by the interpreter.

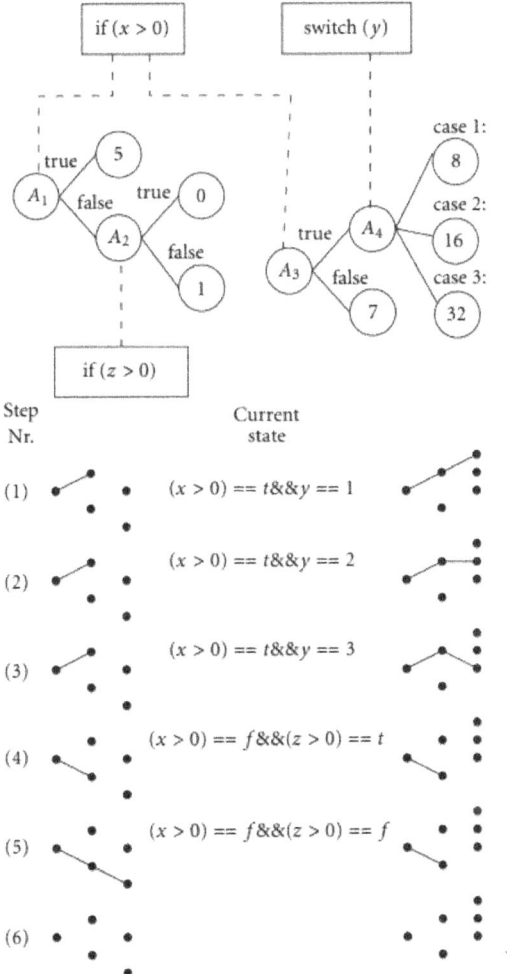

Figure 7: Iterating over ambiguities.

Iterating Over Ambiguities

When abstractly executing statements (Section 5.4) or computing the set of all possible evaluations of an expression,[7] We have to iterate over the alternatives of ambiguities. This is basically done by traversing the corresponding tree. However, the current state is taken into account, that is, only those alternatives are visible, whose conditions are not contradictory to the current state. Furthermore, when selecting an alternative from an ambiguity, the corresponding conditions are—if not yet included—added to the current state. This way, the following is achieved: All data couplings are taken into account, that is, no impossible cases are considered. Alternative executions of statements can be done without further thought about the current state (see Section 5.4). Selecting an alternative from an ambiguity is done by building a path through the corresponding tree. The end of the path is an unambiguous operation. In principle, iterating is performed on all successors of an ambiguity first, until it will be iterated over the alternatives of the ambiguity itself (depth first). When establishing a path through an ambiguity, two basic cases have to be considered: (1) The current state contains a condition respective to the control flow fork that is associated with the ambiguity. In this case, the path must follow the link that corresponds to the condition and may not be altered. The node would be considered a slave node. (2) The current state does not yet contain a condition respective to the control flow branch that is associated with the ambiguity. In this case, a possible branch is selected and the path is extended by the corresponding link. The corresponding condition is added to the current state. The node would be considered a master node. During further iteration, the path will switch to all other links successively. When this is done, the respective condition has to be updated accordingly. After that, the condition is removed from the current state. The trees in Figure 7 show the contents of variables c (left-hand side) and d (right-hand side) connected to line 24 in the code. Figure 7 also illustrates how to iterate over all possible combinations of contents of both variables. Note how building a path through an ambiguity affects the current state and how the current state masks the visible alternatives of ambiguities. First of all value 5 is selected from ambiguity A_1. The corresponding condition $((x > 0) == true)$ is added to the current state. Thus A_1 becomes a master node. When building the path through A_3, A_3 becomes a slave node, because the current state already makes an assumption about the control flow ambiguity that is associated with A_3 $((x > 0))$. Therefore, the path must follow the link from A_3 to A_4. Nodes A_2 and A_4 are associated with different control flow forks, respectively. They always become master nodes and never affect any other ambiguities. Steps 2 and 3 iterate over the remaining visible alternatives of the right-hand tree. Step 4 switches to the second alternative of master node A_1 (false). This affects the slave A_3 in this way as long as the path in the left-hand tree goes from A_1 to A_2 (steps 4 and 5), the only visible alternative of the right-hand tree is 7. In step 6 the iteration has been completed.

Execution of a Program

Figure 8 shows how statements are abstractly executed. The solid lines represent the control flow of a concrete execution. Abstract execution also follows that control flow. However, statements that depend on ambiguous data are executed multiple times (dashed lines), once for every possible vector of the involved ambiguities. The vectors are iterated over as described in Section 5.3. Thus every execution is performed in a different current state, such that changes in memory together with their corresponding states are stored in ambiguity trees. This algorithm is applied recursively for nested statements. Any code constructs can be executed this way. Although a possibly large number of execution states exists, we found that the run time and the memory consumption of the analysis were remarkably low for typical signal processing algorithms. In most cases the control and data

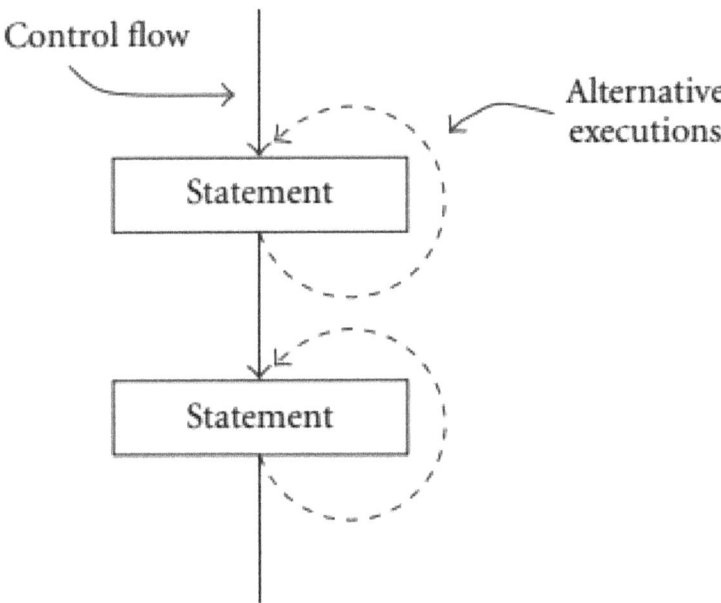

Figure 8: Abstract executions of sequential statements

flow analysis was performed in less than one second on a 800 MHz PC. The information gained during abstract execution is stored in the intermediate representation of the algorithm. The FRIDGE back ends, which will be introduced in the next sections, access this information to perform several code transformation steps.

FAST BIT-TRUE SIMULATION

As pointed out in Section 1, transforming a signal processing algorithm from a floating-point to a fixed-point requires extensive simulations due to the nonlinear nature of the quantization process. The available C++-based fixed-point libraries [10, 11] offer a high modeling efficiency but the simulation speed of these libraries on the other hand is rather poor. This makes simulation speed a major bottleneck in the fixed-point design process. Utilizing C-based fixed-point libraries like the ETSI basic arithmetic operations [24] does not overcome this problem as the simulation speed still has a considerable overhead compared to an equivalent floating-point implementation. Existing C++-based simulation libraries model the fixedpoint operands as objects. In order to offer generic fixedpoint data types without word length restrictions, data container types are used as an internal representation. Bit-true operations are performed by operator overloading. Range checking, the choice of cast modes and many other decisions necessary for correct bit-true behavior are done at simulation time. The price for this flexibility and ease of modeling is slow execution speed as the generic fixed-point data types modeled by extensive C++ constructs cannot be efficiently mapped to the architecture of the host machine by today's C++ compilers. A simulation speedup can be achieved by mapping the fixed-point operands to the mantissa of the floating-point hardware of the host machine and bit level manipulations to maintain bit-true behavior. This restricts the maximum word length of the fixed-point operands to the word length of the mantissa. This approach has been described by Kim et al. [25] and it is also implemented in the SystemC library [11].

Another mean of speeding up fixed-point simulations is the use of a hardware accelerator, for example, an FPGA to perform computationally expensive operations. The acceleration can be achieved either by utilizing configurable logic or by combining configurable logic with a processor. This approach has been described by De Coster [26]. The mapping of the algorithm to the different hardware units and the data transfer between the units make additional transformation steps necessary. The work described in this article proposes a mapping of fixed-point algorithm in SystemC to an integer-based ANSI C algorithm that directly addresses the built-in integer ALU of the host machine. An efficient mapping includes an embedding of all fixed-point operands into the host machine registers, a cast mode optimization and many other aspects, and requires a detailed control and data flow analysis of the algorithm. Independently from the authors' work, De Coster [26] proposed a similar method, using DFL [27] as input language and targeting directly a Motorola DSP65000. Our work presented here represents a continuation of the research results published by Keding et al. [6] and Willems [14] and introduces improved concepts for the mapping process that result in a considerable simulation acceleration. For the fast simulation back end we assume that fixedpoint attributes are assigned to every operation. The back end also requires the information collected during the control and data flow analysis stored in the IR. After a number of IR refinements, an ANSI C representation of the algorithm using only integral data types can be derived from the IR. It is important to note that the transformation in the back end, in contrast to the float-to-fixed transformation in the IR, does not change the behavior of the algorithm. The fully quantized algorithm coded in SystemC and the integer-only ANSI C algorithm yield bit-by-bit identical results, making the fast simulation back end output ideally suited for fast bittrue simulation on a workstation or PC.

TRANSFORMATION TO ANSI C

The LBP Alignment

For the embedding of a fixed-point operand specified by a triple (wl, iwl, sign) into a register of the host machine with the machine word length (mwl) the minimum requirement is

$$mwl \geq wl = iwl + fwl. \tag{9}$$

Figure 9 illustrates different options for embedding an operand with a word length of 5 bit into a given mwl of 8. Obviously, for mwl > wl, a degree of freedom for choosing the location of binary point (lbp) exists:

$$mwl - iwl \geq lbp \geq wl - iwl = fwl. \tag{10}$$

Beside this degree of freedom, there are also a number of constraints for the selection of the lbp: (i) Interface constraints. For interface elements, such as, function parameters or global variables, the lbp must be defined identically for a function and all calls to this function. Otherwise, the data written to or read from these data elements will be misinterpreted.

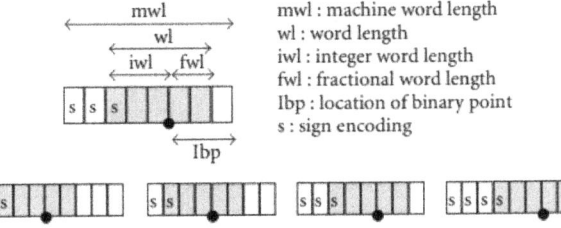

Figure 9: Embedding a 5-bit word into an 8-bit register.

(ii) Operation constraints. Each operation has an lbp syntax. This lbp syntax may include constraints on the lbp of the operand(s) of the operation and/or rules for the calculation of the lbp of the result. For example, the operands and the result of and addition must have the same lbp. (iii) Control and data flow constraints. Generally, a read access to a storage element must use the same lbp as the preceding write access to the storage element. This implies that if a write operation to a memory location occurs in alternative control-flow branches, the lbp must be at the same position in both write operations, as no run time information about the lbp is available in a following read operation. The same applies to ambiguous write operations to arrays and write operations via pointers.

The LBP Alignment Algorithm

The lbp alignment algorithm implemented in the fast simulation back end is designed to take advantage of the degree of freedom described by (10), while meeting the constraints specified above. Meeting these constraints and maintaining the consistency of the lbps require precise information about the control and data flow of the algorithm. To obtain this information we used the data flow analysis method described in Section 5. The data flow information is represented basically as define-use (du) chains and use-define (ud) chains [17, 18], with additional and more accurate information about ambiguous control flow. Initially, for all operands lbp = fwl is chosen. Thus all operands are right aligned. In a first step we set the lbps of all interface elements according to the interface constraints. Then, in an iterative process, the data flow information is used to adjust the lbps by insertion of shift operations to meet the operation constraints and the control and data flow constraints. The algorithm terminates when all conditions are fulfilled and the lbps did not change during the last iteration. The operation constraint lbp alignment algorithm basically consists of an iteration over all operations and an adjustment of the operand and result lbps according to the operation's lbp syntax. The control and data flow constraint lbp alignment algorithm searches for all read accesses from a data element the associated previous write accesses to the same data element, that is, finding all defines for a use of a data element (udchains). According to the control and data flow constraints the lbp of operands linked by such ud-chains are set to the same value. Finally, the embedding of constants can be done in a way that the required shift operations when using the constant are minimized. Unlike described by Kum et al. [28], we do not use a shift operation minimizing approach here, but using the degree of freedom in choosing a suited lbp (10) and the accurate data flow information, we found that there is not sufficient potential for this optimization to justify the effort.

Data type Selection

The next step in the transformation process is the selection of suitable integral data types for fixed-point variables. The FRIDGE internal bit-true specification of the algorithm features arbitrary word lengths. With the SystemC back end this does not represent a problem, since the SystemC data types are generic and may be of any bit length required. With the fast-simulation back end, on the other hand, we only have the limited pool of the built-in data types of the host machine, that is, integral data types like char, short, int, long

Basic Constraints for Any Data Element

A matching data type for every fixed-point variable has to be chosen. The minimum requirement for the data type chosen is that it can be embedded into the host machine data type with word length mwl at the correct location, (see Figure 9 for illustration) iwl + lbp $\leq$ mwl.

Structural Constraints

Additionally, the requirements introduced by data structures that force each of their elements to be of the same data type have to be met. An example for this behavior are arrays. The target data type for the N elements of an array must fulfill the following condition: $\max_{i=0}^{N-1}(\mathrm{iwl}_{\mathrm{array}}[i] + \mathrm{lbp}_{\mathrm{array}}[i]) \leq \mathrm{mwl}$.

Semantical Constraints

Another constraint becomes important if aliasing of data elements, for example, by pointers occurs: a pointer may point to different data elements. For syntax and semantics reasons all aliased data elements and the base type of the pointer must be identical [13]. This only causes a problem if data types are changed like it is done in fixed-point optimizations or the floating-point to fixed-point transformation process described in Section 2: initially, most numerical data types are floating-point types but after the transformation there are various different fixed-point data formats. Hence special care must be taken during the code generation process to ensure that the types are consistent. A detailed description of the data type selection algorithm used can be found in [29].

Cast Mode Transformation

Cast operations can reduce or limit the word length on the MSB side of a word (overflow handling) or at the LSB side of a word (quantization handling). They are used either to prevent indeterministic behavior of fixed-point systems8 or to model a data path that is different from the host machine. This is often the case when algorithms for DSP systems are developed. Fixed-point libraries like in SystemC offer various generic overflow and quantization handling modes, which makes SystemC an efficient means of modeling fixed-point systems. For fast fixed-point simulation, on the other hand, the use of these generic casting modes are simply ruled out for performance reasons.

Overflow Handling

Overflow handling is required if it is necessary to reduce the wl at the MSB side of the word or if the carry bit is set for the MSB. Examples for frequently used overflow handling modes in digital signal processing algorithms are wrap-around and saturation [30].

Saturation

In SystemC, a cast of an expression expr to a wl-bit tc data type with integer word length iwl applying saturation as over- flow mode can be modeled as follows:

```
result = sc_fix(expr,wl,iwl,...,SC_SAT);
```

The fast simulation code generation on the other hand translates this into plain C code that first tests if the range of data type is exceeded, and if so it sets the resulting value to the minimum or maximum of this type, which is

$$\underset{wl,iwl,lbp,tc}{MAX} = 2^{iwl+lbp-1} - 2^{lbl-fwl},$$

$$\underset{wl,iwl,lbp,tc}{MIN} = -2^{iwl+lbp-1} + 2^{lbl-fwl} - 1.$$

$$(11)$$

Thus the fast simulation code construct generated is the following:[9]

```
int tmp;
result=((tmp=expr)>MAX)?MAX:(tmp<MIN)?MIN:tmp;
```

Introducing an additional temporary variable avoids multiple evaluations of expr. Wrap-Around The SystemC way of casting an expression expr to a wl-bit tc data type with integer word length iwl applying wrap-around as overflow mode is shown here,

```
result = sc_fix(expr,wl,iwl,...,SC_WRAP);
```

For the bit-true ANSI C equivalent of this operation several options exist. An example for a code construct for wrap around assuming two's complement arithmetic and a machine word length of mwl is

```
result = (expr << SHIFT) >> SHIFT;
```

The amount of shifts computes to SHIFT = mwl − iwl − lpb. The shift left eliminates the MSBs whereas the arithmetic shift right provides a sign extension for the new MSB.

Quantization Handling

If the word length of an operand is reduced at the LSB side, we can apply different quantization handling modes. The most frequently encountered are rounding and truncation. Rounding In SystemC the method for casting an expression expr to a wl-bit two's complement data type with integer word length iwl applying rounding as quantization mode is

```
result = sc_fix(expr,wl,iwl,SC_RND,...);
```

Rounding is defined by adding DELTA = LSB/2 to the operand and eliminating the LSBs, for example, by shifting it right SHIFT = lbp − fwl bits. Thus the rounding operation can be realized in the fast simulation code by

```
result = ((expr + DELTA)>>SHIFT)<<SHIFT;
```

Truncation

The truncation operation, given in SystemC by

```
result = sc_fix(expr,wl,iwl,SC_TRN,...);
```

can be implemented efficiently by a bit mask operation,

```
result = expr & (~MASK);
```

Where MASK is given by $2^{lpb-fwl-1}$. For several combinations of cast modes, for example, wrap-around combined with rounding or truncation, more efficient joint quantization and overflow handling C code constructs are generated. The shift operations introduced by the cast code constructs are also utilized to adjust the lbp of the expression, eliminating the need for additional scaling shifts.

EXPERIMENTAL RESULTS

The code generated by the FRIDGE fast simulation back end has been benchmarked against the fixed-point simulation classes, which are part of the C++-based SystemC language. The simulation classes offer two simulation modes: a mode supporting unlimited fixed-point word lengths based on concatenated data containers and a mode supporting limited precision up to 53 bits based on float-arithmetic and bit manipulations. The benchmarks have been performed on a SUN Ultra 10 workstation running SOLARIS using the GCC compiler version 2.95.2 with the -O3 option. The SystemC library version 1.0 was utilized for the bit-true simulations. The benchmark is based on typical signal processing kernels, FIR 17-tap FIR filter, DCT 8 × 8 JPEG DCT algorithm, Autocorr 25 elements 5th order autocorrelation, IIR 3rd order IIR filter, FFT complex FFT of length 8, Matrix 4 × 4 matrix multiplication. Four different versions of the kernel functions have been benchmarked: (i) Floating-Point. The execution speed of the floatingpoint implementation of the algorithms serve as reference for the benchmarks. (ii) SystemC. The quantized bit-true version of the algorithms utilizing the SystemC fixed-point data types. The algorithms have been quantized using the FRIDGE design environment. (iii) SystemC limited precision. The quantized bit-true code has been compiled with the limited precision option to speed up SystemC fixed-point operations. (iv) Fast simulation code. The fast fixed-point simulation code based on integral data types has been generated by the FRIDGE back end applying the transformation techniques described in the previous sections. The code yields bit-by-bit the same results as the code utilizing the SystemC data types. The experimental results are presented in Table 1. As the floating-point code has been used as a reference, the experimental data has been scaled relative to the execution speed of the floating-point code. The bit-true SystemC code consumes by a factor of 325 to 1103 more run time than the original floating-point code, making bit-true simulation a major bottleneck in the fixed-point design flow. Utilizing the limited precision mode of the SystemC library, a speedup by a factor of 3.1 ⋯ 5.2 can be achieved, but the fixed-point code is still by a factor of 67 ⋯ 234 slower than the floating-point reference. The fast simulation code runs by a factor of 18.8 ⋯ 90.9 faster compared to the SystemC fixed-point code utilizing the limited precision option. For the unlimited precision the speedup is 91.0 ⋯ 454.2, respectively. Compared to the floating-point reference code, the fast simulation code is by a factor of 2.5 ⋯ 6.9 slower. This is due to the host system's architecture and additional shift and bit mask operations necessary to perform lbp-alignment and cast operations to maintain bit-by-bit consistency with the quantized code. The quantized DCT algorithm contains many cast operations to reduce fixed-point word lengths introduced by the quantization process. As these operations can be modeled ef- ficiently by bit mask operations in the fast simulation code, the highest speedup was achieved for this kernel function.

DSP CODE GENERATION

During the recent years, new architectural approaches for DSP processors have been made. The current generation of high performance DSP processors features a pipelined VLIW architecture (very long instruction word), which offers a very high computing performance if a high degree of software pipelining in combination with instruction level parallelism is used. But programming these processors manually utilizing assembly language is a very tedious task. In awareness of this problem, the modern DSP architectures have been developed using a processor/compiler codesign methodology which led to compiler-efficient processor designs. On the other hand, a significant gap in the system design flow is still evident; there is no direct path from a floatingpoint system level simulation to an optimized fixed-point implementation. Today a manual implementation on the DSP and target specific code optimization is necessary, increasing time-to-market and making design changes very tedious, error prone, and costly. Thus we have developed an optimizing FRIDGE back end to generate target optimized DSP C code. The target specific code generation is necessary for two reasons:

Table 1: Relative execution speed

	Floating-point ANSI C	SystemC	SystemC limited precision	Fast simulation code
FIR	1.0	386.5	102.7	2.8
DCT	1.0	1103.1	233.9	2.5
Autocorr	1.0	694.6	130.6	6.9
IIR	1.0	371.0	120.2	3.1
FFT	1.0	354.7	67.7	2.6
Matrix	1.0	325.9	71.2	3.6

(i) The generic fixed-point data types used for fixedpoint simulations are not suited for DSP implementation, as the currently available DSP compilers do not support C++ fixed-point data types. The upcoming generation of DSP compilers will support C++ language constructs, but compiling the fixed-point libraries for the DSP is no viable alternative as the implementation of the generic data types makes extensive use of operator overloading, templates, and dynamic memory management. This will render fixed-point operations rather inefficient compared to integer arithmetic performed on a DSP. (ii) Compiling the FRIDGE-generated integer ANSI C code on a DSP is also not sufficiently efficient as the generic C code does not exploit the capabilities of the DSP hardware such as built-in saturation and rounding logic or SIMD processing. As a case study, we have chosen the TMS320C62x processor and its C compiler as a target for the FRIDGE design environment. This enables a seamless design-flow from floating-point to optimized C62x C code utilizing integral data types. Generating a C62x optimized version of a signal processing algorithm using a different set of fixed-point parameters becomes a matter of hours instead of days or weeks using the conventional manual techniques. The C62x integer code generated by the design environment yields bit-by-bit the same results as the fixed-point code utilizing C++ simulation classes on the host machine. Thus a comparative simulation to the "golden reference model" gives the designer a high degree of confidence in the generated code. The first objective of our case study was to find out which C code constructs compile into efficient C62x assembly code. Thus we applied the DSPstone benchmarking methodology to the C62x optimizing C compiler. The DSPstone project [31], conducted in 1994 by ISS, Aachen University of Technology established a benchmarking methodology for DSP compilers by comparing the performance of compiled C code to hand optimized assembly code in terms of program/data memory consumption and execution time. As a consequence, it allows to identify a possible mismatch between architecture and compiler. The benchmarking has been done using eleven typical signal processing algorithms (FIR, FFT, DCT, minimum error search, etc.). The benchmarking gives quantitative results for cycle count and program memory consumption. In a second step, we used C62x specific C language extensions (intrinsics) and compiler directives to restructure the off-the-shelf C code while maintaining functional equivalence to the original code. These optimizations led to a considerable improvement in performance in many cases as the compiler was able to utilize software pipelining and instruction level parallelism to speed up the code. It has turned out that software pipelining is the key to achieving a high performance but, on the other hand, requires careful analysis and code restructuring. The evaluation [32] gave quantitative performance data for the C62x compiler and a set of code optimization techniques to generate efficient C62x C code. In a third step, we benchmarked various implementations of the fixed-point quantization and overflow handling modes on the C62x. This led to a set of optimized implementations for the quantization and overflow handling functionality.

DSP Code Transformation

The FRIDGE C62x back end performs similar transformation steps as the fast bit-true simulation code generation presented in Section 6: lbp alignment, cast mode transformation, and data type selection. Additionally, target specific code optimization is performed. The designer has to keep the special requirements of the DSP target in mind to reach a high level of efficiency. Through our experiments we found that, for example, the number of cast statements and shift operations has a strong influence on the efficiency of the generated code. Thus if the designer chooses settings for the global annotations and the default cast mode during the early stages of the transformation which do not represent the properties of the target architecture properly, the code optimization and the DSP compiler are not able to generate efficient assembly code. The optimizations performed in the FRIDGE C62x back end are source level transformations to supply the C62x compiler with the best C code possible. The amount of analysis done in an optimizing compiler is usually limited due to constraints of the time used for compilation. In the FRIDGE design environment, control and data flow analysis is performed with the maximum possible accuracy utilizing the techniques presented in Section 5. The information gained during this analysis is available for the back end code transformation as well. Thus we are able to perform code restructuring techniques, which are usually beyond the scope of an optimizing compiler

The Lbp Alignment

As the TI C6000 processor family has an integer multiplication mode, the right alignment strategy of the lbp alignment algorithm can also be applied in the C62x back end. This algorithm implicitly minimizes the number of scaling shifts. In contrast to the fast bit-true simulation, the number of scaling shifts generated is important for the C62x code generation. For the fast simulation code generation we found the potential of shift minimization limited to a performance improvement of 3% ⋯ 13% [29]. This is different for the C62x code generation. As the C62x can perform two scaling shift operations per cycle, a shortage of functional units limits the performance in highly software pipelined loops. Thus "shift poisoning" of loops must be avoided, for example, by choosing suitable fixed-point data types for function parameters and central data structures.

Data type Selection

As the properties of the data paths of the C62x processor and the width of the integral data types supported by the C62x C compiler are known, the design environment can utilize this information during the transformation process. A set of global annotations for the C62x guides the interpolation process and a set of integral data types with a given bit length is supplied to the C62x back end.

Cast Mode Transformation

The generic overflow- and quantization handling modes offered by SystemC have to be mapped to the target hardware in an efficient manner. The C62x offers built-in saturation hardware which can be used by the back end. This is illustrated by the following example.

Cast Mode: Saturation

A cast of an expression to a wl-bit two's complement data type with integer word length iwl applying saturation as over- flow mode is modeled in SystemC as follows:

```
result=sc_fix(expr,wl,iwl,...,SC_SAT);
```

An implementation of this code construct in generic ANSI C is

```
int tmp;
result=((tmp=expr)>MAX)?MAX:(tmp<MIN)?MIN:tmp;
```

On the C62x the sshl intrinsic (saturating shift left) can be used to perform the saturation operation:

```
result=(signed)_sshl(expr,SHIFT)>>SHIFT;
```

where SHIFT is given by mwl−(iwl+lbp). Utilizing the builtin saturation hardware of the C62x via the sshl intrinsic allows the generation of code with linear control flow in contrast to the forked control flow in the ANSI C implementation. This significantly speeds up the code.

Loop Optimizations

The key to high execution speed on the C62x is software pipelining and instruction level parallelism. This is especially important for loops, where most of the execution time is spent for most digital signal processing algorithms. The latest version of the C62x C compiler is able to perform quite sophisticated loop optimizations to achieve high performance. This can be further improved by restructuring the loops at source level, applying techniques like loop unrolling, scalar expansion and splitting data paths. By introducing SIMD (single instruction multiple data) intrinsics it is possible to reduce the required number of load/store operations significantly. The C62x back end utilizes the data- and control flow information and the code transformation infrastructure to identify possible loop optimizations and to perform the necessary loop restructuring. The design environment maintains the consistency of generated code.

EXPERIMENTAL RESULTS

We have benchmarked the cycle count performance of the generated C62x integer C code using two sets of typical signal processing kernel functions: The first set consists of six off-the-shelf kernels which have been initially coded without DSP specific code optimization. The second set of kernels has been extracted from TI's C6000 compiler benchmarking suite.

Off-the-Shelf Kernels

This set of kernels consists of six signal processing functions, which also have been used for the benchmarks in Section 8: FIR, DCT, Autocorr, IIR, Matrix, Dotprod. The code has been translated using TI's C6x compiler version 4.0

[33] and the performance has been compared with three reference codes: (i) C67x floating-point C code. The C67x floating-point DSP is code-compatible to the C62x and its C compiler is mostly identical to the C62x C compiler, thus the performance of the generated fixed-point C code can be compared to the original floating-point C code. (ii) C62x floating-point emulation. The floating-point emulation library which is part of the C62x compiler's run time library allows the user to perform floating-point arithmetic on the C62x processor. The floating-point operations are executed as function calls. (iii) C62x integer ANSI C code. The FRIDGE back end allows the designer to generate ANSI C fixed-point code without C62x specific optimization. This code can also be compiled and executed on the C62x processor. The efficiency of the target specific code optimization can be benchmarked using this code.

Table 2: Cycle count

Device	Floating-point C67x	Float emulation C62x	Generic ANSI C C62x	Target specific C C62x
FIR	132	1304	523	234
DCT	331	34163	1509	622
Autocorr	564	6581	3057	1041
IIR	73	708	82	81
Matrix	108	4999	1600	233
Dotprod	95	9436	1300	406

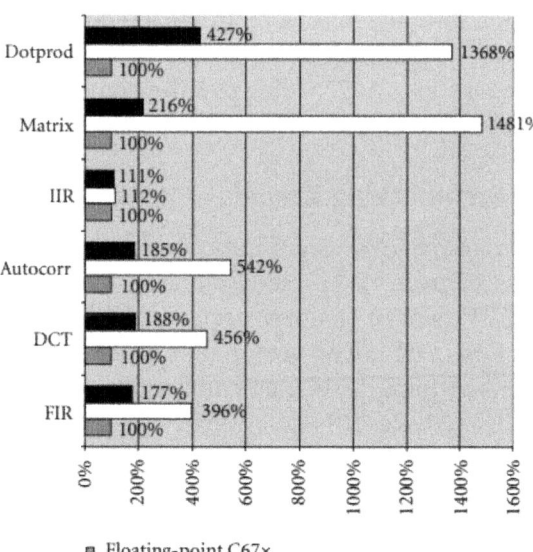

Figure 10: Cycle count relative to floating-point code.

Table 2 presents the benchmarking results for the six kernel functions. Figure 10 illustrates the relative cycle count. As the C67x floating-point code has been used as a reference, it was scaled to 100%. For readability the results of the floatingpoint emulation have been omitted in the bar graph. As depicted in Table 2 the C62x floating-point software emulation has a cycle count which is by a factor of 9.7 to 103 higher than the cycle count of the same code compiled for the floating-point processor. The generic ANSI C integer code without C62x specific language extensions is by a factor of 1.1 to 14.8 slower than the floating-point code. The integer code performs additional shift- and bit-masking operations to ensure the bittrue behavior. Some of the cast-operations cannot easily be modeled in generic ANSI C. Thus a significant overhead is introduced for kernel functions where many cast operations are inserted by the interpolation (e.g., the DCT). The performance can be improved by matching the generated code to the target architecture. For example, utilizing the sshl intrinsic is a convenient way to access the C62x saturation hardware directly. This reduces the overhead introduced by the additional shift and cast operations to a factor of 1.1 to 4.3 compared to the floating-point code. For the floating-point code of the Dotprod kernel function, the compiler was able to generate efficient code using 95 cycles for 64 vector elements. For the fixed-point code, the additional operations needed for cast operations in the inner loop prevent the compiler from achieving similar efficiency. Removing all scaling shifts and overflow protection from the inner loop of the fixed-point code for this kernel yields a cycle count of 83. Introducing a single scaling shift in the inner loop brings the cycle count up to 147, adding overflow protection yields 406 cycles. Similar effects appear in the Matrix kernel benchmark.

TI Compiler Benchmarking Kernels

This set of kernels consists of six signal processing functions: IIR 16-coefficient IIR filter, IIR cas biquads 10 cascaded biquads, FIR 10-tap 40 sample FIR filter, MAC VSELP two 40 samples vectors, VQ MSE MSE between two 256 element vectors, VEC SUM vector sum of two 44 sample vectors. For these kernels hand-optimized C62x assembly code and C62x integer C code is available on TI's website. It is noteworthy that neither the C code nor the assembly code was coded with overflow protection. For the embedding of input and output operands, implicit assumptions were made which reduced the number of scaling shifts in the kernel functions. Thus the hand-optimized C62x assembly code can serve as an "upper bound" for the efficiency of the FRIDGE C62x design flow. We derived the floating-point code from the integer C code. The function interfaces in the floating-point code were manually annotated with fixed-point specifications to get hybrid code. The hybrid code was used as input to generate optimized C62x integer code from the FRIDGE C62x environment. The FRIDGE generated C62x code features full over- flow protection and maintains consistency for the "location of binary point" for input and output operands. The code has been translated using TI's C6x compiler version 4.0 [33] and the performance has been compared to the reference codes: (i) C67x floating-point C code. This is the floating-point code compiled for the C67x processor. (ii) C62x hand-optimized integer C code. This is the original hand-optimized code from the benchmarking suite.

Table 3: Cycle count

	Floating-point	Assembly	Hand optimized ANSI C	FRIDGE
Device	C67x	C62x	C62x	C62x
IIR	85	42	38	72
IIR BIQUAD	149	70	82	108
FIR	315	237	278	373
MAC VSELP	175	61	59	207
VQ MSE	559	279	275	275
VEC SUM	63	48	51	127

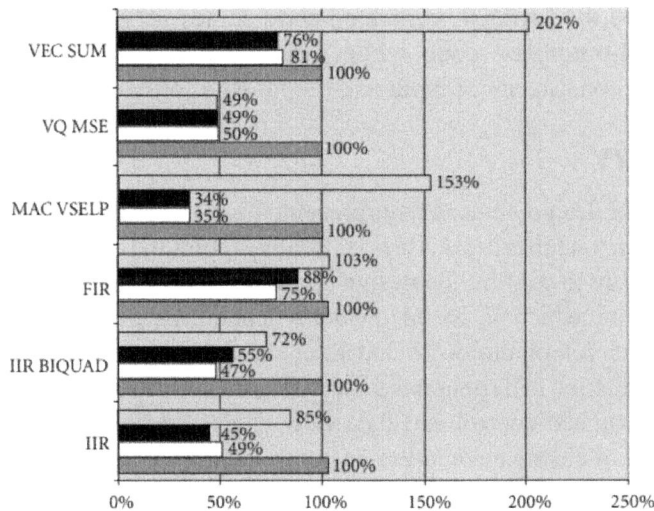

Figure 11: Cycle count relative to floating point code

(iii) C62x hand-optimized assembly code. The handoptimized assembly code served as a reference for the benchmarks. Table 3 presents the benchmarking results for the six kernel functions. Figure 11 illustrates the relative cycle count. For consistency, the floating-point code has been used as a reference, it was scaled to 100%. For these kernels, the C6x compiler was obviously able to generate very efficient code. For consistency we have measured the cycle count including the function call. This causes the hand-optimized C code to be faster than the handoptimized assembly code for some kernels. The floatingpoint code is slower than the hand-optimized assembly and C code in all cases as the floating-point instructions need more execution stages than their integer counterparts. For this set of kernel functions the FRIDGE generated code consumes more cycles than the hand-optimized code as additional shift and cast operations for overflow protection are performed. For some kernels, such as, the MAC VSELP and the VEC SUM, this leads to a significant overhead as the hand-optimized code uses the processor's functional units in a very efficient manner. Introducing additional shift and bit mask operations in the innermost loop slows down the code, as no unused functional units are available in the very tight loop pipelining schedule. Especially the s-unit which performs shift operations is heavily used and becomes the performance bottleneck. Nevertheless, the FRIDGE generated code comes very close in performance to the hand-optimized code while offering full overflow protection and maintaining consistency of input and output data formats.

SUMMARY

The FRIDGE design environment presented in this article allows the designer to concentrate on the critical issues of floating-point to fixed-point design flow. Thus he is able to explore the design space more efficiently. The interpolative transformation which is based on analytical range propagation enables an accelerated development cycle and in consequence a shorter time-to-market. The fast simulation code generation as well as the DSP back end benefits directly from the advanced control and data flow analysis techniques we developed. The concept of abstract execution, in combination with a state-driven memory model and coupled iterators, yields results with the precision necessary for the back end transformation steps. The verification of the fixed-point algorithm has to be performed by means of simulation. Existing C++-based fixed-point libraries increase simulation-time by up to two orders of magnitude compared to the corresponding floating-point simulation. The FRIDGE fast simulation back end applies advanced compile-time analysis concepts, analyzes necessary casting operations, and selects the appropriate built-in data type on the host machine, thus a speedup by a factor of 20 to 400 compared to the SystemC code while maintaining bit-by-bit equivalence was achieved. The target specific C code generation provides a direct link from a floating-point code to C62x C code using integral data types. The generated code yields bit-by-bit the same results as the bit-true SystemC code for host simulation, enabling comparative simulation to the reference model. As proven by the experimental data, the generated C62x C code comes very close to hand-optimized C- and assembly code. These features make FRIDGE a powerful design environment for the specification, evaluation, and implementation of fixed-point algorithms.

REFERENCES

1. Synopsys Inc., "CoCentric System Studio—User's Manual," Mountain View, Calif, USA.

2. Mathworks Inc., "Simulink Reference Manual," March 1996.

3. Cadence Design Systems, 919 E. Hillsdale Blvd., "SPW User's Manual," Foster City, Calif, USA.

4. T. Grotker, E. Multhaup, and O. Mauss, "Evaluation of¨ HW/SW tradeoffs using behavioral synthesis," in Proc. Int. Conf. on Signal Processing Application and Technology, Boston, Mass, USA, October 1996.

5. B. Liu, "Effect of finite word length on the accuracy of digital filters—a review," IEEE Trans. on Circuit Theory, vol. 18, no. 6, pp. 670–677, 1971.

6. H. Keding, M. Willems, M. Coors, and H. Meyr, "FRIDGE: A fixed-point design and simulation environment," in Proc. European Conference on Design, Automation and Test, pp. 429– 435, Paris, France, February 1998.

7. M. Willems, V. Bursgens, and H. Meyr, "FRIDGE: Floating-¨ point programming of fixed-point digital signal processors," in Proc. Int. Conf. on Signal Processing Application and Technology, pp. 1000–1005, San Diego, Calif, USA, September 1997.

8. M. Willems, V. Bursgens, H. Keding, T. Gr¨ otker, and H. Meyr,¨ "System level fixed-point design based on an interpolative approach," in Proc. Design Automation Conference, pp. 293–298, Anaheim, Calif, USA, June 1997.

9. S. Kim, K. Kum, and W. Sung, "Fixed-point optimization utility for C and C++ based digital signal processing programs," in Workshop on VLSI and Signal Processing '95, pp. 197–206, Osaka, Japan, November 1995.

10. Frontier Design Inc., "A|RT Library User's and Reference Documentation," Danville, Calif, USA, 1998.

11. Synopsys Inc., CoWare Inc., Frontier Design Inc., "SystemC User's Guide, Version 2.0," 2001.

12. W. Sung and K. Kum, "Word-length determination and scaling software for a signal flow block diagram," in Proc. IEEE Int. Conf. Acoustics, Speech, Signal Processing, pp. 457–460, Adelaide, Australia, April 1994.

13. B. W. Kernighan and D. M. Ritchie, The C Programming Language, Prentice-Hall, Englewood Cliffs, NJ, USA, 2nd edition, 1988.

14. M. Willems, A methodology for the efficient design of fixedpoint systems, Ph.D. thesis, Aachen University of Technology, 1998.

15. Mentor Graphics, "DSP Station User's Manual," San Jose, Calif, USA.

16. C. Hankin, "Program analysis tools," International Journal on Software Tools for Technology Transfer, vol. 2, no. 1, pp. 6–12, 1998.

17. A. Aho, R. Sethi, and J. Ullman, Compilers, Principles, Techniques and Tools, Addison-Wesley, Reading, Mass, USA, 1986.

18. M. J. Wolfe, High Performance Compilers for Parallel Computing, Addison-Wesley, Redwood City, Calif, USA, 1996.

19. C. Hankin, F. Nielson, and H. R. Nielson, Principles of Program Analysis, Springer, Heidelberg, Germany, 1999.

20. F. Martin, "PAG—an efficient program analyzer generator," International Journal on Software Tools for Technology Transfer, vol. 2, no. 1, pp. 46–67, 1998.

21. MIPS Computer Systems, "UMIPS-V Reference Manual (Pixie and Pixstats)," Sunnyvale, Calif, USA, 1990.

22. T. Ball and J. R. Larus, "Optimally profiling and tracing programs," ACM Transactions on Programming Languages and Systems (TOPLAS), vol. 16, no. 4, pp. 1319–1360, 1994.

23. S. B. Akers, "Binary decision diagrams," IEEE Trans. on Computers, vol. 27, no. 6, pp. 509–516, 1978.

24. European Telecommunication Standard Institute, "GSM full rate speech transcoding," GSM recommendation 06.10, February 1992.

25. S. Kim, K. Kum, and W. Sung, "Fixed-point optimization utility for C and C++ based digital signal processing programs," IEEE Trans. on Circuits and Systems II: Analog and Digital Signal Processing, vol. 45, no. 11, pp. 1455–1464, 1998.

26. L. De Coster, Bit-true simulation of digital signal processing applications, Ph.D. thesis, KU Leuven, 1999.

27. Mentor Graphics, "DSP Architect, DFL User's and Reference Manual," 1994.

28. K. Kum, J. Kang, and W. Sung, "A floating-point to integer C converter with shift reduction for fixed-point digital signal processors," in Proc. IEEE Int. Conf. Acoustics, Speech, Signal Processing, vol. 4, pp. 2163–2166, Phoenix, Ariz, USA, March 1999.

29. H. Keding, M. Coors, O. Luthje, and H. Meyr, "Fast bit-true " simulation," in Proc. the Design Automation Conference, pp. 708–713, Las Vegas, Nev, USA, June 2001.

30. S. K. Mitra, Digital Signal Processing: A Computer-Based Approach, McGraw-Hill, New York, NY, USA, 1998.

31. V. Zivojnovi ˇ c, J. Mart ´ ´ınez, C. Schlager, and H. Meyr, "DSP- ¨ stone: A DSP-oriented benchmarking methodology," in Proc. International Conference on Signal Processing Applications and Technology, Dallas, Tex, USA, October 1994.

Chapter 6

A HYBRID SYSTEM APPROACH FOR HIGH CONSUMPTION INDUSTRIAL FURNACE CONTROL

Goran Stojanovski, Mile Stankovski

Department of Automation and System Engineering, Faculty of Electrical Engineering and Information Technologies, Ss. Cyril and Methodius University, Skopje, Macedonia

ABSTRACT

In this paper we describe a hybrid system approach for high consumption industrial furnace control. The problem is observed in systematic way starting from the need for modeling this system as hybrid. For description of this behavior we use the Hybrid System Description Language. After that, we design an optimal controller for the furnace and we simulate and compare the controller with other relevant predictive controllers. We have shown that using the hybrid approach for control of industrial furnaces leads to significant improvement of the control system performances.

INTRODUCTION

Processes and plant constructions of thermal systems and industrial furnaces, kilns and ovens in particular, have been subject to both scientific and technological research for long time [1] . This is mainly due to the process complexity of energy conversion and transfer into thermal systems, however, their control and supervision have recently become topics of extensive research.

The overall control task is to drive the process to the desired thermodynamic equilibrium and to regulate the temperature profile through the plant. In industrial operating environment, technical control specifications involve goal and task description of aims and procedures of supervision functions. From the general systems theoretical standpoint, it is the thermal systems where it became apparent that controlled processes in the real-world plants constitute a non-separable, unique interplay of the three fundamental natural quantities: energy, mass and information.

From control point of view, in thermal systems the essential impact occurs due to time delay and natural I/O operating modes. These modes interact with the controlling infrastructure in the process real-time, provide the way the complexity of sensor-actuator problem be properly resolved by natural ordering of I/O modes and respective input-output variable pairing [2,3].

In this paper we present a hybrid model for a high consumption industrial furnace that should represent the real plant more accurately. On the basis of this model, we will design controller(s) that will lead to increasing of the control system performance.

The paper is organized as follows. At the beginning we explain the principles of hybrid systems and we elaborate on the need of using hybrid modeling for control of high consumption industrial furnaces. In the third section we present linearized model of the furnace and we derive hybrid model for designing of the predictive controller that is explained in section four. In section five the simulation results are presented. At the end we give concluding remarks and possible future work.

THE NEED FOR HYBRID MODEL FOR HIGH CONSUMPTION INDUSTRIAL FURNACE

A hybrid system denotes in general a system composed of two unlike components. A hybrid control system is a control system with both continuous and binary/integer signals. Such a system generates a mixture of continuous and discrete signals, which take values in a continuum (such as the real numbers $\mathbb{R}$) and a finite set (such as a, b, c), respectively.

In the last decade several modeling formalisms have been developed to describe hybrid systems. One of the most popular and widely used is the class of discrete hybrid automata (DHA) introduced in [4] . DHA result from the interconnection of a finite state machine (FSM), which is the discrete dynamics of the hybrid system, with a switched affine system (SAS), which is the continuous dynamics, see **Figure 1**. as presented in [5] .

The information exchange of the two basic elements of the hybrid automaton is based on the event generator (EG) and the mode selector (MS). The EG is responsible for activating the logic variables based on the continuous

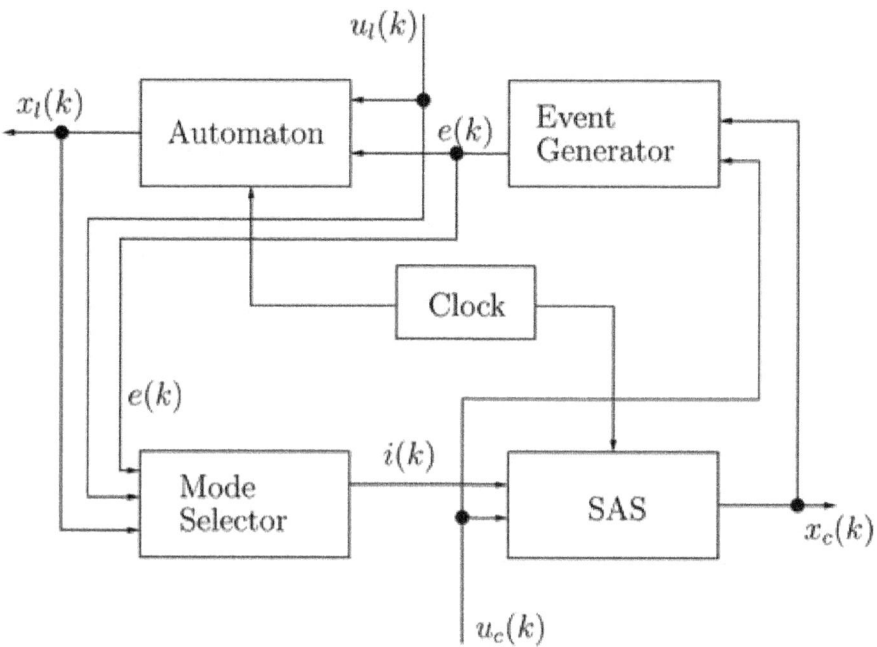

Figure 1: Discrete time hybrid automata.

state and input signals. These logic events and other exogenous logic inputs affect the logic states of the FSM. After that the MS combines all the logic variables to choose the "mode" of the continuous dynamics of the SAS. Continuous dynamics are expressed as linear affine difference equations.

Industrial furnaces are usually represented in control science as nonlinear mathematical models. In order to simplify the model of the furnace in the near surrounding of the operating point, researchers usually use linearization. Nevertheless, it is not unusual for a system to have more than one operating points so linearization must be done in all of them. On the other hand, these systems are usually subject to logic and integer variables inputs (furnace empty/full; door open/closed; line speed: 1, 2, 3, or 4; and so on). In these cases, there are several possible solutions:

a). Linearization in one operating point and neglecting the logic and integer variables. In this case the designer of the control system should carefully choose one operating point to linearize the plant, and design the controller as robust as possible. For the integer and logic variables the most common value must be assumed. Reduced efficiency of the controller is expected due to the neglecting of the variables and when the operating point is different from the one of linearization.

b). Hybrid approach. Here the designer should model the system in one of the popular hybrid modeling languages, and design a controller for the hybrid model. This approach is very similar to the Switching control, but in addition it allows the user to incorporate logical rules in the mode selection of the system, as we have previously described. In order to implement hybrid multi model system, the nonlinear function, must be linearized in several operating points before the hybrid approach is implemented.

c). Nonlinear control techniques. These techniques are the method that can achieve best control performance if no logic and integer variables are involved in the system. Additionally, nonlinear control is too demanding in means of computational power, and in other cases this kind of design is not feasible. That is why this method is rarely implemented in industry.

We can summarize that the furnace has complex model that consists of continuous dynamics over different segments and integer/boolean logic variables that significantly affect the transfer function. That is why a hybrid approach for modeling of the furnace is the most appropriate choice. As one of the possible representations, the authors in [6] have proposed a class of hybrid systems definition of the form of Equations -.

$$x(k+1) = Ax(k) + B_1 u(k) + B_2 \delta(k) + B_3 z(k)$$

$$y(k) = Cx(k) + D_1 u(k) + D_2 \delta(k) + D_3 z(k)$$

$$E_2 \delta(k) + E_3 z(k) \leq E_1 u(k) + E_4 x(k) + E_5$$

where $x(k) = [x_c(k)/x_l(k)]$

is the state vector $x_c(k) \in \mathbb{R}^{n_c}$ and $x_l(k) \in \{0,1\}^{n_l}$, the output vector is

$y(k) = [y_c(k)/y_l(k)] \in \mathbb{R}^{p_c} \times \{0,1\}^{p_l}$

and the input vector is

$$u(k) = [u_c(k)/u_l(k)] \in \mathbb{R}^{m_c} \times 0,1^{m_l},$$

$z(k) \in \mathbb{R}^{r_c}$ and $\delta(k) \in \{0,1\}^{r_l}$ are auxiliary variables. A, B_i, C_i, D and E_i denote real constant matrices, E_5 is real vector, $n_c > 0$, and $p_c, m_c, r_c, n_l, p_l, m_l, r_l \geq 0$. Without loss of generality, we assumed that the continuous components of a mixed-integer vector are always the first. Inequalities must be interpreted component wise. Systems that can be described by model - are called Mixed Logical Dynamical (MLD) systems.

In this paper we will elaborate the hybrid system design approach in modeling and control of high consumption industrial furnace. The operating point of the furnace depends on the temperature profile that we want to achieve. The most used temperature profile in this furnace is when we need to regulate the temperature near 1000° Celsius. Nevertheless, if we want to use the furnace for heating other types of pipes, we need to stabilize the output temperature to different operating points e.g. 500, 700 or 1200 degrees etc. Additionally there are several discrete parameters that significantly influence the furnace behavior. The state that represents the presence of pipe in each of the furnace zones has great impact on the coefficients for the increase/decrease of the temperature in the respective zone. Also the state of the doors (open/closed) at the beginning and at the end of the furnace impacts the cooling of the furnace.

MODEL OF THE FURNACE

In our previous work, we have presented different types of model predictive control implemented on linearized models of industrial furnaces [7,8]. In order to elaborate our results we are going to compare our work with previous related work. In this section we present the linearized model of the furnace and the discrete time hybrid model.

Continuous Time Model

The modeling of an industrial furnace is not an easy task. That is why, in this paper we decided to use the previous work [3] where a complete identification of a high consumption industrial furnace in the factory "FZC 11 Oktomvri" in Kumanovo, in the Republic of Macedonia has been done.

Structural, non-parametric and parameter identification has been carried out using step and PRBS (PseudoRandom Binary Sequence) response techniques in the operational environment of the plant as well as the derivation of equivalent state realization. With regard to heating regulation, furnace process is represented by its 3×3 system model. The families of 3×3 models have 9 controlled and 9 disturbing transfer paths in the steady and transient states. The structural model of the furnace is depicted in **Figure 2**.

Experiments involved the recorded outputs (special thermocouples): temperature changes in the three zones in response to input signal change solely in one of the zones. Firstly, only the burners at the first zone were excited and data for the temperatures in all three zones is collected; the temperature T_j and the corresponding fuel flow Q_i for each input-output process channel (transfer path) were recorded. After collecting the data, the parameter modeling of the furnace was conducted and the system's state space model presented in Equations and was derived. This model represents a linearization of the furnace model near the operating point.

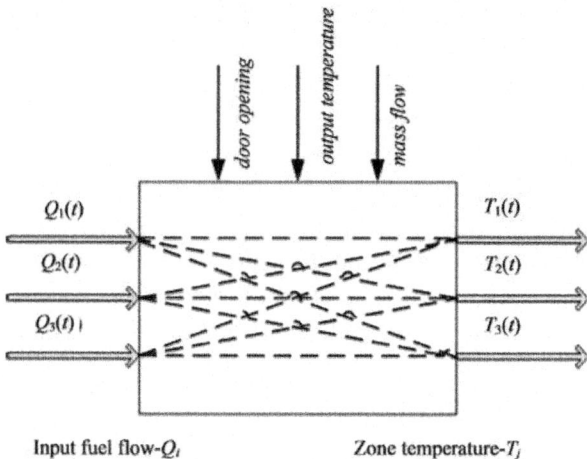

Figure 2: Diagram of the conceptual MIMO system model for gas-fired furnace in FZC "11 Oktomvri".

$$\dot{x} = Ax + Bu$$

$$y = Cx + Du$$

where $\dot{x}$ is the state vector of the system, y is the output vector of the system and the values of matrix A are defined in Equation , the values of matrix B are defined in Equation , and the values of matrices C and D are defined in Equation .

$$A = \mathrm{diag}\,(P_i), i = 1, 2, \cdots, 9;$$

$$P_{ij} = \begin{bmatrix} -1/T_1 & -1/T_1 \\ 0 & -1/T_2 \end{bmatrix}$$

$$B = [S, S, S],$$

$$S = \begin{bmatrix} 0 & 1.93 & 0 & 0 & 0 & 0 \\ 0 & 0 & 0 & 1.29 & 0 & 0 \\ 0 & 0 & 0 & 0 & 0 & 0.2 \end{bmatrix}$$

$$C = \begin{bmatrix} V & \bar{0} & \bar{0} \\ \bar{0} & V & \bar{0} \\ \bar{0} & \bar{0} & V \end{bmatrix}$$

$$V = [1, 0, 1, 0, 1, 0]$$
$$\bar{0} = [0, 0, 0, 0, 0, 0], D = 0$$

The time constants are $T_1 = 6.22$ min and $T_2 = 0.7$ min.

Discrete Time Hybrid Model

Before we introduce the hybrid model we need to elaborate the furnace dynamics. In this paper we are dealing with 3-input 3-output gas fired furnace. The maximum temperature that can be achieved is $1300°$ Celsius when operating at full power (the valves for the burners are opened 100%). The model for the temperature is discrete and it is represented with the Equations -.

$$T_i[k+1] = T_{out} - (T_i[k] - T_{out})\{0.5/(T_{max} - T_{out}) - \alpha_i\}$$
$$+0.05(T_i[k-1] - T_{out})$$
$$+\theta_i F_i(u)(T_{max} - T_i[k])/\in$$

$$\alpha_1 = 0.945 - hc_F - hc_B/5$$

$$\alpha_2 = 0.945 - hc_F/3 - hc_B/3$$

$$\alpha_3 = 0.945 - hc_F/5 - hc_B$$

$$F_1(u) = n_{S1} \cdot U1_{k-3} + n_{S2} \cdot U1_{k-4} + n_{F1} \cdot U2_{k-4}$$
$$+ n_{F2} \cdot U2_{k-5} + n_{D1} \cdot U3_{k-6} + n_{D2} \cdot U3_{k-7}$$
$$F_2(u) = n_{S1} \cdot U2_{k-3} + n_{S2} \cdot U2_{k-4} + n_{F1} \cdot U1_{k-4}$$
$$+ n_{F2} \cdot U1_{k-5} + n_{F1} \cdot U3_{k-4} + n_{F2} \cdot U3_{k-5}$$
$$F_3(u) = n_{S1} \cdot U3_{k-3} + n_{S2} \cdot U3_{k-4} + n_{F1} \cdot U2_{k-4}$$
$$+ n_{F2} \cdot U2_{k-5} + n_{D1} \cdot U1_{k-6} + n_{D2} \cdot U1_{k-7}$$

$$\theta_i = \begin{cases} 1 & \text{if pipe}_i = 0 \\ 0.95 & \text{if pipe}_i = 1 \end{cases}$$

$$hc_F = \begin{cases} 0 & \text{if Frontdoor is closed "=0"} \\ 0.005 & \text{if Frontdoor is open "=1"} \end{cases}$$

$$hc_B = \begin{cases} 0 & \text{if Backdoor is closed "=0"} \\ 0.005 & \text{if Backdoor is open "=1"} \end{cases}$$

where $n_{S1} = 1.195$; $n_{S2} = 0.6232$; $n_{F1} = 0.07968$; $n_{F2} = 0.04155$; $n_{D1} = 0.01245$; $n_{D2} = 0.006492$. T_{max} represents the maximum temperature that can be achieved in this furnace and is equal to $1300°$ Celsius. Signals hc_F, hc_B and θ_i are logic signals that can change their value according to the process dynamics and represent disturbances of the system. The outdoor temperature around the furnace T_{out} is continuous state disturbance to this system.

It is obvious that the system is both discrete and nonlinear by nature but cannot be implemented as a discrete control system because of the logical conditions in the transfer function and the interconnection between the states and variables that combine a non-affine set for synthesis of the control system.

In order to overcome this problem we are going to propose 2 solutions.

a).   The model could be linearized near the best-fit operating point, and after that to be represented as a hybrid system, by decomposing this model to sections according to the logic rules.

b).   The model will be divided into several sections and linearized in each of them. In this way we will derive a multi-model of the furnace that can be later decomposed to more section according to the logic rules. This way we will have more complex model and controller, but improvements in the results are expected.

Both of the solution will be compared between each other, and with a standard MPC controller on a linearized model of the furnace.

CONTROLLER SYNTHESIS

Model Predictive Control (MPC) has become the accepted standard for complex constrained multi variable control problems in the process industries. Here at each sampling time, starting at the current state, an open-loop optimal control problem is solved over a finite horizon [9] . Only the first computed control value in the sequence is implemented. At the next time step the computation is repeated starting from the new state and over a shifted horizon, leading to a moving horizon policy [10] .

Controlling a system means to calculate input signals in a manner that when the calculated sequence is applied to the system it will eliminate the difference between the referent signal and the measured output of the system. In this paper we will compare three different MPC techniques to a highly complex nonlinear model of an Industrial furnace. As mentioned in the previous section the controllers to be compared are linear MPC, hybrid MPC on a model linearized in one operating point, and hybrid multi-model MPC.

The optimization problem of linear MPC is known for a long time and it is not a subject of this paper. The reader can find detailed explanations on MPC in [11] , [12] and many other books. Regarding the hybrid optimization problem, it should be of the form

$$\min_{\{u,\delta,z\}_0^{N-1}} J\left(\{u,\delta,z\}_0^{N-1}, x(t)\right)$$

$$\triangleq \left\|Q_{xN}\left(x(N|t)-x\right)_r\right\|_p + \sum_{k=1}^{N-1}\left\|Q_x\left(x(k)-x_r\right)\right\|_p$$

$$+ \sum_{k=0}^{N-1}\left\|Q_u\left(u(k)-u_r\right)\right\|_p + \left\|Q_z\left(z(k|t)-z_r\right)\right\|_p$$

$$+ \left\|Q_y\left(y(k|t)-y_r\right)\right\|_p$$

$$s.t. \begin{cases} x(0|t) = x(t) \\ x(k+1|t) = Ax(k|t) + B_1 u(k|t) + \\ \qquad\qquad + B_2\delta(k|t) + B_3 z(k|t) \\ y(k|t) = Cx(k|t) + D_1 u(k|t) + \\ \qquad\qquad + D_2\delta(k|t) + D_3 z(k|t) \\ E_2\delta(k|t) + E_3 z(k|t) \le E_1 u(k|t) + \\ \qquad\qquad + E_4 x(k|t) + E_5 \\ u_{min} \le u(t+k) \le u_{max}, k \in [0, N-1] \\ x_{min} \le x(t+k|t) \le x_{max}, k \in [0, N] \\ y_{min} \le y(t+k) \le y_{max}, k \in [0, N-1] \\ S_x x(N|t) \le T_x \end{cases}$$

where the explanation of the elements is the same as in Equations -.

We use the Hybrid Toolbox for Matlab [13] as a design tool for the controller for the high consumption industrial furnace. This toolbox can work with several different types of hybrid system models (e.g. Mixed Logical Dynamical Systems, Piecewise Affine Systems and Discrete-time Hybrid Automata) and presents a formal mathematical equivalence between these models. We use HYSDEL to represent the model of the furnace.

In order to achieve better results we have divided the temperature domain of the furnace in five sections as presented here:

$section_1 = T_2 \in [-10, 260]$; $T_2 \in [260, 520]$ $section_2 = $; $section_3 = T_2 \in [520, 780]$;

$T_2 \in [780, 1040]$ $section_4 = $; $section_5 = T_2 \in [1040, 1300]$.

For each of the section a linearized model for the furnace was derived near the midpoint of the respective section (e.g. for $section_4$ the model was linearized near $T_2 = 910°$ Celsius).

MLD hybrid model generated from the HYSDEL file for the multi-model linearized problem has 25 continuous states, 9 inputs (4 continuous, 5 binary) and 3 continuous outputs. The HYSDEL model has 22 continuous auxiliary and 15 binary auxiliary variables. The optimization problem to be solved has 118 mixed-integer linear inequalities. The sampling time of the system is 0.5 minutes. If comparison to the hybrid model of the furnace linearized in one operating point whose HYSDEL representation has only 38 mixed-integer linear inequalities, it is obvious that the complexity of the optimization problem is significantly increased with the introduction of multi-model linearization. This affects the computation time of the optimization algorithm and favors the one point linearization method for implementation if it has satisfactory behavior.

SIMULATION RESULTS

To verify the hybrid approach for control of high consumption industrial furnace the authors have conducted series of simulations. The Disturbance signals from the front and the back door, and the timing of the pipe entering in the first zone of the furnace are graphically represented on Figure 3. On Figure 3 the logic variable for pipes entering zone 1 is resented. The logic variables for zone 2 and 3 have deterministic dependence on this value with fixed delay. In reality this delay is represented through the line speed of the conveyor driving the pipes in the furnace, but this is to be done in near future. During this simulation a fixed delay time of 10 minutes between zones is adopted. During the simulation the continuous disturbance signal T_{out} has value of 15° Celsius.

The main results are presented on Figures 4-6 where the temperatures in the respective zones of the furnace are presented long with the reference signal. The control signals applied to the three control valves respective are presented on Figures 7-9.

From the presented results it is obvious that introduceing the hybrid control approach for high consumption industrial furnace improves the quality of the control. The controller leads the system faster to the referent setpoint and the steady state error is acceptable. The hybrid MPC—one linearized model method, has also satisfactory results. Nevertheless we must point out that the tracking of the referent trajectory is best when it is near the linearization point (s), and as the referent trajectory moves from this point we have bigger error in the control algorithm. This is more expressed in the hybrid controller with only linearization point, which is linearized near 800° degrees. In this case it is obvious that output tracks the reference without any problem near this region, but if we have work plans that require a lot of temperature

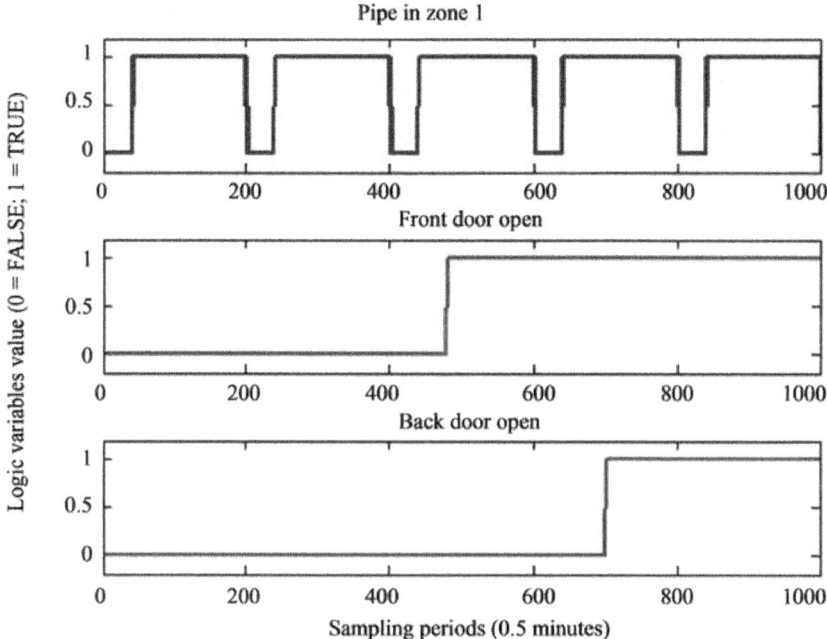

Figure 3: Timing of the logic variable disturbances of the furnace during the simulation.

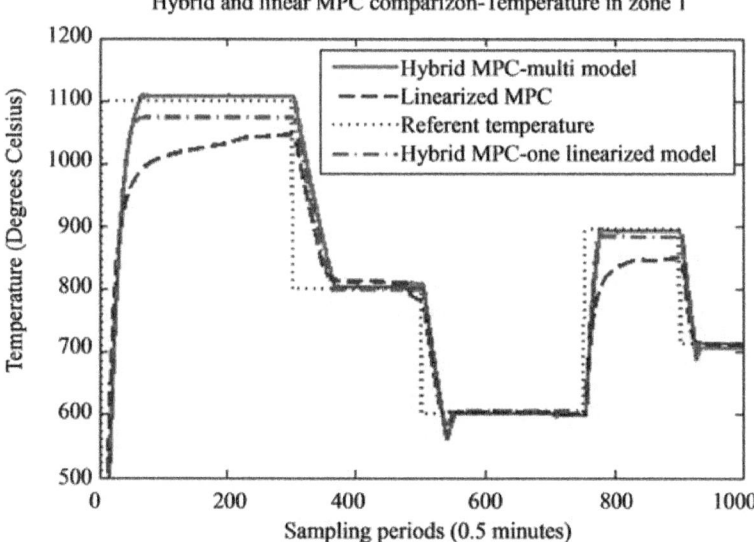

Figure 4: Temperature in the first zone of the furnace during the simulation.

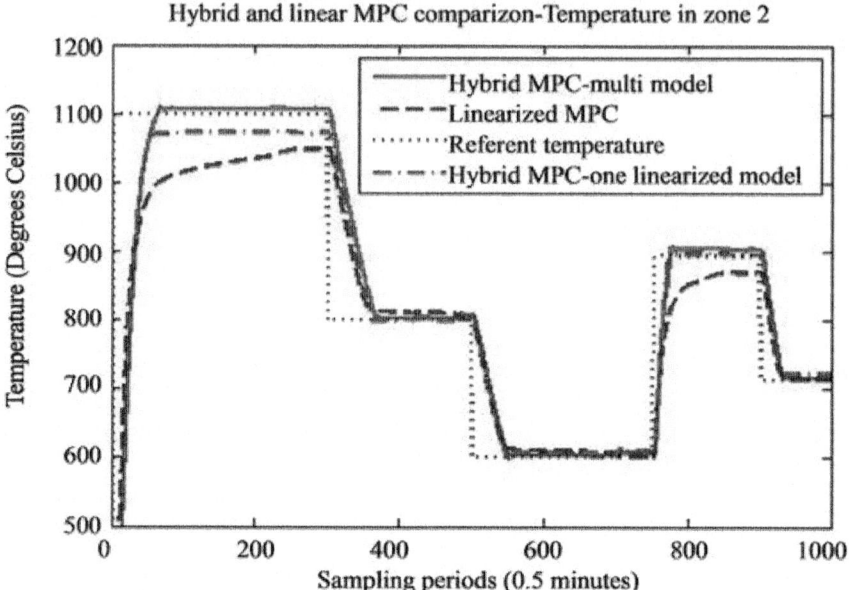

Figure 5: Temperature in the second zone of the furnace during the simulation.

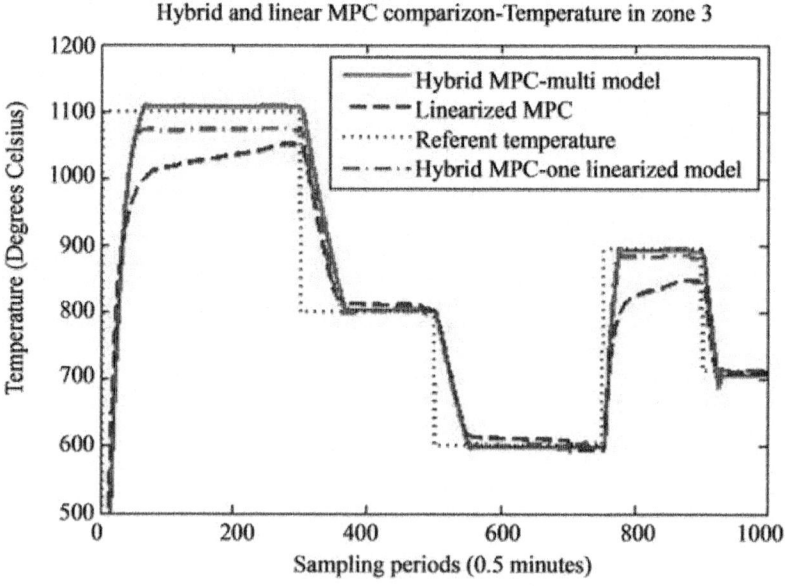

Figure 6: Temperature in the third zone of the furnace during the simulation.

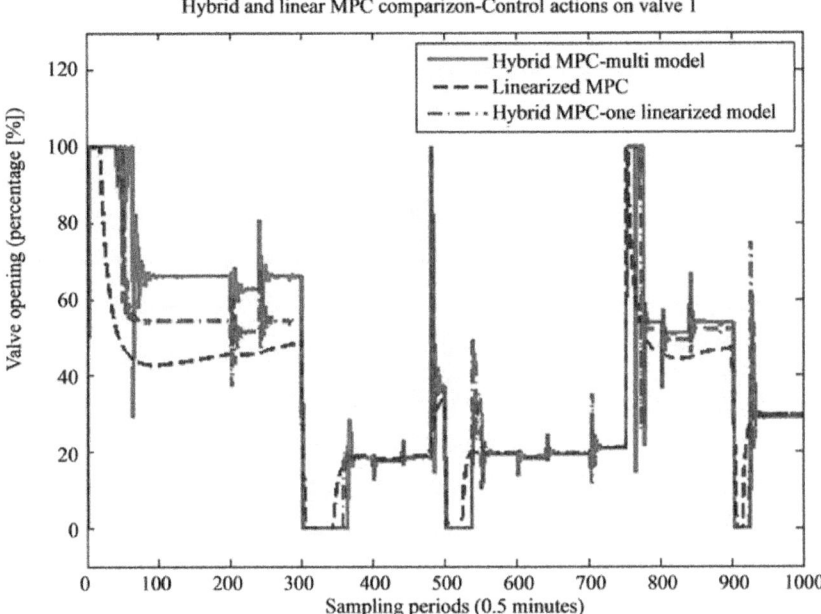

Figure 7: Valve openings on the first valve of the furnace during the simulation.

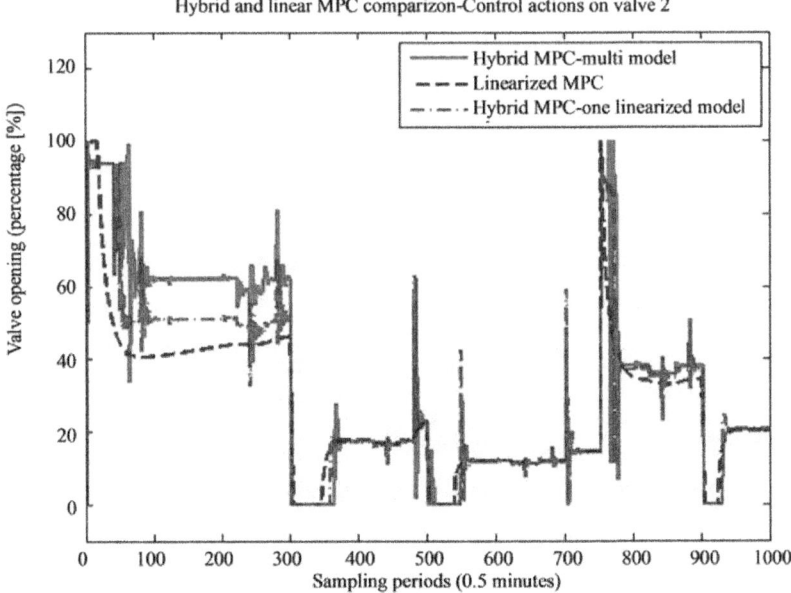

Figure 8: Valve openings on the second valve of the furnace during the simulation.

changes throughout the temperature domain of the furnace, the multi-model hybrid approach is to be considered. The previous remark, regarding the performance of the controller near the linearization point also stands for the multi-model hybrid approach. The difference here is that we have several models and the difference between the set-point and the active model cannot be very big. Logically if we introduce more models linearized in different operating point we will increase the performance of the controller, but also we will increase the complexity and the time necessary to perform the optimization.

Regarding the control signals, on all three figures (Figures 7-9) we can note that the hybrid controllers have fast reaction time to the disturbances. When there is new pipe entering in the one of the zones of the furnace, the control signal in the respective zone, acts towards

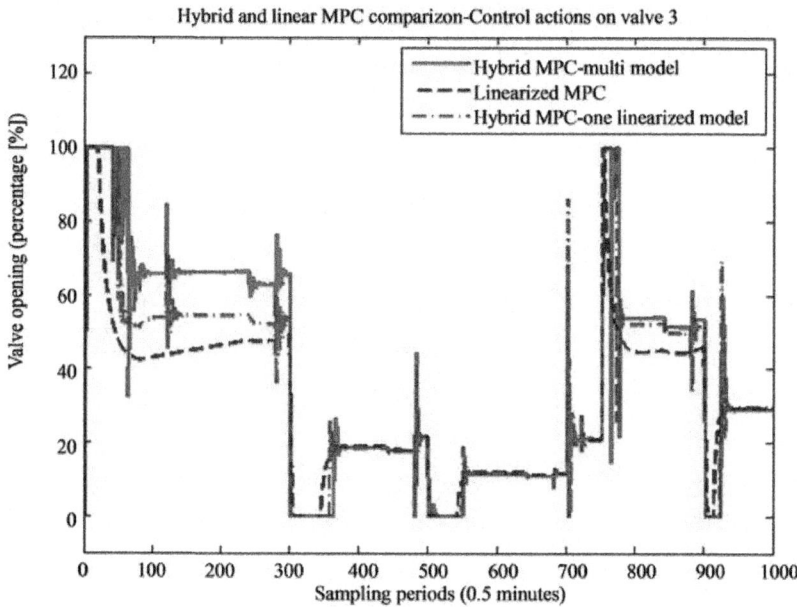

Figure 9: Valve openings on the third valve of the furnace during the simulation.

stabilization of the temperature. Also we can note that when the furnace is operating near 800° degrees, all three controller generate the same control value, but if we move far from this central linearization point, the calculated values for the control action differ a lot.

CONCLUSIONS

In this paper, a hybrid model of the high consumption industrial furnace in the factory "FZC 11 Oktomvri" in Kumanovo, R. Macedonia was presented. This approach resulted with significant improvements regarding to the linearized model that have been used before. Also we have shown that increasing the complexity of the model is not always necessary and depends on the specifics of the problem.

The new model incorporates the logic signals that act as disturbances to the furnace (new pipe entering in the zone, opening of the back and the front cooling door). Also in order to improve the performance of the furnace, multi point linearization was implemented on five characteristic points in the temperature domain. These results are confirmed with the presented simulation results.

We are also currently working to extend the results of this paper towards implementation of the line speed control of the furnace in the model and practical implementtation of the controller to the furnace.

ACKNOWLEDGEMENTS

This work was partially supported by the Faculty of Electrical Engineering and Information Technologies in Skopje, project: DEPAMPC—Development of Probability Algorithms for Model Predictive Control.

REFERENCES

1. J. Rhine and R. Tucker, "Modelling of Gas-Fired Furnaces and Boilers," McGraw-Hill, Boston, 1991.

2. G. Dimirovski, A. Dourado, N. Gough, B. Ribeiro, M. Stankovski, I. Ting and E. Tulunay, "On Learning Control in Industrial Furnaces and Boilers," Proceedings of the IEEE International Symposium on Intelligence Control, Patras, 17-19 July 2000, pp. 67-72.

3. M. Stankovski, "Non-Conventional Control of Industrial Energy Processes in Large Heating Furnaces," Ph.D. Dissertation, Ss. Cyril and Methodius University, Skopje, 1997.

4. F. Torrisi and A. Bemporad, "HYSDEL—A Tool for Generating Computational Hybrid Models for Analysis and Synthesis Problems," IEEE Transactions on Control Systems Technology, Vol. 12, 2004, pp. 235-249. doi:10.1109/TCST.2004.824309

5. Bemporad, S. Di Cairano and N. Giorgetti, "Model Predictive Control of Hybrid Systems with Applications to Supply Chain Management," Congress of ANIPLA Associazione Nazionale per LAutomazione, Napoli, 23-24 November 2005, pp. 1-15.

6. Bemporad and M. Morari, "Control of Systems Integrating Logic,

Dynamics, and Constraints," Automatica, Vol. 35, No. 3, 1999, pp. 407-427. doi:10.1016/S0005-1098(98)00178-2

7.   G. Stojanovski, M. Stankovski and G. Dimirovski, "Multiple-Model Model Predictive Control for High Consumption Industrial Furnaces," Facta Universitatis Series: Automatic Control and Robotics, Vol. 9, No. 1, 2010, pp. 131-139.

8.   G. Stojanovski and M. Stankovski, "Advanced Industrial Control Using Fuzzy-Model Predictive Control on a Tunnel Klin Brick Production," Proceedings of the 18th World Congress, The International Federation of Automatic Control, Milano 2011, pp. 10733-10738.

9.   Bemporad, W. Heemels and B. De Schutter, "On Hybrid Systems and Closed-Loop MPC Systems," IEEE Transactions on Automatic Control, Vol. 47, No. 5, 2002, pp. 863-869.doi:10.1109/TAC.2002.1000287

10.   D. Q. Mayne, J. B. Rawlings, C. V. Rao and P. O. M. Scokaert, "Constrained Model Predictive Control: Stability and Optimality," Automatica, Vol. 36, No. 6, 200, pp. 789-814. doi:10.1016/S0005-1098(99)00214-9

11.   E. Camacho and C. Bordons, "Model Predictive Control," Springer-Verlag, London, 2004.

12.   J. Maciejowski, "Predictive Control with Constraints," Prentice Hall, Upper Saddle River, 2002.

13.   Bemporad, "Hybrid Toolbox—User Guide," 2004. http://cse.lab. imtlucca.it/ bemporad/hybrid/toolbox

Chapter 7

TOWARDS A MODEL-DRIVEN IEC 61131-BASED DEVELOPMENT PROCESS IN INDUSTRIAL AUTOMATION

Kleanthis Thramboulidis[1,2], Georg Frey[2]

[1] Electrical and Computer Engineering, University of Patras, Patras, Greece

[2] Saarland University, Saarbrucken, Germany

ABSTRACT

The IEC 61131-3 standard defines a model and a set of programming languages for the development of industrial automation software. It is widely accepted by industry and most of the commercial tool vendors advertise compliance with it. On the other side, Model Driven Development (MDD) has been proved as a quite successful paradigm in general-purpose computing. This was the motivation for exploiting the benefits of MDD in the industrial automation domain. With the emerging IEC 61131 specification that defines an object-oriented (OO) extension to the function block model, there will be a push to the industry to better exploit the benefits of MDD in automation systems development. This work discusses possible alternatives to integrate the current but also the emerging specification of IEC 61131 in the model driven development process of automation systems. IEC 61499, UML and SysML are considered as possible alternatives to allow the developer to work in higher layers of abstraction than the one supported by IEC 61131 and to more effectively move from requirement specifications into the implementation model of the system.

INTRODUCTION

The IEC 61131-3 standard [1] has been adopted by the industry and is widely used by control engineers in specifying the software part of systems in the industrial automation domain. However, it imposes several restrictions for the development of today's complex systems. There is a trend to exploit best practices from the desktop application domain. Object orientation, component based development and model driven engineering are among these widely accepted best practices. Several research groups are already working to this direction, e.g., [2-5]. Standardization is also following this trend. The IEC 61499 standard [6] is considered as an extension of IEC 61131, to address among others object oriented (OO) concepts and the IEC 61131 working group is currently discussing an Object-Oriented extension to the standard [7]. Model Driven Development (MDD) was widely accepted as a successful paradigm in the desktop domain. The key issue in MDD is that models have become primary artifacts of software design, shifting much of the focus away from program code. Thus MDD refers to "a set of approaches in which code is automatically or semi automatically generated from more abstract models, and which employs standard specification languages for describing those models and the transformations between them" [8]. According to this definition the use of the Function Block Diagram (FBD) graphical programming language of IEC 611131-3 in the specification of the design model of the controller's software and its subsequent automatic translation to executable code for the target PLC platform, characterizes the development process as model driven, at least partially. Moreover, existing tools provide support for code generation for several execution platforms. This leads to the following question: what is all this research about defining MDD approaches based on IEC 61131-3 and 61499, in the industrial automation domain? The drawback of the IEC 61131 FBD is that it provides only one kind of diagram that can be used to construct the model of the application software. This diagram is composed of FB instances and their interconnections. A similar diagram is also found in IEC 61499. In this paper, we refer to this diagram with the term Function Block Network (FBN). The FBN is used in the FBD language to model the structure and the behavior of Programs or FBs. Taking into consideration now that the Unified Modeling Language (UML) and the Systems Modeling Language (SysML) use several diagrams for the definition of the application's structure and behavior, it is clear that the one diagram used by IEC 61131 is not enough to construct a reliable and expressive model for the application software. This means that all discussion about an MDD approach based on 61131 has to do with the use of more diagrams to allow: 1) more abstract models to be constructed, and 2) more aspects of the system to be captured in order to have a more complete and comprehensible model of the system. In this paper, IEC 61499, UML and SysML are considered as candidate notations to address this challenge. In a first step the IEC 61131 FBD notation is analyzed from the viewpoint of the object oriented paradigm. This is a prerequisite in order to have a sound and clear understanding of the OO concepts supported by the 61131 FBD notation. It will also simplify the process of mapping this notation to the three other notations that are examined in this paper. Instead of what is widely believed,

IEC 61131 has already introduced in the industrial automation domain, at least at the specification level, basic concepts of the OO paradigm. The Festo MPS laboratory system, a well documented system used by many universities for research and education purposes, is used as a running example in this paper. It is composed of three units. The Distribution unit, which is composed of a pneumatic feeder and a converter, forwards cylindrical work pieces from a Stack to the Testing unit. The Testing unit performs checks on work pieces for height, material type and color. Work pieces that successfully pass this check are forwarded to the rotating disk of the Processing unit, where the drilling of the work piece is performed as the primary processing of this system. The result of the drilling operation is next checked by the checking machine and the work piece is forwarded for further processing to another mechanical unit. A detailed description of the system, as well as a design of the control application based on the IEC 61499 can be found in [9]. The remainder of this paper is organized as follows: Section 2, presents the related work on using UML and SysML. In section 3, the IEC 61131 FBD notation is examined in detail from the object oriented viewpoint. The focus is on structure and behavior definition of Function Block type and Function Block network. In Section 4, we consider the potentials of using IEC 61499, UML and SysML towards a more effective MDD process based on the IEC 61131 FBD notation. The paper is concluded in the last section.

RELATED WORK

There are already several works that try to integrate IEC 61131 with UML and SysML. Vogel-Heuser, et al. in [10] describe the automatic generation of IEC 61131 code expressed in Structured Text (ST) and Sequential Function Chart (SFC) from UML 1.4 diagrams. Class diagrams are used to capture the structure of the application, while state charts are considered similar to SFCs and they are used to capture the behavior. The authors do not map the UML class to FB directly but instead they propose a complex implementation of the class concept using nested FBs. Furthermore, the object oriented view of the FB construct is not exploited and the FB diagram is considered as a diagram to capture behavior. Ramos et al., in [11] describe an OO environment they have developed for the development and implementation of distributed process control systems. This environment is based on the integration of UML with IEC 61131 and Simulink. They use class diagrams and statechart diagrams to create the requirements model of the application and they implement a mapping of SFC to UML statecharts. In fact they use UML as an intermediate model in their transformation from Simulink and IEC 61131 to Java code. Sacha in [12] describes an approach that allows the developer to model the behavior using UML state charts which are verified using UPPAAL, a model checking tool for timed automata. These state charts are then automatically transformed to an IEC 61131 program for STEP 7. This work does not give any focus on the structural aspects of the application and does not use scenario and activity diagrams that provide a more effective modeling process for the behavior, when used in combination with state charts. A UML specialization for process automation (UML-PA) is presented by Katzke, et al., in [13]. The authors propose the use of only six UML diagrams for the modeling of the system and the use of IEC 61131 languages to describe the actions in UML state charts. However, a specific mapping between these six UML diagrams and the corresponding IEC 61131 code is not given. SysML is evaluated in [14] by Chiron et al., as a modeling notation for programmable logical controllers. Even though the authors propose the use of block definition diagram (bdd) of SysML to represent structure, they consider the internal block diagram (ibd) of SysML as the proper diagram to represent the IEC 61131 FBN. They use atomic flow ports to represent the interconnection points of the FB with the environment, and the activity diagram to represent the behavior. In the decomposition process of the program they use the term task and this is confusing for those that are familiar with IEC 61131. In general the authors use the SysML to represent the application design at the same level of abstraction as the one presented using IEC 61131 FBD. In all the above works, the authors do not exploit the OO aspects of IEC 61131 which results into inefficient mappings of UML and SysML to IEC 61131. We are not aware of any work that exploits the OO view point of IEC 61131 to propose an effective MDD process based on UML or SysML.

IEC 61131: THE OO POINT OF VIEW

The FBD language of IEC 61131-3 has already introduced a few basic concepts of the object oriented programming paradigm, in the automation domain. The FB concept can be used to capture the structure and behavior of a collection of objects (instances) that may be used in an automation project. In the Festo MPS case study, the Feeder FB is the software representative of the real world feeder of the mechanical part of the laboratory system into the software domain. Figure 1(a) presents the Feeder FB that captures the interface of the corresponding software module, while Figure 1(b) presents its behavior using SFC. Input variables are used to get the sensor values; output variables are used to affect the feeder's actuators; and internal variables are used to store the state of the object.

The Function Block

The FB icon is a graphical representation, i.e., a view element, of the FB concept and it can be considered as a design time construct. There is also a textual construct to represent this concept. The graphical representation of the FB may be compared with the UML class design construct and the textual one with the class construct of OO languages such as Java and C++. It should be noted that the IEC 61131-3 allows a mixing of various textual and graphical languages to be used for the program specification which is not common in software engineering notations. Common programming languages use mainly only a textual representation, while UML or similar graphical notations for modeling, are based mainly on graphical symbols. In general, the Function Block (FB), (we use the term FB for the function block type and the term instance when we refer to an FB instance) may be considered as a special kind of class with several restrictions but also extensions. In a similar way to the class, it has a name; it defines the state of its instances using a set of local variables, declared either textually or graphically; and it defines the behavior of its instances through its body. However, there are several differentiations regarding: 1) Behavior There is a restriction in the behavior definition; only one method can be defined1 However, this method usually captures all the different behaviors exhibited by the FB instance in response to the various messages an instance receives from the environment. This means that there are no method signatures as in common OO languages; actually there is no signature even for this one method defined by the FB body. This method is executed when the FB instance is called. The call of the FB instance depends on the language used, but in all the cases at least the name of the FB instance followed by a list of actual parameters is used. Based on this description of the FB concept, the FB can be considered as a special type of class that defines the behavior of its instances by only one method. The specification of this method can be given either in textual or in graphical notation. One or more FBNs can be used to graphically specify the behavior. The SFC is more convenient in the case that the FB directly implements a state machine. In practice, commercial tools do not support parallel branches of SFC in FB specification, even though this is not defined by the standard. If parallel branches of SFC are supported, the IEC 61131 FB will be much more powerful construct compared to the IEC 61499 FB. A comparison of the semantics of SFC with those of the state chart is given in [4]. A textual representation can also be used to specify the behavior. 2) Structure There are no formal parameters defined for the method of the FB, but instead input and output variables are used for this purpose. This means that input and output variables have semantics similar to the method formal parameters, even though they are defined along with local variables as part of the structure of the FB instances. Input and output variables may be considered analogous to the flow ports defined by SysML, which constitute part of the structure of the classifier. This means that UML aggregation of type composite is supported by FBs.

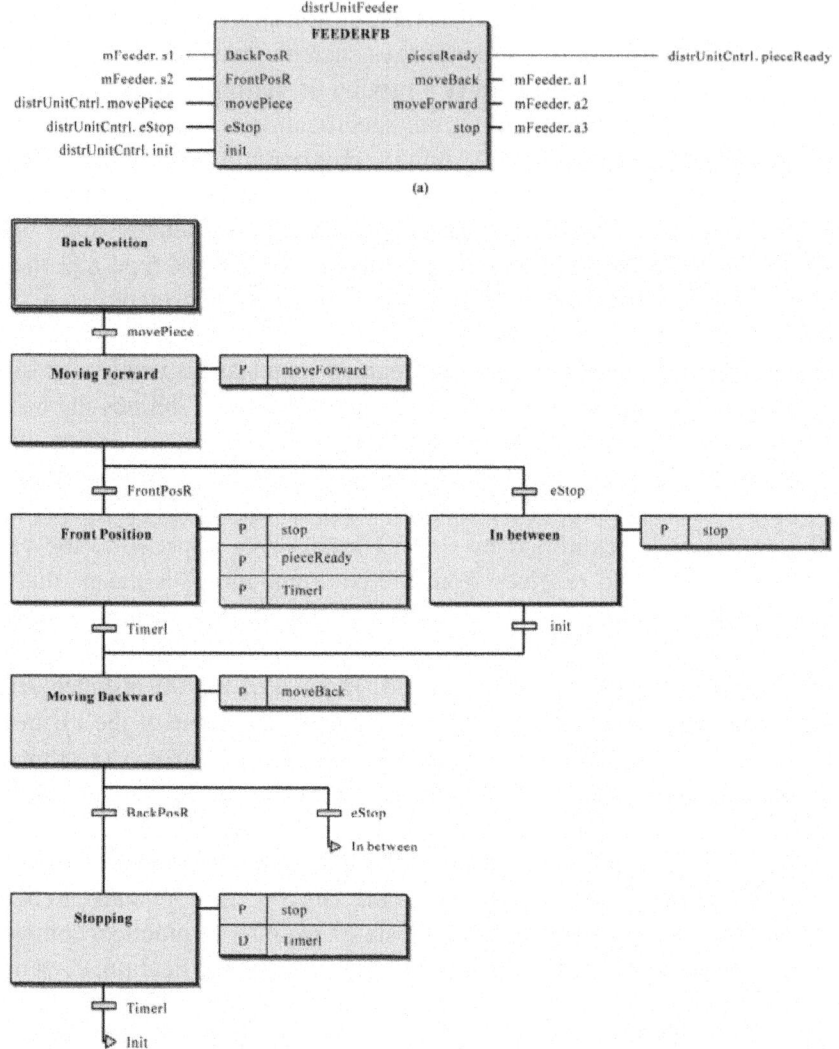

Figure 1: IEC 61131 FB for the Festo MPS feeder: (a) interface, (b) behavior specification using SFC.

It should be noted that it is not possible to define an FB variable as public, so there is no direct access to the FB state variables from outside of the FB. Output variables may be used to export state information. 3) Instantiation

Instantiation is not permitted during run time; all the instantiations should take place during the deployment of the part of the application that corresponds to the specific FB diagram. An instantiation is forced on every variable declaration of FB type. This means that only the composite kind of aggregation, which is called PartAssociation in SysML, is implemented. This kind of association implies that the composite object is responsible for the existence and storage of the composed objects, i.e., its parts. 4) Inheritance Inheritance between FB types is not supported. 5) Interface Interface definition is not supported.

The Function Block Network

The FBN is a design time artifact. Even though the FBN can be considered as a structural specification diagram like the composite structure diagram of UML 2.0 or the internal block diagram (ibd) of SysML, it is actually used, according to the standard, to capture the behavior of Programs and FBs. The closest UML diagrams that can be compared with the FBN are the activity and collaboration diagrams. However, the execution semantics of the FBN, and more specifically those related to the execution order and flow of control are quite different from the ones of the above diagrams. Moreover, the level of abstraction applied in FBN is lower compared to the level of abstraction of the UML diagrams. In the activity and collaboration diagrams of UML, the execution order and flow of control may be captured on the diagrams and explicitly specified by the designer. On the other hand, there are predefined execution semantics in FBN, such as the order of calling functions or FB instances. It is the responsibility of the designer to ensure that the proposed design exploits properly the predefined execution semantics in order to get the desired behavior. The FBN is analogous to the abstract syntax tree generated by a C compiler to calculate the value of an expression. Like the Activity diagram, the FBN diagram captures the activities that have to be performed and the flow of information between these activities. An FBN that includes only FB instances is more close to a collaboration diagram with the restriction that every object has one method. Message passing between the nodes of an FBN is not realized in terms of method call that is the common mechanism for message passing in OO languages. In fact, there are no method signatures, but just one FB method which is activated when the FB instance is called. An FB instance can be called by any Program Organization Unit (POU) that has visibility access to it. The FB method is executed on the specific instances structure, i.e., inputs, state and output. The result is normally: 1) a new instance state that is captured in the instance's internal variables, in the case of state depended behavior, and 2) a response to the environment that is captured by the instance's output variables and the value of the FB call. Message passing between FB instances allocated in different resources in the same configuration is implemented by global variables (VAR_GLOBAL) of the configuration, while between those allocated in different configurations is implemented through access paths defined in configurations with VAR_ACCESS.

TOWARDS A MORE EFFECTIVE MODELING PROCESS FOR IEC 61131

The Need for More Diagrams and for Higher Levels of Abstraction in Application Modeling

As it was stated above, the IEC 61131-3 FBD language can be considered as a realization of the MDD paradigm in the industrial automation domain. However, it can be used to model the application to a very low level of abstraction, very close to the executable code. It does not allow the designer to capture in a graphical way all the aspects of the application, as for example its structure. Moreover, even though the execution semantics are well defined and common for all the IEC 61131-3 compliant execution environments, it is more informative for the control engineer to have a graphical representation of this information using sequence or activity diagrams. In this case, these diagrams may be used even before the assignment of the application activities to function blocks, allowing a more flexible transition from the system requirements to the application design specifications. In order to increase the effectiveness of this MDD process, more models should be exploited, e.g., to capture the behavior, and also new ones more abstract should be used, e.g., to support the definition of the structure of the application. These models, if properly defined, would be automatically transformed to the IEC 61131 FBD notation and finally to executable code for existing run-time environments. In this section, three notations are considered as possible candidates for such an extension of the IEC 61131-based modeling process: 1) the IEC 61499 function block model, 2) the UML, and 3) the systems modeling language SysML. Moreover, it should be noted that the device centric approach, which is currently used with the IEC 61131, does not provide a strong request for higher layers of abstraction in the development process. However, this approach does not allow for the adoption of a synergistic development process of the constituent parts of the automation system that is the current trend in Metchatronic systems development [15], such as the industrial automation systems. The application centric paradigm, that is proposed to better fit with the synergistic development of the constituent parts of the automation system, is almost impossible to be applied without special emphasis on the architecture of the system. This is why UML and SysML may be used as early specifications of the automation system, during the development process, before the 61131 one. In [16] a detailed discussion, including examples, is given on the limitations of IEC 61499 to be used as notation for architectural specification. On the other hand UML and SysML are widely accepted as architecture specification languages.

Using IEC 61499

The IEC 61499 FB model has been proposed by IEC as an extension of the IEC 61131 FB one to exploit OO concepts and address several other challenges in industrial automation systems, such as interoperability, portabi-lity, run-time reconfigurability, etc. However, even though several researchers have published many articles on the applicability of the new standard, e.g. [17], and also on the migration from IEC 61131 to IEC 61449, such as Gerber et al., in [18] and Hussain et al. in [19], the industry has not yet made serious steps towards its adoption [16]. When the first proposal for the ObjectOriented extension of IEC 61131 FB model was presented by CODESYS, a debate has raised between the two communities [20]. Does the industry need both standards or the new version of IEC 61131 will officially denote the abandonment of IEC 61499? Or, in other words, what is the value added of IEC 61499 compared to IEC 61131? In this subsection we assess the potential of using IEC 61499 to enhance the IEC 61131 development process. Due to our assumption to be compliant with the execution semantics of IEC 61131 that includes execution of the program based on cyclic or periodic mode, the events of IEC 61499 FB, at least those that interconnect the FBN with the controlled system, are useless. Based on this, the main differences between IEC 61131 and IEC 61499 FB are: 1) The IEC 61499 uses the ECC to capture the dynamic behavior of its instances. However, this is nothing more than a graphical way of representing the IEC 61131 FB body that can also be done using SFC, as shown in Figure 1. 2) The ability to specify in IEC 61499 the behavior to incoming events as independent distinct algorithms (methods). This allows for a more modular FB body but since there is no support for inheritance there are no direct benefits regarding reusability. It should be noted that the restriction of IEC 61499 to disallow a direct method call bypassing the ECC is considered very positive for this kind of systems. However, the introduction of the algorithm scheduling function imposes a very heavy constraint on the Safety Integrity Level that can be obtained using IEC 61499. The differences are more important when we consider the FBNs. As already mentioned the IEC 61131 FBN is used to model the behavior. The IEC 61499 FBN is quite similar to the UML composite diagram or the SysML internal block definition (ibd) diagram. It represents the structure of the application or the composite FB and all the possible interactions among them. Figure 2 presents an FBN for the distribution unit of the Festo MPS example application. As argued by Thramboulidis et al. in [21], the FBN is not considered appropriate for behavior modeling. This means that the IEC 61499 does not provide a better way to capture the behavior of the application compared to the IEC 61131. It is only the FBN that has to be evaluated as a possible diagram to enhance the IEC 61131 development process. But as argued in [19], the IEC 61499 is not considered as an effective architecture specification language. Based on the above analysis, the authors do not see any valuable benefit in using IEC 61499 to enhance the IEC 61131 development process. It does not provide diagrams for higher level of abstractions models, nor even diagrams to effectively model the other aspects of the application.

Using UML

It is evident that a specific UML class may be used to represent the IEC 61131 FB. Moreover, several UML diagrams can be used to create more abstract models of the application, compared to the one supported by the IEC 61131 FBD. More specifically: 1) The UML composite diagram can be used to capture the structure of programs and FBs. 2) The state diagram can be used to capture the behavior of programs and FBs. 3) The sequence and/or activity diagrams may be used to model the behavior in terms of interactions between FB instances, and to model the behavior of an enclosing FB. 4) The class diagram may be used to represent the structure of the PLC infrastructure. 5) The deployment diagram may be used as graphical representation of a configuration. To have a better exploitation of the above UML diagrams in the domain of IEC 61131, a UML profile may be defined. This profile will contain stereotypes for the main key constructs of IEC 61131. For example the specific UML class that will be used to represent the IEC 61131 FB will be a stereotype of this profile. This profile will allow the industrial engineer to build the models using already known constructs and at the same time to work in the UML abstraction layer. This means that it is possible to create the model of the application in more abstract format compared to IEC 61131, while still using the IEC 61131 basic terminology. Specific model-to-model transformers are required to have an automatic transformation of the UML application specification to an IEC 61131 based one, which will be subsequently translated using existing tools to executable code for commercial run-time environments. The definition of the meta-models of the source and target domains, as well as the set of mapping rules between them, is a prerequisite for this approach to be effectively implemented. A problem with UML is that the class concept is specifically defined to support the OO programming paradigm, so UML does not provide inherent support for the procedural paradigm that is also supported by IEC 61131. This is why SysML is considered a better choice for building a more effective higher layer of abstraction on top of IEC 61131.

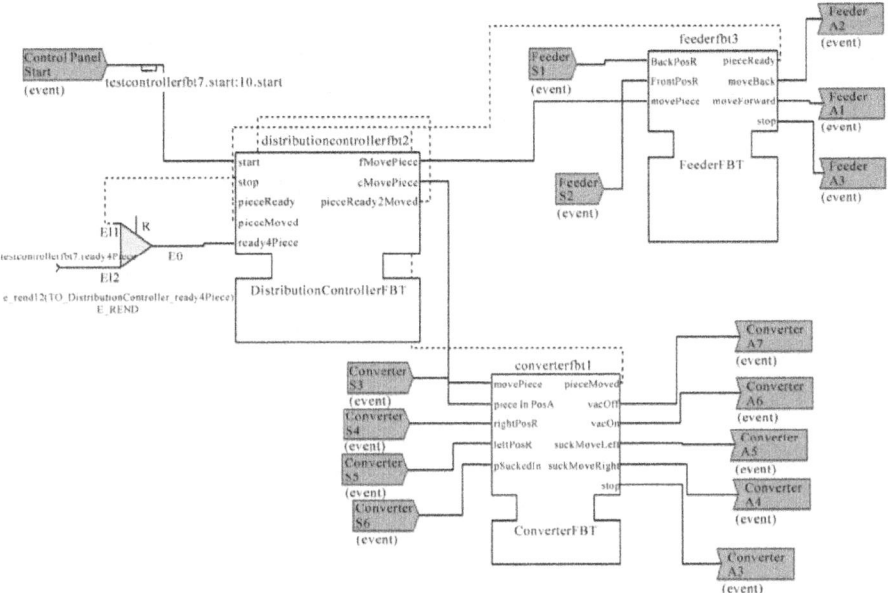

Figure 2: IEC 61499 function block network for the Festo MPS distribution unit.

Using SysML

The construct of block that is the basic static, structural construct of SysML, is broader than the UML class. The block can be used to represent the modular units of system description. A block can define the "type of logical or conceptual entity, a physical entity, a hardware, software or data component, a person, a facility, an entity that flows in the system or even entities from the environment" [22]. This means that the block construct of SysML may be used to represent not only an FB, but also a function and a configuration. The ⬦ stereotype will be used to represent the IEC 61131 FB and the ⬦ stereotype will be used to represent the IEC 61131 function. In a similar way ⬦ and ⬦ will be defined to represent a program and a configuration respectively. The SysML block definition diagram (bdd) can be used to capture the structure of configurations, programs and also FBs that may accept FB instances as input variables. Figure 3 shows the bdd for the Distribution Unit block of the Festo MPS modeled in SysML. It consists of a Feeder FB instance, a Converter FB instance and an FB instance of type DistributionUnit, that coordinates these. The same diagram presents also the interfaces of Feeder and Converter, as well as the flow specifications for the interactions of real world feeder and converter with the corresponding FB instances. The SysML internal block diagram (ibd) can be used to capture the interconnections among the constituent parts of constructs, such as configurations, programs and also FBs that may accept FB instances as input variables. The sequence diagram can be used to capture the interactions of the system components in the context of a particular operation of the system. The SysML sequence diagram, shown in Figure 4, captures the interaction of the system components in the context of the reaction of

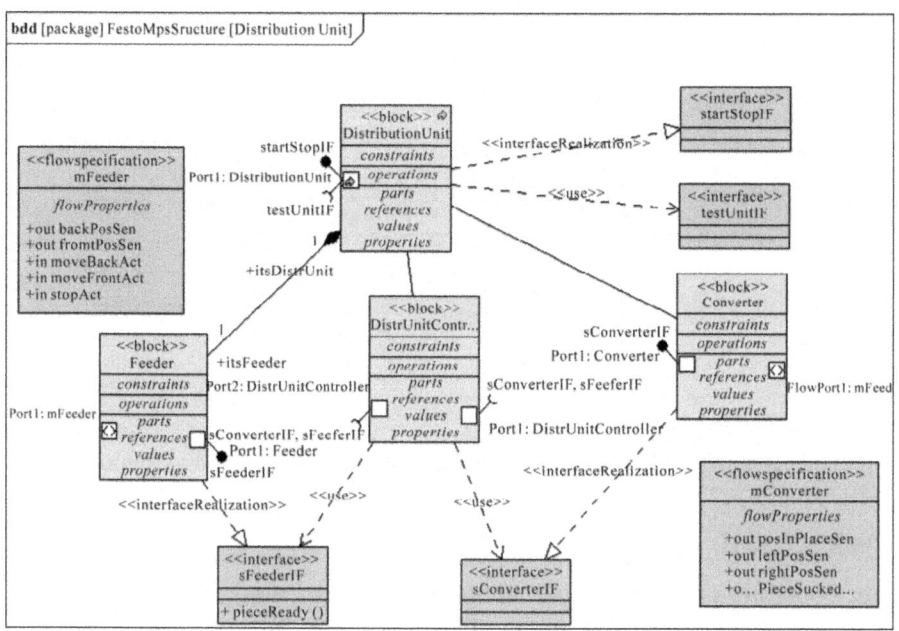

Figure 3: SysML block definition diagram (bdd) for the festo MPS distribution unit.

the control application to the sensor event frontPosReached that comes from the corresponding sensor of the mechanical Feeder. From SysML design diagrams as the ones shown in Figures 3 and 4, using proper model-to-model transformers, we may get the corresponding IEC 61131 design diagrams for the application. In this way SysML can be used to provide a completely graphical environment for modeling control applications in a higher and also more expressive layer than the one already supported by IEC 61131 market tools. This profile can be developed from scratch or more effectively based on an existing profile in the embedded real-time systems domain, such as the MARTE profile [23]. Assuming that the corresponding profile will be standardized, this approach will result in a uniform way of modeling industrial automation software that will hide the proprietary individual characteristics of IEC 61131 tools at the modeling layer.

CONCLUSIONS

The IEC 61131-3 standard has already introduced in the industrial automation domain basic concepts of the OO paradigm, even though these are not widely exploited in practice by existing development tools. It has also introduced the use of graphical notations, i.e., the use of models, in the development process. The FBD language is an attempt to use graphical models in the development process that is one of the basics of model driven development paradigm. However, the FBD notation does not support modeling at high levels of abstraction, that is required when the size and complexity of the application is increasing. Moreover, it does not allow the modeling of all aspects of the application. We have discussed, in this paper, three notations as candidates to address the increasing size and complexity of industrial applications. IEC 61499, UML and SysML were examined for possible integration with IEC 61131 to increase the effectiveness of the IEC 61131-based development processes. IEC 61499 does not provide any valuable benefit in enhancing the IEC 61131 development process. UML may be successfully used, but SysML seems to better match the semantics of IEC 61131. With a proper profile, SysML can be used to define a completely graphical environment that will allow the industrial automation developer to create the model of the application and automatically transform it to IEC 61131 specification. This integration may also hide the proprietary individual characteristics of IEC 61131 market tools at the modeling layer and result in a uniform modeling notation.

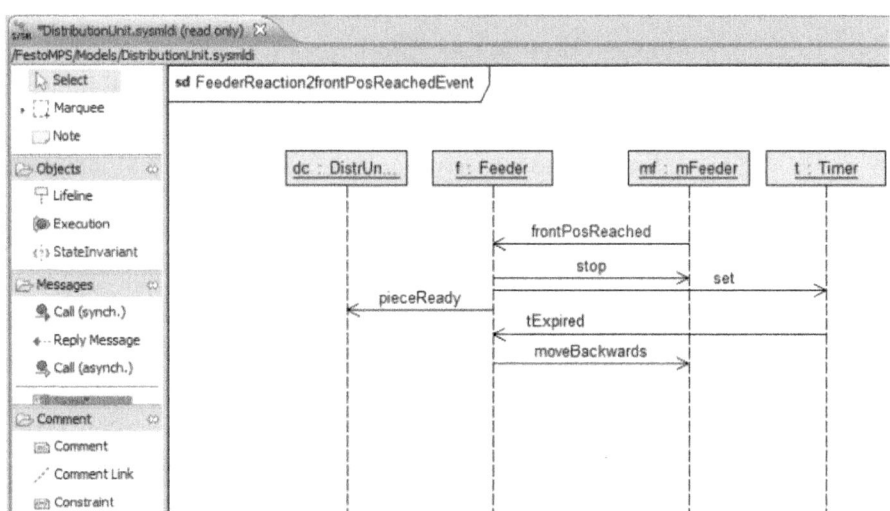

Figure 4: SysML Collaboration diagram for the reaction of the Feeder unit to the frontPosReached signal of the corresponding Feeder sensor.

REFERENCES

1. International Electrotechnical Commission, "IEC International Standard IEC 61131-3: Programmable Controllers, Part 3: Programming Languages," IEC, 2003.

2. G. Doukas and K. Thramboulidis, "A Real-Time Linux Based Framework for Model-Driven Engineering in Control and Automation," IEEE Transactions on Industrial Electronics, Vol. 58, No. 3, March 2011, pp. 914-924.

3. D. Streitferdt, G. Wendt, P. Nenninger, A. Nyssen and H. Lichter, "Model Driven Development Challenges in the Automation Domain," 32nd Annual IEEE International Conference on Computer Software and Applications, Turku, 28 July-1 August 2008, pp. 1372-1375.

4. S. Panjaitan and G. Frey, "Combination of UML Modeling and the IEC 61499 Function Block Concept for the development of Distributed Automation Systems," 11th IEEE International Conference on Emerging Technologies and Factory Automation, Prague, September 2006, pp. 766-773.

5. K. Thramboulidis, D. Perdikis and S. Kantas, "Model Driven Development of Distributed Control Applications," The International Journal of Advanced Manufacturing Technology, Vol. 33, No. 3-4, 2007, pp. 233-242.

6. International Electrotechnical Commission, "International Standard IEC61499, Function Blocks, Part 1-Part 4," IEC, 2005.

7. B. Werner, "Object-Oriented Extensions for IEC 61131," IEEE Industrial Electronics Magazine, Vol. 3, No. 4, 2009, pp. 36-39.

8. B. Selic, "From Model-Driven Development to ModelDriven Engineering," 19th Euromicro Conference on Real-Time Systems, Pisa, 4-6 July 2007.

9. K. Thramboulidis, "Design Alternatives in the IEC 61499 Function Block Model," 11th IEEE International Conference on Emerging Technologies and Factory Automation, Prague, September 2006, pp. 1309-1316.

10. B. Vogel-Heuser, D. Witsch and U. Katzke, "Automatic Code Generation from a UML model to JEC 61131-3 and System Configuration Tools," International Conference on Control and Automation, Budapest, 27-29 June 2005, pp. 1034-1039.

11. D. N. Ramos-Hernandez, P. J. Fleming and J. M. Bass, "A Novel Object-Oriented Environment for Distributed Process Control Systems," Control Engineering Practice, Vol. 13, No. 2, 2005, pp. 213-230.

12. K. Sacha, "Verification and Implementation of Dependable Controllers," 3rd International Conference on Dependability of Computer Systems DepCoS-RELCOMEX, Szklarska Poreba, Poland, June 2008, pp. 143-151.

13. U. Katzke and B. Vogel-Heuser, "Combining UML with IEC 61131-3 Languages to Preserve the Usability of Graphical Notations in the Software Development of Complex Automation Systems," 10th IFAC, IFIP, IFORS, IEA Symposium on Analysis, Design, and Evaluation of Human-Machine Systems, Seoul, September 2007.

14. F. Chiron and K. Kouiss, "Design of IEC 61131-3 Function Blocks Using SysML," Mediterranean Conference on Control & Automation, Athens, 2007.

15. K. Thramboulidis, "The 3 + 1 SysML View-Model in Model Integrated Mechatronics," Journal of Software Engineering and Applications, Vol. 3, No. 2, 2010, pp. 109-118.

16. K. Thramboulidis, "IEC61499 Function Block Model: Facts and Fallacies," IEEE Industrial Electronics Magazine, Vol. 3, No. 4, December 2009, pp. 7-26.

17. A. Zoitl and V. Vyatkin, "IEC 61499 Architecture forDistributed Automation: The 'Glass Half Full' View'," IEEE Industrial Electronics Magazine, Vol. 3, No. 4, 2009, pp. 7-23.

18. C. Gerber, H. M. Hanisch and S. Ebbinghaus, "From IEC 61131 to IEC 61499 for Distributed Systems: A Case Study," EURASIP Journal on Embedded Systems, Vol. 2008, Article ID 231630, p. 8.

19. T. Hussain and G. Frey, "Migration of a PLC Controller to an IEC 61499 Compliant Distributed Control System: Hands-on Experiences," 21st IEEE International Conference on Robotics and Automation, Barcelona, 18-22 April 2005, pp. 3984-3989.

20. K. Thramboulidis, "The Function Block Model in Embedded Control and Automation: From IEC61131 to IEC61499," WSEAS Transactions on Computers, Vol. 8, No. 9, September 2009, pp. 1597-1609.

21. K. Thramboulidis, G. Doukas and A. Frantzis, "Towards an Implementation Model for FB-Based Reconfigurable Distributed Control Applications," 7th International Symposium on Object-Oriented Real-Time Distributed Computing, Viena, 2004, pp. 193-200.

22. OMG, "OMG Systems Modeling Language," V1.0, September 2007.

23. OMG, "A UML Profile for MARTE: Modeling and Analysis of Real-Time Embeded systems," Beta 2, June 2008.

Chapter 8

A SECURE, INTELLIGENT, AND SMART-SENSING APPROACH FOR INDUSTRIAL SYSTEM AUTOMATION AND TRANSMISSION OVER UNSECURED WIRELESS NETWORKS

Aamir Shahzad [1], Malrey Lee [1] Neal Naixue Xiong [2,3], Gisung Jeong [4], Young-Keun Lee [5], Jae-Young Choi [6], Abdul Wheed Mahesar [7] **and** Iftikhar Ahmad [8]

[1]Center for Advanced Image and Information Technology, School of Electronics & Information Engineering, Chon Buk National University, 664-14, 1Ga, Deokjin-Dong, Jeonju 561-756, Korea

[2]School of Information Technology, Jiangxi University of Finance and Economics, Nanchang 330013, China

[3]Department of Business and Computer Science, Southwestern Oklahoma State University, Oklahoma, OK 73096, USA

[4]Department of Fire Service Administration, WonKwang University, Iksan 570-749, Korea

[5]Department of Orthopedic Surgery, Chonbuk National University Hospital, Jeonju 561-756, Korea

[6]College of Information and Communication Engineering, Sungkyunkwan University, Suwon 16419, Korea

[7]Department of Computer Science, International Islamic University Malaysia, Kuala Lumpur 53100, Malaysia

[8]Department of Software Engineering, College of Computer and Information Sciences, King Saud University, Riyadh 11543, Saudi Arabia

ABSTRACT

In Industrial systems, Supervisory control and data acquisition (SCADA) system, the pseudo-transport layer of the distributed network protocol (DNP3) performs the functions of the transport layer and network layer of the open systems interconnection (OSI) model. This study used a simulation design of water pumping system, in-which the network nodes are directly and wirelessly connected with sensors, and are monitored by the main controller, as part of the wireless SCADA system. This study also intends to focus on the security issues inherent in the pseudo-transport layer of the DNP3 protocol.

During disassembly and reassembling processes, the pseudo-transport layer keeps track of the bytes sequence. However, no mechanism is available that can verify the message or maintain the integrity of the bytes in the bytes received/transmitted from/to the data link layer or in the send/respond from the main controller/sensors. To properly and sequentially keep track of the bytes, a mechanism is required that can perform verification while bytes are received/transmitted from/to the lower layer of the DNP3 protocol or the send/respond to/from field sensors. For security and byte verification purposes, a mechanism needs to be proposed for the pseudo-transport layer, by employing cryptography algorithm. A dynamic choice security buffer (SB) is designed and employed during the security development. To achieve the desired goals of the proposed study, a pseudo-transport layer stack model is designed using the DNP3 protocol open library and the security is deployed and tested, without changing the original design.

INTRODUCTION

In the last two decades, a number of enhancements have been made in industrial sectors such as Water, Oil, Gas, and Electric. The communication methods have been changed from standalone systems to network based systems; furthermore, to fulfill the current communication demands of industries, there is also a requirement to connect the industrial remote (geographical) located stations to one or more centralized station. To connect several remote located stations, the best, most efficient and cost effective way is to use the wireless technologies (*i.e.*, cellular, satellite, and others). Through the deployment of wireless technology, the industrial systems or the SCADA systems are able to access, monitor, and control their remotely located networked stations from a control center in minimal time; the wireless technology also overcomes the costs required with wired technology [1,2].

Human machine interface (HMI) is part of the SCADA system that provides interaction between supervisory control and data acquisition (SCADA) operators and devices. SCADA/HMIs are highly designed graphics-based interfaces in which a SCADA monitor and controlling systems are visualized; SCADA system operators use the HMI to manage the overall network structure and communication is usually displayed in the form of text (or text stream) and graphic symbols. SCADA systems are highly distributed network systems in which a number of field devices are located graphically and controlled for the main center; multimedia audio/video contents are employed at a remote location, embedded with processes, and monitored from the control center in case there is no operator available at that site [3,4]. Typically, SCADA field devices are designed for low bandwidth transmission over serial channels and the SCADA communication links and employed protocols are also limited for low bandwidth access, and are inadequate to accomplish the requirements of advance multimedia components such as audio/video. Thus, the SCADA system components need to be integrated with advanced communication networks to enable the employment or integration of advanced multimedia applications into the SCADA systems [1,4,5,6,7].

With the arrival of new technology, SCADA systems are also connected to numbers of advanced networks such as LAN/WAN and cellular networks; significant changes have also been observed in SCADA protocol designs and network supported specifications, which are connected to the internet for faster multimedia information delivery via non-proprietary transport control protocol (TCP). This means that the SCADA messages are constructed then passed to lower layered protocols (e.g., TCP/IP and UDP), which are usually designed to manage and meet the needs of a high bandwidth corresponding to the transmission of SCADA multimedia applications [1,8,9,10,11]. For example, the delivery time of SCADA messaging may typically be in the range of 10 ms to 100 ms, while the reliable and sequential delivery of messages (or packets) is also an important factor in the SCADA system. If we transmit the packets using SCADA serial protocols (e.g., DNP3 protocol and Modbus protocol) over Ethernet LAN using 100 Mbps transmission link, then a SCADA message (or multimedia message) can be sent in less than or within the range of 10 ms. However, converters and/or gateways such as 5201-DFNT-DNPM, Moxa NPort 6110, VLINX MODBUS, and PLX31-MBTCP-MBS are used during the transmitting of a message from SCADA serial protocols to TCP/IP protocols and vice versa. Nowadays, SCADA predominated protocols such as the DNP3 protocol and Modbus protocol are also available in TCP/IP versions as the DNP3 TCP/IP protocol and Modbus TCP/IP protocol, respectively; the frames are constructed according to protocol specifications and directly encapsulated into TCP/IP packets to ensure reliable transmission over LAN/WAN [1,11,12,13]. Moreover, the network architecture of the SCADA system is illustrated in Figure 1.

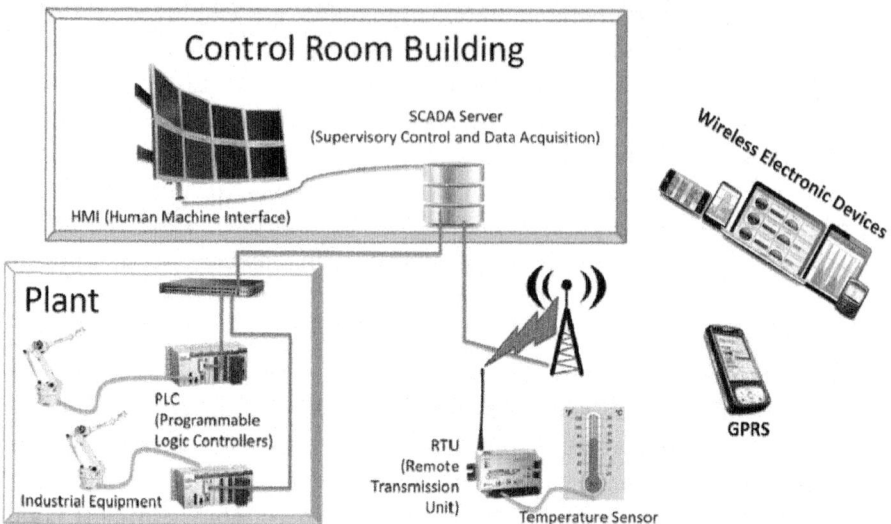

Figure 1: SCADA system and network components.

In SCADA system, DNP3 protocol has been considered as an major protocol due to its functionalities and reliable communication over the physical channels, but at the same time it has several security issues while travelling over the open networks and/or Internet [11,13]. As a consequence, the proposed study focuses on the DNP3 pseudo-transport layer security issues that most often occur during the transmission of fragments and a corresponding security mechanism or hashing function is deployed as a strong security wall that provides protection against adversaries (*i.e.*, integrity attacks); however, this development is also able to test the other cryptography algorithms according to security demands. In addition, attack scenarios are defined in which attackers gain access to the fragments, user defined attacks are launched by employing various built-in tools for performance evaluation purposes, formal proofs are employed for validation purposes, and approximate communication is visualized as part of the multimedia technology.

The remainder of this research paper is organized as follows. Section 2 reviews the related works of study. Simulation Design and Environment is explained in Section 3, Section 4 describes the Pseudo-Transport layer message structure, Section 5 describes the Payload Design and Security Development, and Section 6 explains the Algorithm. In Section 7, Attacking Scenarios are defined, while the Setup and Abnormal Communication, as well as Measurement and Discussion are explained in Section 8 and Section 9. Multimedia Contexts are highlighted in Section 10. The significance of the study is discussed in Section 11 and Section 12 provides the conclusion and suggestions for future research.

RELATED WORK

SCADA system security issues [14,15,16,17,18,19] have been considered as the most prominent and important counter measures of communication [11,12,13]. Therefore, an evaluated potential method is proposed that would be significant to fight against SCADA security challenges; however, security enhancement is limited to specified goals (or security goals). Typically, SCADA system networks and their components are distributed in various locations including in one specific place, in many cities in a country, and around the world. To connect the several networks points, SCADA has been employed in various wired/wireless communication media and the transmission can be accessed over modern technology platforms such as cellular phones using 2G, 3G, 4G, and general packet radio service (GPRS) [20,21,22]. However, overall SCADA communication is carried out by non-proprietary protocols which are ranked above the SCADA proprietary protocols [11,12,13,14].

The larger SCADA system defines the communication structure between the master terminal unit (MTU) and the remote terminal unit (RTU) or/and RTU and MTU. Each station is identified as a master or client/slave station in the SCADA network. However, in a SCADA hierarchical structure, some field devices perform the function of master and slave together. Two terms are defined within the data link layer such as balanced and unbalanced communication. In the application layer, the application protocol control information (APCI) defines data/message that is requested/responded; and response header differs by two additional bytes designated as internal indications (IIN). In the case of an unsolicited response, message is received from the terminal station to the master station and the master station responds to the terminal station. Therefore, different forms of header are added during message construction in the application layer, while the data link layer link protocol data unit (LPDU) bytes remain unchanged in either the message sent from the master station or the terminal station [20,21]. In an unbalanced system, only the master station is able to send the request and will respond according to the request slave station. This means that the master station works as a primary station and other stations work as terminals in an unbalanced system. Whenever the master station sends a request, the substation will then be able to send a response to the master station. However, in a balanced system, each station in the SCADA hierarchical structure acts as a master or slave at the same time. To distinguish between the master and the outstation in the balance system, a direction bit or DIR is set within the message from the master station to the terminal station or from one station to another station. Therefore, any station can initialize or send a request to other stations in the SCADA network. As part of the link layer, a cyclic redundancy check (CRC) is employed which performs the function of detecting errors in the transmission, while the detection mechanism is limited for information authentication and authorization [20,21]. Figure 2 illustrates the DNP3 protocol model and data link layer design [20].

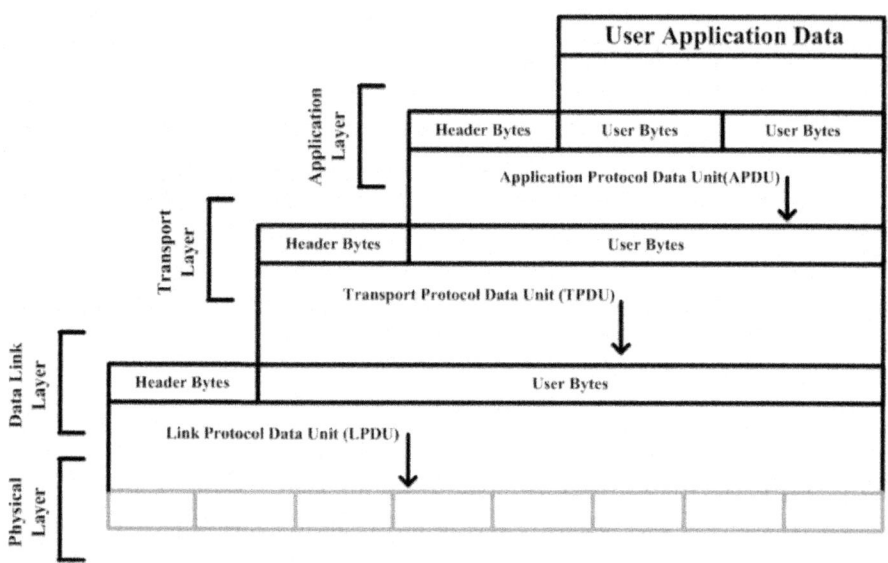

Figure 2: DNP3 protocol model.

Taxonomy of DNP3 protocol attacks is developed, in which attacks are categorized into three main groups: (i) DNP3 specifications attacks; (ii) DNP3 vendor based attacks; and (iii) DNP3 underlying infrastructure attacks [13]. The DNP3 specifications attacks are more prominent and harmful among the other attack groups; The SCADA system is targeted, which then suffers, and in this case, transmission is carried out by the DNP3 protocol [12,13]. The initial DNP3 protocol was designed without considering security; on the other hand, the DNP3 protocol resides in non-proprietary protocols (*i.e.*, TCP and UDP) for the purpose of information delivery on the internet [12,13,18]. As analyzed, three issues of interception, interruption, and modification always interact with the SCADA system and/or its component paths, including the main controller, outstation (or sub-controller), and communication network [11,12,13]. Typically, the DNP3 protocol design does not deploy the potential security mechanisms such as authentication, encryption, and authorization. Due to security limitations, outside attackers can easily interrupt the DNP3 transmission, or directly target the DNP3 layers including the application layer, the pseudo-transport layer, and the data link layer; and the configured DNP3 nodes are also not able to analyze that the incoming message, and its contents are valid, or have not been changed during transmission [11,13,14]. The attack taxonomies for the DNP3 pseudo-transport layer and data link layer are depicted in Table 1 and Table 2 [13].

Table 1. Attack taxonomy for the DNP3 pseudo-transport layer

No.	Attacks	Attacks Instances (Description)
1	Passive Network Reconnaissance	Interception of main station, sub-controller, and network information.
2	Baseline Response Replay	Interruption, modification, and fabrication of the main station and sub-controller.
3	Rogue Interloper	Interruption, modification, and fabrication of the main station, sub-controller, and network information.
4	Fragment Interruption	Interruption of the main station and sub-controller.
5	Sequence Modification	Interception of Main Station, sub-controller, and network information.

Table 2. Attack taxonomy for the DNP3 data link layer

No.	Attacks	Attacks Instances (Description)
1	Passive Network Reconnaissance	Interception of main station, sub-controller, and network information.
2	Baseline Response Replay	Interruption, modification, and fabrication of the main Station and sub-controller.
3	Rogue Interloper	Interruption, modification, and fabrication of the main station, sub-controller, and network information.
4	Length Overflow Attack	Interruption and modification of the main station and sub-controller.
5	Flag Attack	Interruption of sub-controller.
6	Reset Function Attack	Interruption and modification of the main station and sub-controller
7	Unavailable Function Attack	Interruption of the main station
8	Destination Address Alteration	Interruption, modification, and fabrication of the main station, sub-controller, and network information

DNP3 protocol layers such as the application layer and link layers are considered more vulnerable to security threats than the pseudo-transport layer. This is because the pseudo-transport layer provides fewer functionalities than the other layers of the DNP3 protocol; therefore, a limited number of attacks are linked with the pseudo-transport layer [11,13,23,24,25]. However, two potential attacks account for the pseudo-transport layer: first frame (FIR) and final frame (FIN) flags interruption and sequence number modification [11,12,13]. In the DNP3 original design, there is no defined mechanism that detects abnormal entities in the transmission; therefore, DNP3 devices (or nodes) are unaware in cases where unauthorized entities are successful in transmission by attacks such as interruption, modification, and fake reply [11, 12,13,16,17,18,19,20,21,22,23,24,25]. However, the major explained pseudo-transport layer attacks fall under the category of integrity attacks and should be resolved by employing cryptography based integrity functions.

In [26,27,28,29], cryptography based end-to-end security mechanisms are used for SCADA systems, and various cryptography algorithms such as symmetric (*i.e.*, AES and DES), asymmetric (*i.e.*, RSA, Diffie-Hellman, and DSS), and hashing (*i.e.*, MD5 and SHA2) algorithms are deployed to secure the SCADA communication from networks adversaries such as message sniffers, man-in-the-middle attackers, eavesdroppers and password crackers, data interruption, and modification attackers, and others. As a consequence, cryptography based developments are considered more reliable and secure developments for SCADA systems [11,25,26,30,31,32,33,34,35,36,37,38,39, 40]. In symmetric encryption, while the desired message is encrypted, this does not ensure that the message contents are not modified during transmission because a single secret can be shared between the sender and the receiver. Therefore, public key encryptions are considered to be better approaches than symmetric encryptions; in addition, a non-repudiation security service should be achieved while employing the public key encryption with hashing function, or by employing the digital signature technique [11,25]. In [25], an end-to-end security solution was implemented in the transmission of the SCADA system. The SCADA nodes such as the master terminal unit (MTU) and the remote terminal unit (RTU) were installed with DNP3 protocol, and were configured in the SCADA testbed setup. In the testbed, communication is initiated from the MTU and the desired message is treated with a hash algorithm and public key encryption before transmitting to the destination. The message hash digest is computed by employing a hashing function and the computed hash value is then encrypted with a private key for the received message (or RTU). The message does not encrypt itself, and this minimizes the computation time of the encryption process. At the RTU side, the MTU public and RTU private keys are deployed and the MTU/RTU hash values are compared to verify the message contents. In the testbed, each node is installed with a snort tool that monitors the traffic and a snort analyzer is used to detect the intrusions and generate corresponding alerts during communication between the MTU and RTU and vice versa [11,25,37,40].

SIMULATION DESIGN AND ENVIRONMENT

To measure the desired goals of current study, a simulation environment is designed for water pumping system as a part of wireless SCADA system. In wireless SCADA system, the field devices' (or field sensors) are configured and directly connected with the sub-controllers, which are designated to carry the real time information from the sensors, or to monitor the real time information, as required by the main controller. The main controller is superior in the whole system design and network setup and is authorized to send the commands to the field sensors through the sub-controller(s). In water pumping system, as shown in Figure 3, only its two functional parts are considered: pumping for the cooler and pumping for the heater, the heating/cooling points are measured in-accordance to the normal set points that added at the time of configuration; and alarms are generated in-case the abnormal points or critical points will be measured from the field sensors. In wireless SCADA systems, each network node, such as sub-controller and main controller, is installed and configured using of DNP3 protocol as a part of SCADA system. Each time communication has occurrs between the nodes, the message is generated by deploying of DNP3 specified message structure and transmitted between the networked nodes, through employment of WAP (Wireless Application Protocol), the SCADA/DNP3 system would able to made the connection and to communicate wirelessly, to its remote located terminals (or remote field devices). In conclusion, the proposed study uses the SCADA/DNP3 protocol for messaging, the TCP/IP protocols to communicate over the Internet, and WAP for wireless communication; and moreover to secure the communication of wireless SCADA system, the cryptography hashing algorithm is deployed and tested at the pseudo-transport layer of SCADA/DNP3 protocol. The details for: message design, security design, security implementation, and security testing, are described in the below sections, of this study.

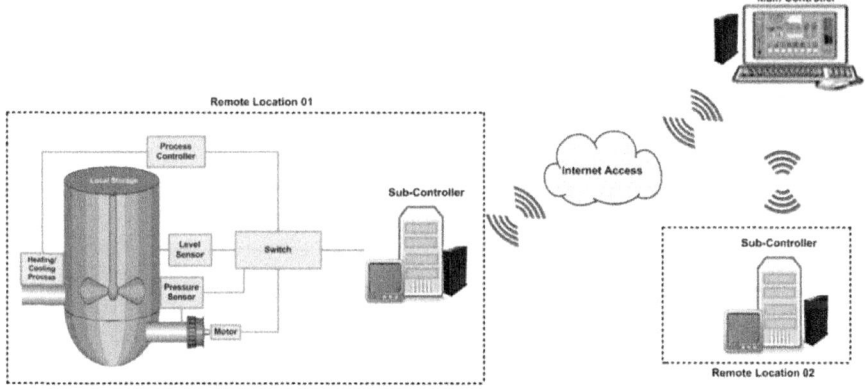

Figure 3: Simulation design and environment.

DNP3 PSEUDO-TRANSPORT LAYER

The pseudo-transport layer is the second layer of DNP3 after the application layer. The pseudo-transport layer takes the application protocol data units (APDUs) from the application layer of the DNP3 protocol and the upper layer bytes are treated as a transport service data unit (TSDU) or user bytes in the lower layer (or in the pseudo-transport layer). The main function performed by the transport layer is the disassembling and reassembling of bytes. The disassembling and reassembling processes allow a larger block of user data from the application layer to be handled easily by a data link layer [20]. In this research, transport protocol data units (TPDUs) are constructed as part of the DNP3 transport layer. Subsequently, control should be passed to the security development process where the hash function is applied using the SHA-2 hashing algorithm, as part of the cryptography mechanism.

Message Structure

The pseudo-transport layer breaks the TSDU into a number of units called transport protocol data unit (TPDUs) and each TPDU is made up of 250 bytes including 1 header byte. In Figure 4, 249 bytes are added with 1 byte of transport header (TH) information; this TH was originally named transport protocol control information (TPCI). In the case where a complete payload (or 2048 bytes information) has been received from the upper layer, the APDUs are then generated according to the payload size. In Figure 5, a total of eight TPDUs are generated and the remaining 56 bytes of the Application Protocol Data Unit (APDU) or 32 bytes of the cyclic redundancy check (CRC) from Link Protocol Data Unit (LPDU) would be employed for especial purposes. The size of each TPDU is fixed to 250 bytes because TPDU block could easily fit within a frame of the data link layer. This study made an alignment of the APDU which could be easily assembled within one segment (or TPDU) of the pseudo-transport layer.

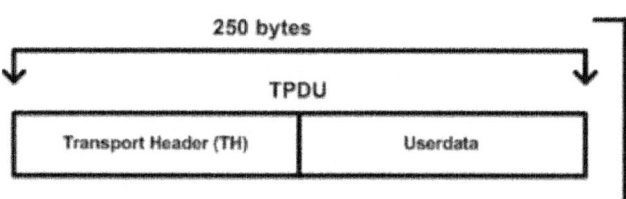

Figure 4: Single TPDU block.

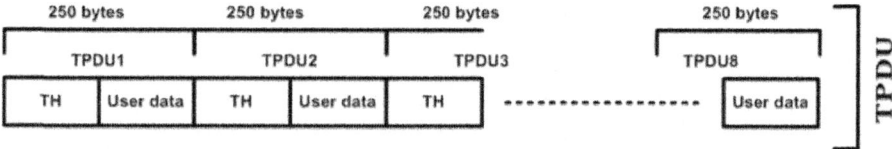

Figure 5: Multiple TPDU blocks.

The transport header is composed of three fields: FIR, FIN, and sequence number. Each TPDU is 250 bytes in length, which easily fits into the data link layer frame, called FT3. In Figure 6, the TH contains one byte of information and each bit has a specific function. The last two bits define the start and end of the TPDU sequence and the remaining six bits define the sequence counter.

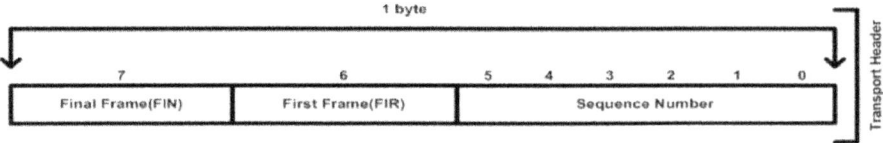

Figure 6: Transport header field structure.

PAYLOAD DESIGN AND SECURITY DEVELOPMENT USING HASHING

Similar to other SCADA protocols, the initial design of the DNP3 protocol was also limited in terms of security, or the security design was associated with the physical parts of the system [11,12,13,14]. To fulfill the requirements of industrial processes and automations, SCADA systems are connected to almost all modern networks [20,21,22]. To minimize the security falls that have been associated with communication of the SCADA system, several studies [25,26,27,28,29,30,31,32] have been conducted that provide node-to-node security protection against various vulnerabilities [12,13,14,41,42,43]. The DNP3 application layer and data link layer security have been analyzed and various cryptography techniques have been suggested to enhance the security of these layers, but are still under development [11,22,23]. As a consequence [12,13,22,23], security issues have seldom been considered for the pseudo-transport layer. The current research therefore emphasizes the pseudo-transport layer security issues and deploys a cryptography mechanism as the best approach to significantly enhance the security of this layer.

While the security development at the pseudo-transport layer is simple and straightforward, fulfilling the requirements of the pseudo-transport layer design, or its functional specifications, is more complex. However, we employed the C# tool to design and construct the transport layer bytes and employed the security development process using the SHA-2 hashing function. The entire development is also validated through proofs and evaluated through computed results.

This section is divided into three Sub-Sections: Section 5.1 Payload Design and Computation; Section 5.2 Security implementation; and Section

5.3 Proof of development. In Section 5.1, the transport layer payload is computed, and further described in Section 5.2 for the purposes of security computation. Section 5.3 demonstrates the proof of development fromSection 5.1 and Section 5.2.

Payload Design and Computation

In the DNP3 stack, the pseudo-transport layer takes the APDU as the user bytes from the application layer, and assembles the upcoming bytes into TSDU (bytes). In the reassembling process, the transport layer receives each TPDU (bytes) from the data link layer, and the TH is then stripped off and the TSDU bytes are recreated (or reformed) from the tripping process of TPCI. The pseudo-transport layer is also responsible for ensuring the sequence of TPDUs during the TSDU reassembling process. Due to the disassembling/ reassembling process of the pseudo-transport layer, the data link layer is able to handle the bulk of the data, but the functionality is finite in the transport layer of the open systems interconnection (OSI) model (as illustrated in Figure 7) [20].

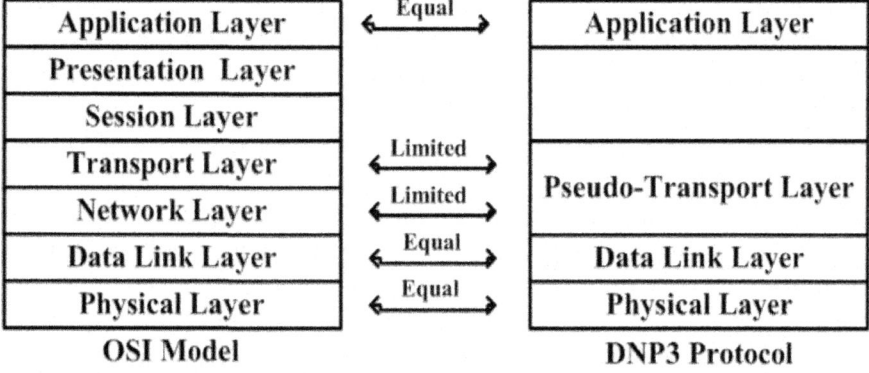

Figure 7: Interconnection between OSI model and DNP3 protocol.

The DNP3 protocol is a proprietary protocol and its design is limited for advanced IP based client/server applications; therefore, TCP/IP protocols are employed instead of the DNP3 physical layer, to communicate over networks such as LAN/WAN and over the internet. Figure 8 shows the pseudo-transport layer interrelation and flow of communication.

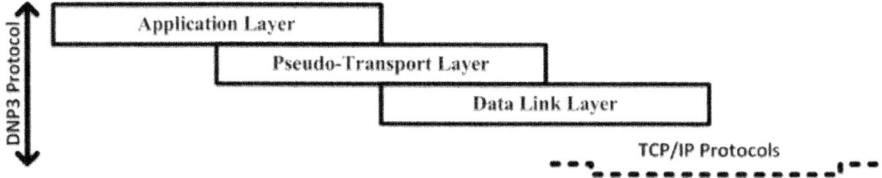

Figure 8: Logical interrelation and communication flow.

As described above, the overall development has been made in the C# platform and in a few available implicit code libraries. Examples are employed as references [44], with user defined codes to validate the approximate and best development, according to the best of our knowledge. The following definitions demonstrated the pseudo-transport layer payload design, and its operations.

Definition 1 (Bytes Assembling):

The number of user bytes "B" is received by the interaction of variable "Q" and "fQ" is an explicit dual non-linear function which assembles the upper layer bytes "B_{APDU}" with the lower layer bytes "B_{TSDU}" and vice versa by the interaction of "Q". However, since "B" is limited, an integer "$Z*$" (*i.e.*, not negative integer) exists if "B" defines the limit as lim←k, such that,

$$\Leftrightarrow B_{APDU} \propto B_{TSDU} \Rightarrow f_Q : B_{APDU} \longmapsto B_{TSDU}$$

Definition 2 (Bytes Dissembling):

Assume b∈B, where "b" refers to the fixed/non-fixed number of user bytes during the disassembling process of "B_{TSDU}". In the case where i=0 or i0 is manipulated, then b∈B∈∅, such that,

$$B_{TSDU} \Rightarrow$$

$$\sum_{B(b,k)}^{lim \leftarrow k} b_{i \in (\varnothing,n)}, \ i = 0,1,2,3, \ldots \ldots, \ n-1, \ n, b \in B \leqslant lim$$

Definition 3 (Payload):

"α" is a variable that counts the number of bytes "b", and the explicit user function "$f\alpha$" is employed to manipulate the transport layer (TL) user bytes $Q_{b \leqslant (lim, \varnothing)}^{TL}$ corresponding to the disassembling process, with header (h) functional bytes $Q_{h,h \neq \varnothing}^{TL}$, where $Q_{(h,b)}^{TL} \leqslant lim$.

Bytes Alignment and Security Computational Bytes

During the payload design and computation, a keyword "limit" (lim) is defined, the purpose of which is twofold: (1) limit the number of bytes in each TSDU; and (2) limit the number of bytes in each TPDU. However, the size of TSDU is directly proportional to the size of APDU, but the size of each TPDU is limited to 249 bytes, plus 1 byte of header [20,21]. In this study, we limited the upper bytes (or APDU) size to 1992 bytes in both cases: request and response payload. This would further align with the TPDUs. For example, if we define the size of APDU as 1992 bytes, then eight equal TPDUs are created, as an addition to the transport protocol control information (APCI). This would also significantly protect information from non-legitimate users; fixed sized data is transmitted rather than variable size data.

For the alignment process of APDU and the fixing of TPDUs, the remaining 56 bytes are employed to keep track of security development and to protect sensitive information from unauthorized users. Hence, all remaining 56 bytes are not employed in this development, but are utilized and considered for other parts of the DNP3 protocol security enhancement purposes [37]. Some functions are deployed by employing the bytes from the total of 56 bytes, while the remaining functions are padded with zeros to be un-padded later for future developments. The functions details are as follows.

- Payload Counter (Two Bytes): Payload (or TPDU) is created, and 250 bytes are counted in the payload counter. In the case where minimal bytes are defined, the remaining bytes are padded to protect the payload from data modification and reply attacks.

- Hash Sequence Counter (One Byte): In the case where the number of TPDUs is defined by a single TSDU, the hashing sequence is counted in the range of 0–63, and should be recycled as 63–0 on the remote side. Two bits are used that designate the first and last hashing sequence in the defined range.

- Security Method (One Byte): In the proposed study, SHA-2 hashing is deployed to protect the sensitive information of the transport layer against integrity attacks. However, this development is also able to test other algorithms such as secret key and public key algorithms. In this case, if multiple algorithms have been deployed, the dynamic selection is made by this functional field.

- Padding Counter (Two Bytes): Initially, two bytes are defined that accumulate the number of padding bytes in the entire development. The size would be changed as required by this functional field by allocating bytes from dynamic storage (DS).

- Acknowledge No (Two Bytes): Acknowledge flags are set at both sides of the communication. Therefore, acknowledgement is required at both sides followed by the acknowledgement number.

- Useful Contents (One Byte): Typically, the payload contents are verified corresponding to the content list before being transmitted to the networks (or remote site).

- Dynamic Storage (10–46 Bytes): Bytes are dynamically allocated to other fields, if required; or these bytes are reserved for future development.

Security Implementation

In the DNP3 protocol, the pseudo-transport layer performs a limited functionality of the transport layer and data link layer of the OSI model. As described, the functionality is fairly limited; therefore, the vulnerabilities are also limited, or a limited number of attacks are linked with the pseudo-transport layer [12,13]. In [13,18], three commonly potential attacks including Interruption, Modification, and Fabrication, with 32 instances, are counted against the pseudo-transport layer in terms of security, two of which are directly linked with the TPDU flags and their sequence in the DNP3 transmission. However, data modification, fake messaging, and byte interruption are considered as part of the current research. The SHA-2 hashing algorithm is deployed, and is considered in order to enhance the security of the pseudo-transport layer as part of the DNP3 protocol; this development is also able to test the other security algorithms [11,25].

In security implementation, the remote terminal station (RTU) is responsible for generating and sending responses according to the main controller request. The proposed work is based on a simulated environment and the scope is limited to pseudo-transport layer security; therefore, we do not give a detailed explanation of the phenomenon of the client/server architecture. The following steps are followed to deploy the SHA-2 algorithm, and to enhance the security of the pseudo-transport layer, while Table 3 summarizes the notations that are employed in the development.

Table 3: Security notations

Notations	Description
$Q^{TL}(B)$	Assembled bytes.
$f_Q : b_i$	Disassembled bytes.
$f_\alpha : (Q_h^{TL})$	Manipulated header bytes.
$f_\alpha : (Q_b^{TL})$	Manipulated data bytes, after disassembling.
$k \rightarrow lim$	K is dual integer that defines the limit (lim).
$\alpha_{TL} = \wedge$	User defined index pointer.
f_H	User defined hashing function.
f_{Comp^H}	Hashing comparison function.
$f_w : w_{h,b}$	User defined relation function.
$f_p : p_{h,b}$	User defined bytes separator function.

i $Q_{h,h\neq\varnothing}^{TL}, Q_{b\leq(lim,\varnothing)}^{TL} \implies Q_{(h,b)}^{TL}$ is the transport layer payload that is being manipulated by security function $H_{digest(s,Q)}^{TL}$ using SHA-2 algorithm. The maximum size of each $Q_{(h,b)}^{TL}$ is 250 bytes, if a number of $Q_{(h,b)}^{TL}$ are created then the hash sequence is counted to keep the track of each $Q_{(h,b)}^{TL}$. The original payload $Sender(s) : Q_{(s,h,b)}^{TL}$ and computed security function $Hash(H) : H_{digest(s,Q)}^{TL}$ are transmitted, while the parameter, which designates the sender information, is added.

ii   Upon receiving at the other side, the receive hash digest $H_{digest(R,Q)}^{TL}$ is computed based on the original payload $Sender(s) : Q_{(s,h,b)}^{TL}$ and compared with $H_{digest(s,Q)}^{TL}$. As a consequence, if $H_{digest(R,Q)}^{TL} = H_{digest(s,Q)}^{TL}$ then the payload would be accepted; otherwise, it is rejected in the case of $H_{digest(R,Q)}^{TL} \neq H_{digest(s,Q)}^{TL}$.

In security development (or in Figure 9), the number of integrity attacks such as data modification, data detection, and data reply could be verified in the transmission and this would also be concluded in the security (or lack of security) of the pseudo-transport layer. More detail is described in Algorithm 1, in Section 6.

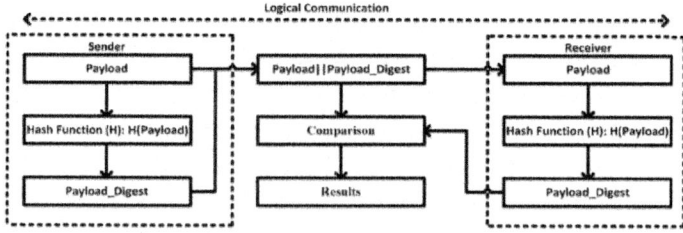

Figure 9: Security development Using hashing function.

ALGORITHM: PSEUDO CODE TRANSPORT LAYER MESSAGE CONSTRUCTION WITH SECURITY DESIGN

Algorithm 1: Transport Layer Security.

Input: The input of Upper Layer Bytes.

Output: Transport protocol data unit (TPDU), Hash digest.

1. The input of "n" bytes is received from upper layer. In pseudo-transport layer, these bytes are assembled as transport service data unit (TSDU) .Hence, the bytes are limited are upper layer therefore, we defined TSDU corresponding to APDU.

TranLayer Header TH = 1 byte, TranLayerUserdate UD = 249 byte, TranLayer TPDU = 250 byte;

Bytes TSDU [ ] = [ ]; Bytes H [ ] = [ ]; Bytes UD [ ] = [ ]; Bytes TPDU [ ] = [ ];

2. TranTPDU ( )

{ Bytes TPDU [ ] = [H [ ] = [ ], UD [ ] = [ ]]; Bytes H [ ] = [Sequence No., FIR, FIN];

}

3. TranTSDU ( )

{ Bytes TSDU [ ] = [TPDU1 [ ], TPDU2 [ ], TPDU2 [ ],..........., TPDUn [ ]];

Where n = 1,2,3,..........., n and maximum size of each TSDU is equal to APDU size.

If TSDU [ ] = APDU [ ]

Print ("Process transport layer Information");

else if TSDU [ ] = APCI [ ]

Print ("Process transport layer Information corresponding to application header or APCI ");

Else Print (Unknown Communication);

4. Cryptography ( )

{ RequestMessage M, j, Hash Digest HD ;

Hash ( ) //Sender function

{ Total Payload = Hash(M)

$i = TPDU_0 + TPDU_1 + TPDU_2 + \cdots + TPDU_{n-1} + TPDU_n)$, $i = 1, 2, 3$ ....limit. Here, $i = TPDU_0$ is defined, if only header bytes are manipulated and limit shows the maximum value, or maximum value of "n".

Bytes M[ ] = [ ];

For ($i = TPDU_1$; $i <= TPDU_n$; i++);

Print ("Add in to Hashing Buffer" +M[ ]);

Sender: Hashing $\Rightarrow SHA2[(TPDU)_i] = (TPDU)_i\_Digest$

5. Hash ( ) //Receiver function

{ Total Payload = Hash(M)

Bytes M[ ] = [ ];

For ($i = TPDU_n$; $i <= TPDU_1$; i--); Print ("Add in to Hashing Buffer" +M[ ]);

Receiver: Hashing $\Rightarrow SHA2[(TPDU)_i] = (TPDU)_i\_Digest$

Comparison: Receiver: $(TPDU)_i\_Digest =\neq$ Sender: $(TPDU)_i\_Digest$

6. Conclusion: At sender side, hash function is deployed on TPDU and computed and compared at remote side (or at receiver side) to verify the integrity of payload (or APDU).

ATTACKING SCENARIOS

In the pseudo-transport layer header or transport protocol control information (TPCI), one byte is designed to represent the header information, six bits define the sequence number counter, and the remaining two bits are employed to designate the FIR and FIN frames of APDU (or fragment) [13,20]. In the transmission, the number of frames of a payload are sent and counted in sequence, where the FIR and FIN terms define the special meaning in the processing of the payload. In the case where the payload is transmitted with the FIR indication flag set, all the existing fragments (or partially-completed fragments) are then wasted, and are no longer considered. In some scenarios, the sensitive information of the pseudo-transport layer is interrupted.

- In the reassembling process, the original payloads are disrupted; if a newer payload enters with the FIR flag set, the fragmented payload transmission subsequently starts.

- The numbers of payload are transmitted and counted in the sequence counter while the sequence should be recycled at the remote side. An interruption is created during the manipulation of the incomplete (or partially completed) payload if the new payload is entered with the FIN indication flag set; as a consequence, the assembling process is closed, as it is untimely.

- The APCI information is sensitive, and needs to be protected from unauthorized entities. The adversary has many chances to delete the payload information during transmission. The attacker uses sensitive information by using various capturing tools [11] and deletes/modifies the flags set such as FIR and FIN, and at the remote side, the receiver assumes that the payload originates from a secure source.

- In APCI, 6 bits are occupied by a sequence number (field) which ensures the transmission of a fragmented payload (APDU) in a sequence order. Each time, the fragment is created and transmitted, and the corresponding number is added to the sequence counter; thus, the transmitted and transmitting fragmented payloads are recorded with a unique sequence number. However, an attacker could have many chances to change the fragment sequence, monitor the traffic, and capture the fragments. Using a sequence number, an attacker employs various inject tools to change the sequence counter value and to inject a new fabricated fragment instead of an original fragmented payload [11]. As a consequence, there are many scenarios in which the sensitive information of the pseudo-transport layer can suffer from internal/external adversaries [12,13,18].

Security approaches [26,27,28,29] have been proposed to hide sensitive information from attackers [1,2,3,4,5,6,7,8,9,10,11,12,13,18], but these security approaches are limited in terms of specification design, protocol dependencies, and transmission requirements [11,25]. To hide the information, cryptography approaches are considered as the best solutions for system security [26,37]; in a few cases, the encrypted information cannot be satisfied at the remote side, especially during decryption of the header [11,37,43]; therefore, the best solution is to encrypt the user bytes, excluding the header bytes [43]. On the other hand, if header information is not secured, there is a chance an adversary [12,13] modifies the header with false information while replying to the message. Therefore, in this research paper, a hashing algorithm was employed that generates a fixed size security code and travels along the original payload that keeps the receiver aware of unauthorized opponents. In the following section, attacking tools are employed to interrupt the normal flow of the pseudo-transport layer as part of the DNP3 protocol, and the corresponding observed measurements are discussed.

SETUP AND ABNORMAL COMMUNICATION

In a SCADA wireless network setup, the number of nodes is configured to exchange information with the main controller, although the total number of SCADA nodes is not discussed here because of unicasting communication; the system is designed according to the terminologies of an unbalanced system in which only the main controller is authorized to initial communication with the remote controller(s) [20]. However, the terms such as balanced and unbalanced, which are defined at the data link layer, are not part of the pseudo-transport layer. Therefore, this study does not specifically emphasize these terms, but we conclude that, to the best of our knowledge, the unbalanced system is more appropriate than the balanced system for this study.

To interrupt the logical normal flow of the pseudo-transport layer, predominated attacking tools such as airpwn, file2air with wireshark, and injection tools, are used which perform traffic monitoring and frames (or fragments) captured as an attacker of the system [11]. However, security development is limited to an integrity security service; thus, the attacks such as fragment injection, payload replay, and payload deletion are considered as corresponding to the proposed security implementation.

MEASUREMENT AND DISCUSSION

DNP3 protocol unitization has been massively increasing (*i.e.*, 70%) in SCADA systems [13]. Due to the lack of security precautions in the initial design of the DNP3 protocol, several potential adversaries take advantage of the DNP3 protocol's vulnerable platform [11,13]. The current study employs a hashing function to enable awareness between the SCADA and DNP3 nodes, if transmission is interrupted by network adversaries. This research paper also deals with various developments of multimedia based security followed by communication requirements, although the study scope is limited to the pseudo-transport layer, as a layer of the DNP3 protocol.

To compute the performance measurements, random size fragment payloads are generated several times and transmitted between the main controller and the remote controller and vice versa; however, each fragment is limited to 1–250 bytes in length. In the case where no TSDU bytes are assembled from the upper layer, only the TPCI is transmitted with the computed hashing code. In the transmission, each fragment hash digest is calculated before transmitting to the remote side; the fixed hash code (or digest) travels along the original payload and is again computed at the receiver side to verify the contents of the payload. Of all the experiments, 200 are selected as the best experiments according to the best of our knowledge, and further performances such as attacks detection and security assessment, are also based on these selected experiments. Figure 10 shows the 200 successful experiments that are tested with random size payloads (or segments) and received at the remote side, whereby the first half of all the experiments are designated for sending to the payload and the remaining half are designated for the response payload. Each half is separated by a line.

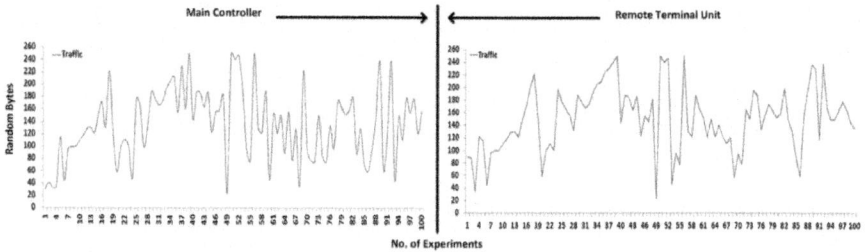

Figure 10: Traffic: (**left**): Main Controller; (**right**) Remote Terminal Unit.

Figure 11 and Figure 12 show the 200 successful attacks experiments that are tested with random size payloads (or segments), whereby 100 successful experiments are designated for sending to the payload and the remaining 100 as shown in Figure 12 are designated for the response payload.

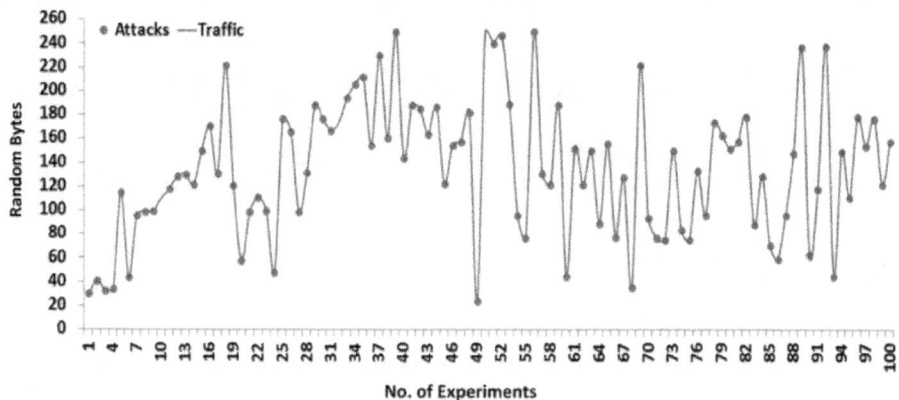

Figure 11: Attacks: Main controller traffic.

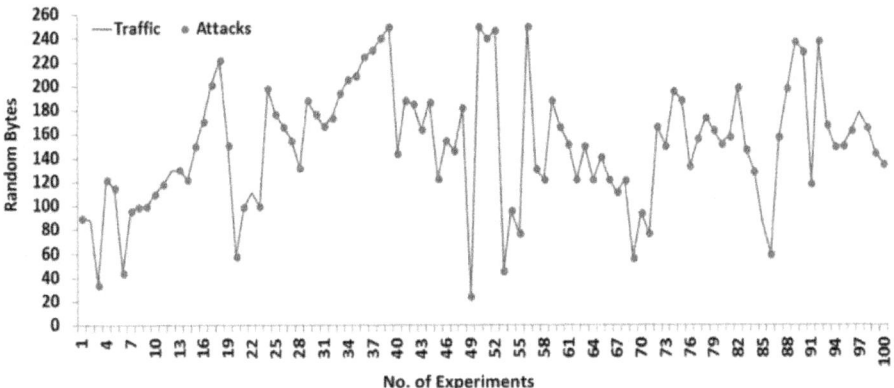

Figure 12: Attacks: Remote terminal unit traffic.

To evaluate the performances, preliminary packet analyzer tools such as wireshark, dSniff, Kismet, ethereal, and ettercap are employed which analyze the packets (or fragments), as a consequence, and approximately 192 times fragments are intercepted in the transmission. Thus, we can conclude that the DNP3 pseudo-transport layer has a lack of security design, or the DNP3 pseudo-transport layer was designed without considering any security. In [11,18,33], the number of attacks is defined and detected as part of the SCADA system, and security mechanisms are also used that protect the SCADA communication against several potential attacks and ensure the SCADA platform is invulnerable [12,13,25,27,29]. However, security is accounted in SCADA systems or/and SCADA protocols, with the exception of the deliberation of pseudo-transport layer security.

During the fragments interception shown in Figure 11 and Figure 12, the SCADA nodes are configured without any security paradigms such as firewalls, demilitarized zone (DMZ), antivirus protection, *etc.*, which determine the approximate security level during the transmission of fragmented payloads. However, if payload security was enhanced, the receiver would also be made aware of adversaries by contents verification. The fragmented payload hash digest is computed 200 times and transmitted along the original payload; upon receiving, the receiver also computes the hash digest of the original payload to compare with the sender hash digest. If two hash digest values are matched, then the receiver assumes that the payload came from an original source; otherwise, the payload contents are discarded and the exception (*i.e.*, payload contents have not been verified, there is chanced of adversary in transmission), is generated against the adversary.

In the existing studies [45,46,47,48], several limitations of SHA-2 hashing function are analyzed and creaking tools are employed; it is also assumed that the hash code is breakable. However, we did not fully succeed in breaking the computed hashing codes, or the results were captured with zero impact. In the case where the computed hashing values are breakable, we propose a method called a digital signature to resolve these issues. In this method, a fragmented payload hash digest is computed HTLdigest(S,Q) and a private key is deployed on the hash digital Pr(k,S)(HTLdigest(S,Q)), which acts as a digital signature. The original payload QTL(S,h,b) and digital signature Pr(k,S)(hTLdigest(S,Q)) are then encrypted with a public key Pu(k,R) of the receiver(R) as Pu(k,R) (QTL(S,h,b), Pr(k,S)(hTLdigest(S,Q))), and transmitted to the remote side. Upon receiving, the receiver uses the sender (S) public key Pu(k,S) and the private key Pr(k,R) of the receiver(R) to open (or decrypt) the original payload QTL(S,h,b) and hash digest HTLdigest(S,Q). Subsequently, the hash digest of QTL(S,h,b) is calculated, and is designated as HTLdigest(R,Q) and compared with HTLdigest(S,Q). The keys such as private keys and public keys are defined and generated using an RSA algorithm; however, the keys are distributed statistically among the network nodes. As a consequence, we concluded that the hash digest is secured and the payload contents are not altered during transmission, even in cases of adversary. In study [49], the attack scenarios were conducted, in which the authentication and confidentiality attacks such as brute force, cryptography key cracking, eavesdropping, and man-in-the-middle are launched 200 times and the numbers of detected attacks are counted and visualized [11,49]. As a result, minimal impact is computed that is so far able to break the hash digest; also, it is very difficult for an adversary to inject, modify, and delete the sensitive information of the payload.

MULTIMEDIA CONTEXTS

In this study, various multimedia contexts are employed in the form of text and images. The human machine interface (HMI) is designed and installed at both sides of the transmission. The basic configuration and setup, including the connection type (*i.e.*, TCP/IP), IP addresses, Port numbers, Channel setting, *etc.* required between the main controller and the remote controller are visualized as part of HMI. During transmission, the total number of bytes fragmented in the case of sending and responding (as part of the pseudo-transport layer), are also visualized which make it convenient for the end user prospective. The fragments flow in sequence and are shown on HMI at both sides of the transmission; the end users or operators can check the flow of fragments during the construction and distribution at both sides using sharing media (*i.e.*, team viewer *etc.*), which also determines the effects of networks adversaries, in the case where abnormal flows are visualized.

In a few cases, the main controller requires exceptional reports and screen shots (or images) of the physical setup, such as sensors, actuators, PLCs, and hardware devices; the information is then secured from the network adversaries, the images are transmitted in compressed form in order to minimize the memory space, and the security using the SHA-2 algorithm is deployed before responding to the main controller. Normally, SCADA systems are designed and used for low bandwidth; therefore, hashing is considered a secure and reliable approach [1,4]. In the case where there are potential adversaries that successfully break the hashing value, a digital signature is considered as the best approach according to the best of our knowledge and according to our measurements.

SIGNIFICANCE OF STUDY

Hashing is a good approach which verifies the payload contents by comparing the computed hashing values of the sender and receiver. In this study, pseudo-transport layer security issues are analyzed and the SHA-2 hashing algorithm is selected and deployed on the fragmented payload; TPDU is made up of user bytes and a header byte, while the FIR, FIN, and sequence number are part of TPCI. In the case where an adversary causes an interruption (*i.e.*, injection, modification, and deletion) by means of FIR, FIN, and sequence number, he/she cannot be successful because the hash digests are computed at both sides and compared at the remote side; if he/she is successful, the digital signature is computed for the hash value(s). As a consequence, the overall transmission is secured from the adversaries. This study also employed various multimedia contexts in the form of text and images, while security development and communication have been demonstrated to make information more convenient and reliable for the user.

CONCLUSIONS AND FUTURE WORK

This study used a simulation based environment of water pumping system, and SCADA wireless sensors based network system to deploy the cryptography mechanism while communicating over unsecured network, or over Internet communication. Furthermore, the main security issues realized during the pseudo-transport layer disassembly and reassembling process are highlighted and a security solution using the SHA-2 hashing function is deployed, which ensures the integrity of bytes received/transmitted from/to the data link layer. Therefore, a DNP3 pseudo-transport layer stack has been designed and evaluated from formal evidence, security implementation is employed, and evidence is given of the protection against byte verification issues.

In future work, the SCADA system information will be accessed and monitored via cellular phones; and the SCADA/DNP3 testbed attack (abnormal) setup will be developed and simulation tools or software will be used to test the integrity attacks such as packet/data injection, packet/data replay, and data (byte) deletion and others. The security percentage will be measured based on the attack impact percentage on the overall system (or at the pseudo-transport layer stack). The other cryptography functions such as asymmetric and symmetric will also be deployed and security results will be validated against attacks such as shared key guessing, brute force, cracking key, man-in-the-middle, and others.

ACKNOWLEDGMENTS

This paper was supported by research funds of Wonkwang University in 2016. This research was supported by Next-Generation Information Computing Development Program through the National Research Foundation of Korea (NRF) funded by the Ministry of Science, ICT & Future Planning (NRF-2014M3C4A7030503).

AUTHOR CONTRIBUTIONS

In this research, Aamir Shahzad, Malrey Lee and Neal Naixue Xiong conceived and designed the experiments; Aamir Shahzad, Gisung Jeong and Young-Keun Lee performed the experiments; Aamir Shahzad, Jae-Young Choi and Abdul Wheed Mahesar analyzed the data; Aamir Shahzad, Malrey Lee and Iftikhar Ahmad Khan contributed materials/analysis tools; Aamir Shahzad, Malrey Lee and Neal Naixue Xiong wrote the paper.

REFERENCES

1. Cheung, R.; Fung, Y. Wireless access to SCADA system. In Proceedings of the 2000 International Conference on Advances in Power System Control, Operation and Management, Hong Kong, China, 30 October–1 November 2000; pp. 553–556.

2. Flammini, A.; Ferrari, P.; Marioli, D.; Sisinni, E.; Taroni, A. Wired and wireless sensor networks for industrial applications. *Microelectron. J.* 2009, *40*, 1322–1336.

3. Escudero, J.I.; Rodriguez, J.A.; Romero, M.C.; Luque, J. IDOLO: Multimedia data deployment on SCADA systems. In Proceedings of the Power Systems Conference and Exposition, New York, NY, USA, 10–13

October 2004; pp. 252–257.

4.   Escudero, J.I.; Rodriguez, J.A.; Romero, M.C.; Diaz, S. Deployment of digital video and Audio Over electrical SCADA networks. *IEEE Trans. Power Deliv.* 2005, *20*, 691–695.

5.   Morsi, I.; el Deeb, M.; El Zwawi, A. SCADA/HMI Development for a Multi Stage Desalination Plant. In Proceedings of the Computation world 09. Computation World Future Computing, Service Computation, Cognitive, Adaptive, Content, Patterns, Athens, Greece, 15–20 November 2009; pp. 67–71.

6.   Laurence, K.; Rémi, B. The synergy between system modelization and HMI modelization: Application on a workbench HMI. In Proceedings of the 2014 Ergonomie et Informatique Avancée Conference—Design, Ergonomie et IHM: Quelle articulation pour la co-conception de l'interaction (Ergo'IA '14), New York, NY, USA, 15 October 2014; pp. 122–129.

7.   Adnan, S.; Vlatko, M.; Zoran, C.; Elvedin, K.; Nina, D. Web based multilayered distributed SCADA/HMI system in refinery application. *Comput. Stand. Interfaces* 2009, *31*, 599–612.

8.   Mander, T.; Nabhani, F.; Wang, L.; Cheung, R. Data Object Based Security for DNP3 Over TCP/IP for Increased Utility Commercial Aspects Security. In Proceedings of the Power Engineering Society General Meeting, Tampa, FL, USA, 24–28 June 2007; pp. 1–8.

9.   Ortega, A.; Akira Shinoda, A. Simulation in NS-2 of DNP3 protocol encapsulated over TCP/IP in smart grid applications. In Proceedings of the 2013 IEEE PES Conference on Innovative Smart Grid Technologies Latin America (ISGT LA), Chicago, IL, USA, 15–17 April 2013; pp. 1–8.

10.   Haydn, A.T. Wireless and Internet communications technologies for monitoring and control. *Control Eng. Pract.* 2004, *12*, 781–791.

11.   Musa, S.; Aborujilah, A. Secure security model implementation for security services and related attacks base on end-to-end, application layer and data link layer security. In Proceedings of the 7th International Conference on Ubiquitous Information Management and Communication, Kota Kinabalu, Malaysia, 17–19 January 2013.

12.   Huitsing, P.; Chandia, R.; Papa, M.; Shenoi, S. Attack taxonomies for the Modbus protocols. *Int. J. Crit. Infrastruct. Prot.* 2008, *1*, 37–44.

13.   East, S.; Butts, J.; Papa, M.; Shenoi, S. A Taxonomy of Attacks on the DNP3 Protocol. *Crit. Infrastruct. Prot.* 2009, *3*, 67–81.

14. Shahzad, A.; Lee, M.; Kim, S.; Kim, K.; Choi, J.-Y.; Cho, Y.; Lee, K.-K. Design and Development of Layered Security: Future Enhancements and Directions in Transmission. *Sensors* 2016, *16*.

15. Sugwon, H.; Lee, M. Challenges and Direction toward Secure Communication in the SCADA System. In Proceedings of the Communication Networks and Services Research Conference (CNSR), Montreal, QC, Canada, 11–14 May 2010.

16. Gao, J.; Liu, J.; Rajan, B.; Nori, R.; Fu, B.; Xiao, Y.; Liang, W.; Chen, P. SCADA communication and security issues.*Secur. Commun. Netw.* 2014, *7*, 175–194.

17. Kim, H. Security and Vulnerability of SCADA Systems over IP-Based Wireless Sensor Networks. *Int. J. Distrib. Sens. Netw.* 2012, *2012*.

18. Lee, D.; HakJu, K.; Kwangjo, K.; Yoo, P.D. Simulated Attack on DNP3 Protocol in SCADA System. In Proceedings of the 31th Symposium on Cryptography and Information Security, Kagoshima, Japan, 21–24 January 2014.

19. Willig, A.; Matheus, K.; Wolisz, A. Wireless Technology in Industrial Networks. *IEEE Proc.* 2005.

20. Gordon, C.; Deon, R.; Edwin, W. *Practical Modern SCADA Protocols: DNP3, 60870.5 and Related Systems*; Elsevier: New York, NY, USA, 2004; pp. 73–129.

21. Susanto, I.; Jackson, R.; Paul, D.L. Industrial Process Control System Security. In *Wiley Handbook of Science and Technology for Homeland Security*; John Wiley & Sons: Hoboken, NJ, USA, 2009; pp. 1–15.

22. Jeffrey, L.; Hieb, J.H.; Graham, S.C. *Cyber Security Enhancements for SCADA and DCS Systems. ISRL-TR-07-02, Intelligent Systems Research Laboratory*; Technical Report TR-ISRL-07-02; University of Louisville: Louisville, KY, USA, 2007.

23. Steve, G. The SCADA challenge: Securing critical infrastructure. *Netw. Secur.* 2009, *2009*, 18–20.

24. Igor Nai, F.; Andrea, C.; Marcelo, M.; Alberto, T. An experimental investigation of malware attacks on SCADA systems. *Int. J. Crit. Infrastruct. Prot.* 2009, *2*, 139–145.

25. Patel, S.C.; Bhatt, G.D.; Graham, J.H. Improving the cyber security of SCADA communication networks. *Commun ACM* 2009, *52*, 139–142.

26. Hieb, J.; Graham, J.; Patel, S. Security Enhancements for Distributed Control Systems, Critical Infrastructure Protection. *IFIP Int. Fed. Inf. Process.* 2008, *253*, 133–146.

27. Kim, H.M. A Proposal for Key Policy of Symmetric Encryption Application to Cyber Security of KEPCO SCADA Network. In Proceedings of the Future Generation Communication and Networking, Jeju-Island, Korea, 6–8 December 2007.

28. Azeem, I.; Muhammad, S.; Muhammad, S.F. A secure authentication scheme for session initiation protocol by using ECC on the basis of the Tang and Liu scheme. *Secur. Commun. Netw.* 2014, *7*, 1210–1218.

29. Seongan, L.; Eunjeong, L.; Cheol-Min, P. Equivalent public keys and a key substitution attack on the schemes from vector decomposition. *Secur. Commun. Netw.* 2014, *7*, 1274–1282.

30. Vyas, P. Wireless Sensor Networks for Industrial Process Monitoring and Control with Security Architecture: A survey for Research Issues. *IJESRT* 2013, *2*, 930–936.

31. Anupam, S.; Om, P.; Zia, S. Public Key Cryptography Based Approach for Securing SCADA Communications, Computer Networks and Information Technologies. *Commun. Comput. Inf. Sci.* 2011, *142*, 56–62.

32. Martin, D.; Maricel, B. Cipher for Internet-based Supervisory Control and Data Acquisition Architecture. *J. Secur. Eng.* 2011, *8*, 337–348.

33. Shahzad, A.; Musa, S.; Irfan, M. N-Secure Cryptography Solution for SCADA Security Enhancement. *Trends Appl. Sci. Res.* 2014, *9*, 381–395.

34. Fujisaki, E.; Okamoto, T. Secure integration of asymmetric and symmetric metric encryption schemes. In *Advances in Cryptology—CRYPTO'99*; LNCS; Spring-Verlag: Berlin, Germany, 1999; pp. 537–554.

35. He, D.; Chen, J.; Chen, Y. A secure mutual authentication scheme for session initiation protocol using elliptic curve cryptography. *Secur. Commun. Netw.* 2012, *5*, 1423–1429.

36. Robles, R.-J.; Balitanas, J. Comparison of Encryption Schemes as Used in Communication between SCADA Components. *Ubiquitous Comput. Mult. Appl.* 2011.

37. Shahzad, A.; Musa, S.; Irfan, M.; Asadullah, S. Deployment of New Dynamic Cryptography Buffer for SCADA Security Enhancement. *J. Appl. Sci.* 2014, *14*, 2487–2497.

38. Chen, Y.; Dong, Q. RCCA security for KEM + DEM style hybrid encryptions and a general hybrid paradigm from RCCA-secure KEMs to CCA-secure encryptions. *Secur. Commun. Netw.* 2014, *7*, 1219–1231.

39. Rosslin, J.R.; Maricel, B.; Tai-hoon, K. Security Encryption Schemes for Internet SCADA: Comparison of the Solutions. *Commun. Comput. Inf. Sci.* 2011, *223*, 19–27.

40.   Sandip, C.P. Secure Internet-Based Communication Protocol for Scada Networks. Ph.D. Thesis, University of Louisville, Louisville, KY, USA, 2006.

41.   Ralston, P.A.S.; Graham, J.H.; Hieb, J.L. Cyber security risk assessment for SCADA and DCS networks. *ISA Trans.*2007, *46*, 583–594.

42.   Gilchrist, G. Secure authentication for DNP3. In Proceedings of the Power and Energy Society General Meeting—Conversion and Delivery of Electrical Energy in the 21st Century, 2008 IEEE, Pittsburgh, PA, USA, 20–24 July 2008; pp. 1–3.

43.   Majdalawieh, M.; Parisi-Presicce, F.; Wijesekera, D. DNPSec: Distributed Network Protocol Version 3 (DNP3) Security Framework. *Adv. Comput. Inf. Syst. Sci. Eng.* 2006, *3*, 227–234.

44.   Shahzad, A.; Lee, M. The Protocol Design and New Approach for SCADA Security Enhancement during Sensors Broadcasting System. *Multimed. Tools Appl. Springerlink* 2015.

45.   Helena, H.; Henri, G. Evaluation Report, Security Level of Cryptography—SHA-256, 2002. Available online: http://www.ipa.go.jp/security/enc/CRYPTREC/fy15/doc/1045_IPA-SHA256.pdf (accessed on 1 October 2015).

46.   Somitra, K.S.; Palash, S. A new hash family obtained by modifying the SHA-2 family. In Proceedings of the 4th International Symposium on Information, Computer, and Communications Security, Sydney, NSW, Australia, 10–12 March 2009.

47.   Florian, M.; Tomislav, N.; Martin, S. Finding SHA-2 characteristics: Searching through a minefield of contradictions. In Proceedings of the 17th International Conference on the Theory and Application of Cryptology and Information Security (ASIACRYPT'11), Seoul, Korea, 4–8 December 2011; pp. 288–307.

48.   Henri, G.; Helena, H. Security Analysis of SHA-256 and Sisters. *Sel. Areas Cryptogr. Lect. Notes Comput. Sci.* 2004, *3006*, 175–193.

49.   Shahzad, A.; Lee, M.; Lee, Y.; Kim, S.; Xiong, K.; Choi, J.; Cho, Y. Real Time MODBUS Transmissions and Cryptography Security Designs and Enhancements of Protocol Sensitive Information. *Symmetry* 2015.

Chapter 9

SURVEY ON WIRELESS SENSOR NETWORK TECHNOLOGIES FOR INDUSTRIAL AUTOMATION: THE SECURITY AND QUALITY OF SERVICE PERSPECTIVES

Delphine Christin [1], Parag S. Mogre [2] and Matthias Hollick [1]

[1]Secure Mobile Networking Lab, Center for Advanced Security Research Darmstadt, Department of Computer Science, Technische Universität Darmstadt, Mornewegstr. 32, 64293 Darmstadt, Germany

[2]Multimedia Communications Lab, Department of Computer Science, Technische Universität Darmstadt, Rundeturmstr. 10, 64283 Darmstadt, Germany

ABSTRACT

Wireless Sensor Networks (WSNs) are gradually adopted in the industrial world due to their advantages over wired networks. In addition to saving cabling costs, WSNs widen the realm of environments feasible for monitoring. They thus add sensing and acting capabilities to objects in the physical world and allow for communication among these objects or with services in the future Internet. However, the acceptance of WSNs by the industrial automation community is impeded by open issues, such as security guarantees and provision of Quality of Service (QoS). To examine both of these perspectives, we select and survey relevant WSN technologies dedicated to industrial automation. We determine QoS requirements and carry out a threat analysis, which act as basis of our evaluation of the current state-of-the-art. According to the results of this evaluation, we identify and discuss open research issues.

INTRODUCTION

Industrial automation has been successfully introduced in a countless amount of industries ranging from food to energy industries. Even if the products differ from one industry to another, the automated processes can be classified according to three main layers as proposed by [1]: the plant-floor automation layer, the manufacturing execution system layer and the enterprise resource planning layer. Internet technology can be considered as the link that interconnects all these layers and allows for information exchange. For example, it serves as a backbone to interconnect different production locations within one enterprise, to transfer production control data in near real-time to the headquarters or to integrate suppliers into a production workflow.

Within the scope of this survey, we focus on the plant-floor automation layer including sensors, switches, programmable controllers and motor starters (Fig. 1) that ensure the correct operation of machines and execution of processes, while the remaining layers are dedicated to the optimization of the production by managing resource allocation and operation scheduling for example. In addition to productivity gain and precision improvement, the automation of processes at the plant-floor automation layer allows the replacement of workers in harsh and hazardous environments or assigned to tedious tasks [2]. The WSNs are part of this layer and can be used for multiple purposes, such as monitoring synchronous or asynchronous events that require periodic data collection or detecting exceptional events, respectively [3]. For example, vibration, heat or thermal sensors can be deployed in proximity of machines to monitor their health. The analysis of the measured parameters can allow the detection of abnormal operating conditions and aids therefore in preventing potential machine failure. In addition to machine monitoring, WSNs can be deployed to measure basic physical quantities such as pressure, temperature, flow or more complex events such as process quality or automotive performance in industrial environments [4].

Although wired sensor networks can also be deployed for such monitoring scenarios, WSNs present additional advantages. In fact, their wireless capability allows deployments in hostile environments, where vibrations or moving parts may prevent the use of cables that would be damaged or even broken. In addition to reduce cabling costs, the WSNs provide network flexibility, as the sensor nodes may be relocated quickly without necessitating time-consuming cable installation and maintenance. However, the nature of the wireless medium opens up security and QoS issues. For example, potential attackers may easily eavesdrop or manipulate wireless communication in absence of security mechanisms and the wireless channel has to be efficiently allocated between the different devices to provide the required QoS.

Our contributions are as follows:

- We first select relevant WSN technologies dedicated to industrial automation and we provide an exhaustive survey of their characteristics.
- We then analyze industrial automation applications to determine QoS requirements and evaluate the selected standards according to each identified QoS requirement. Open issues are discussed based on the results of this evaluation.

- We carry out a threat analysis to identify pertinent security requirements and we investigate if and how the selected standards fulfill the previously identified security requirements. Related open issues are finally highlighted and discussed.

The paper is structured as follows. Section 2 provides a detailed overview of the following WSN technologies: Wireless Interface for Sensor and Actuators (WISA), WirelessHART, ISA100.11a, ZigBee, ZigBee PRO, and 802.15.4e Factory Automation MAC Layer. Section 3 and Section 4 focus on QoS and security respectively. Both sections are composed of an analysis of the respective requirements and an evaluation of the selected specifications. Open issues are listed and discussed at the end of both sections. Section 5 concludes our work.

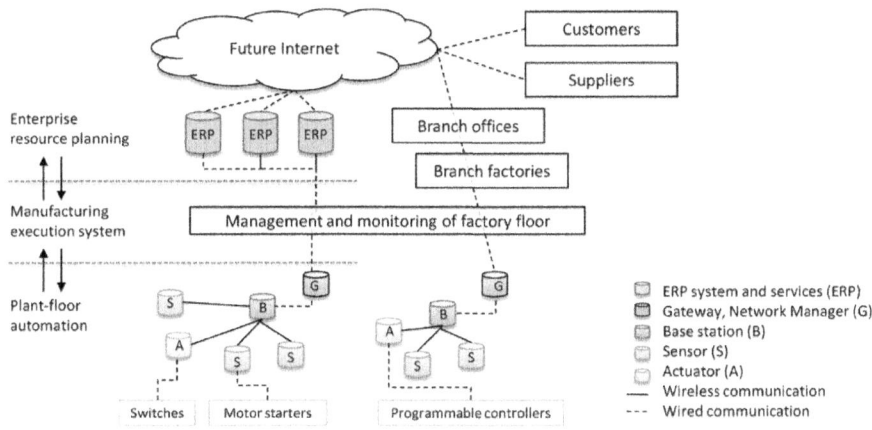

Figure 1. From the sensors to the customers.

SELECTED WIRELESS SENSOR NETWORKS STANDARDS: STATE-OF-THE-ART

Wireless communication in industrial automation is mostly based on standardized technologies, such as the IEEE 802.11 [5] and IEEE 802.15 standard families [6], also designated as *Wireless Local Area Networks* (WLAN) and *Wireless Personal Area Networks* (WPAN). Both of these standard families were conceived for application purposes different than industrial automation. In fact, the IEEE 802.11-based standards offer high data rates in the order of tens of Mbit/s and ranges up to tens/hundreds of meters, while the IEEE 802.15-based standards only supports data rates of hundreds of kbit/s to several Mbits/s with ranges from a few meters up to hundreds of meters.

However, to provide greater data rate and range, IEEE 802.11 technology consumes a greater energy budget that can limit the benefits obtained by wireless communications. Indeed, the sensor nodes are either powered by cables or batteries. In the former case, the advantages provided by wireless communication are partially negated, whereas in the latter case, the scarce energy resource has to be parsimoniously consumed in order to avoid frequent human interventions to recharge the batteries. Energy is thus a major concern in both previous cases and we therefore focus on the IEEE 802.15-based standards, and particularly on the IEEE 802.15.1 [7] and IEEE 802.15.4 [8] standard, within the scope of this work.

IEEE 802.15.1-based Standards

The IEEE 802.15.1 standard, also known as Bluetooth®, can be classified to fall between the IEEE 802.11 and IEEE 802.15.4 standards in terms of energy consumption and data rates. With medium data rates and lower energy consumption than the IEEE 802.11 standard, IEEE 802.15.1 offers an interesting compromise between energy consumption and data rate, and is therefore particularly suited for high-end applications requiring high data rates as well as applications with strong real-time requirements such as *factory automation*. The Wireless Interface for Sensor and Actuators (WISA) has been selected as a representative 802.15.1-based specification for further discussion.

Wireless Interface for Sensor and Actuators (WISA)

Released by ABB and presented in [9], the proprietary *Wireless Interface for Sensors and Actuators* (WISA) specification is based on the IEEE 802.15.1 physical layer and targets factory automation WSNs with packet error rate less than 1−9 and cycle time of 2ms [10].

Network Elements

WISA networks [11] can be deployed in cellular topology with up to three cells (Fig. 2). Each cell uses a different transmission frequency, and is composed of a *base station* and up to 120 *end devices* including sensors and/or actuators organized in a star topology. The end devices communicate wirelessly via standard Bluetooth transceivers, while the base station is equipped with a specific transceiver, which is able to receive up to four channels in parallel [12]. Additionally, the base stations exchange information with the network manager via wired fieldbus such as DeviceNet [13] and Modbus [14].

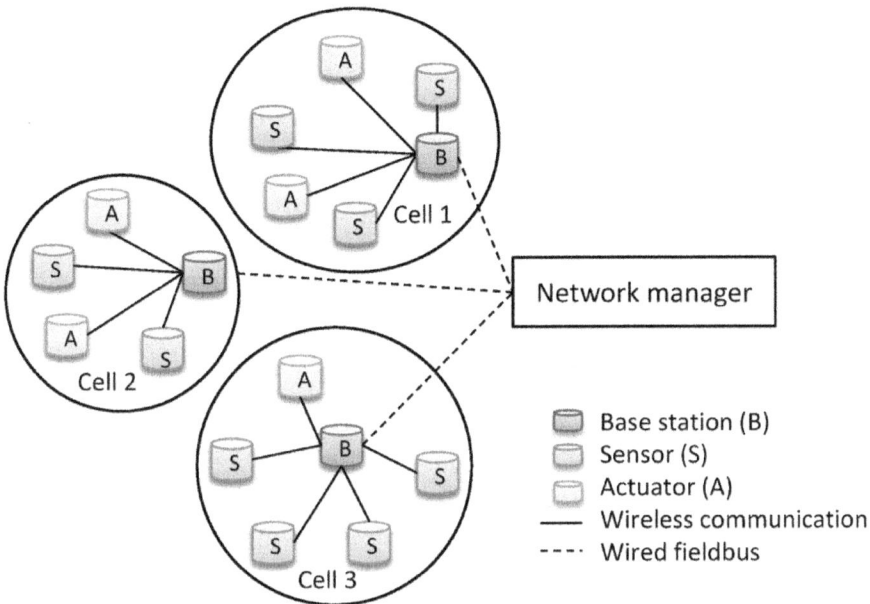

Figure 2. WISA network elements.

Architecture

The WISA architecture is limited to the physical and MAC layers, as sensors and actuators communicate exclusively with a central base station in a star topology within each cell. As mentioned previously, the WISA specification relies on the physical layer of the IEEE 802.15.1 standard operating in the 2.4 GHz frequency band at a data rate of 1 Mb/s. The WISA specification [15] is based on *Time Division Multiple Access* (TDMA) and *Frequency Division Duplex* (FDD), meaning that communication from base station to sensors, defined as *downlink* direction, and from sensors to base station, defined as *uplink* direction, occur at different frequencies. The downlink direction is exclusively reserved for the base station that remains continuously active and manages the TDMA scheme occurring in the uplink direction. The four uplink channels are divided into superframes of 2048 μs, which are composed of 30 timeslots able to support packets up to 64 bit length (Fig. 3). These time slots are allocated by the base station to each sensor willing to transmit data. In case of successful transmission of the sensor data, an acknowledgement is sent by the base station in the downlink. Otherwise the sensor retransmits the data in the next frame. To avoid interference and improve the reliability, frequency hopping is additionally applied after each superframe with a carrier spacing of 1 MHz.

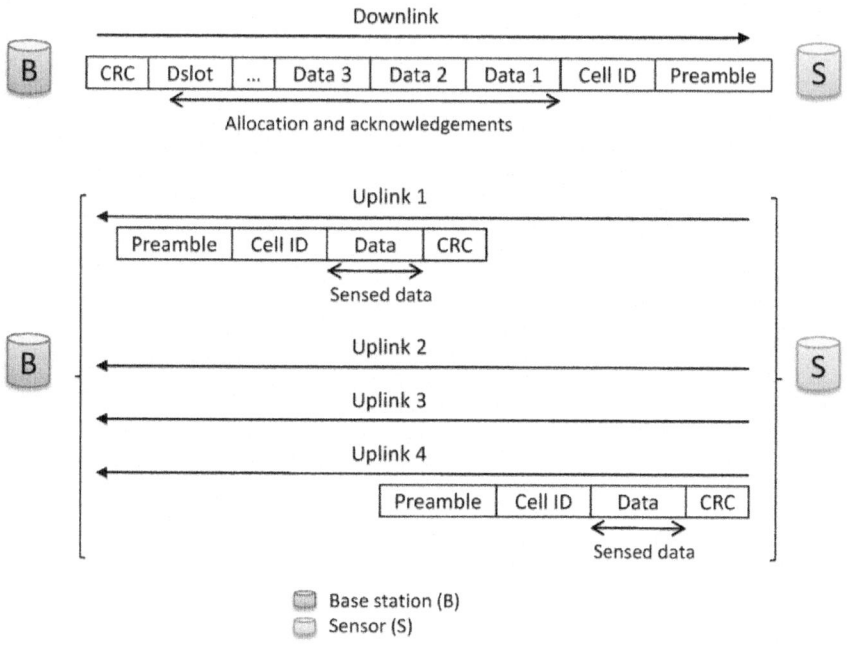

Figure 3. WISA superframe structure [16].

IEEE 802.15.4-based Standards

In comparison with Bluetooth®, the IEEE 802.15.4 standard presents lower data rates, but requires also lower energy budgets. According to [6], the standard is appropriate for infrequent exchanges of small packets, when power consumption is an important issue. The standard is therefore suited for *process automation applications*, where continuous production streams are monitored.

The IEEE 802.15.4 physical layer is common to all the following standards and operates in the 2.4 GHz frequency band as well as the 868 MHz and 915 MHz bands in Europe and North America, respectively. The 2.4 GHz frequency band is divided into 16 channels with a maximal data rate of 250 Kbits/s per channel and separated by a 5 MHz gap, while the 915 MHz band is divided into 10 channels with a maximal data rate of 40 Kbit/s each. The single channel in the 868 MHz frequency band presents a data rate of 20 Kbit/s. However, the effective data rates are smaller than the announced nominal values in reality, as mentioned in [6]. In addition to WISA, we have selected the WirelessHART, ISA100.11a, ZigBee, ZigBee PRO, and 802.15.4e Factory Automation MAC

Layer technologies. These technologies address the complete protocol stack from the physical layer to the application layer (Table 1) and thus provide a complete system, except for the 802.15.4e FA MAC and the WISA technology that mainly focus on the data link layer.

Table 1. Scope of the selected technologies

	WISA	WirelessHART	ISA100.11a	ZigBee	802.15.4e MAC
PHY/MAC layers	802.15.1-based	802.15.4-based	802.15.4-based	802.15.4-based	802.15.4-based
NET/TRANS APP layers	unspecified	specified	specified	specified	unspecified

WirelessHART

The HART Communication Foundation is an independent and not-for-profit organization that ensures the development of the HART Protocol. As technology owner and central authority, the foundation released the open WirelessHART™ standard in 2007, considered by [17] as the only released open wireless standard suitable for process measurement and control applications.

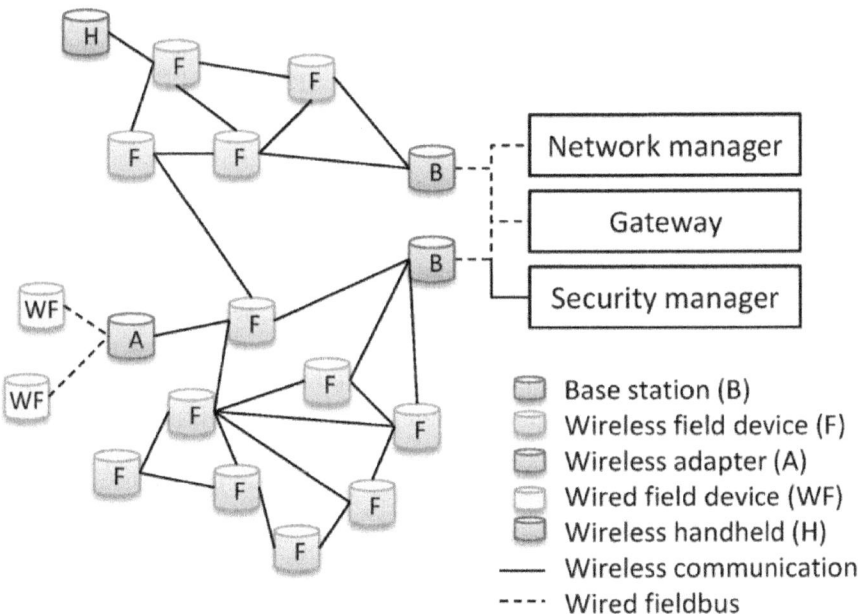

Figure 4. WirelessHART network elements [19].

Network Elements

WirelessHART networks are composed of different devices as illustrated in Fig. 4, including field devices, gateways, network and security managers. The *field devices* are organized in either star or mesh topology. However, the star topology is not recommended by [18]. The *gateway* is a bridge between the field device network and the host application. The gateway is configured by the *network manager* using HART commands and allows buffering large sensor data, event notifications, diagnostics, and command responses. In addition to the gateway configuration, the network manager configures the remaining devices and maintains the whole network. WirelessHART networks may include several network managers for redundancy reasons, however only one should be active at a time. The active network manager also schedules communication, manages routing tables, and monitors network health. In addition, the network manager can receive input from the host application and also queries the field devices about particular information via the gateway. The *security manager* collaborates with the network manager and prevents intrusion and attacks against the network by generating session keys, joint keys and network keys. Each security manager can collaborate with several network managers responsible for the key distribution to the concerned field devices. Additionally to these devices required by the standard, *adapters* and *handheld devices* may be added. The adapters connect HART devices to WirelessHART networks, whereas handheld devices configure and maintain WirelessHART compliant devices.

Architecture

The WirelessHART standard is presented by the HART Communication Foundation, as a reliable, secure, and robust standard. We consider successively the data link, network as well as transport and application layers that compose WirelessHART's protocol stack. At the data link layer, the WirelessHART standard coordinates and manages each device's transmission time by using TDMA with timeslots of 10 ms. Each time slot may be allocated to one source or may be shared between several sources using the *Carrier Sense Multiple Access with Collision Avoidance* (CSMA/CA) mechanism. In the former case, minimal latency can be reached, whereas the latter case supports efficient bandwidth utilization. Possible collisions due to multiple sources sending simultaneously are reduced by setting random back-off intervals for each source. During the timeslot assignment, the data to transmit are first prioritized by the network manager. The data originating from the network manager have the highest priority, followed by sensed data and event information. The timeslot allocation is then communicated by the network manager in the superframe to each device. At least one superframe (Fig. 5) is continuously repeated at fixed rate and further superframes can be added to support additional traffic [20]. The length of the superframes can be adapted to the needs of the application. However, the length is fixed once the superframe becomes active. In addition to the timeslot allocation, the network manager indicates the transmission channel in the superframe. This frequency hopping helps in reducing multipath fading and interferences. Moreover, the faulty channels are eliminated by blacklisting.

The WirelessHART standard uses two possible routing protocols [22] at the network layer: graph and source routing. In *graph routing*, the network manager determines the different paths forming the graph. The paths are stored by each device and are then used to identify the next node to forward the packet. In case of *source routing*, the packet header contains the list of the devices from the source to the destination. In addition to the paths stored for the graph routing, each device maintains tables about its communication statistics and neighbor activities. Furthermore, the WirelessHART network layer offers broadcast, multicast and unicast transmissions.

At the transport layer, the WirelessHART standard supports connection-oriented as well as connectionless communication. The *connection-oriented* communications are set up for applications requiring a reliable transfer of data between the host application and the field device for example. The connection

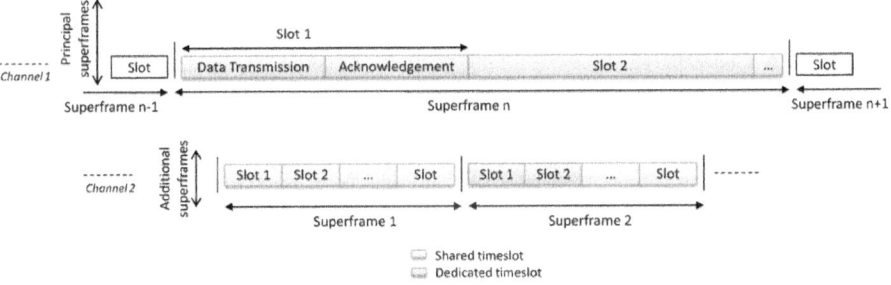

Figure 5: WirelessHART superframes [21].

set up starts by opening a dedicated port on the targeted field device with a specific HART command and the transmission data rate is then negotiated with the network manager, before the data transmission between the both entities can begin. Once all packets have been transmitted, the connection established between both entities is removed. The reliability of the data transfer is ensured by the acknowledgement and the order of the packets is maintained from the source to the destination. Nevertheless, these guarantees introduce additional overhead and a trade-off between reliability and overhead has to be found. For applications supporting out-of-order packet delivery, *connectionless* communication can be sufficient. Each datagram contains the full destination address and is routed independently from the others through the network [23]. Depending on the required level of reliability and the tolerance of packet loss, end-to-end acknowledgement can be introduced, but would also cause additional overhead. In connection-oriented and connectionless communication, the transport layer is also responsible for the segmentation of the data blocks at the source and their reassembling at the destination in order to allow transparent transfers for the upper layers. The WirelessHART application layer is based on the HART commands and extended by additional features allowing data publishing only when required for example.

ISA100.11a

The ISA100.11a-2009 standard [24] has been developed by the ISA100 standards committee, part of the non-profit International Society of Automation (ISA) organization, and approved by the ISA Standards and Practices Board in September 2009. This first release focuses only on process applications that tolerate delays up to 100ms [12]. However, further releases addressing factory and building automation applications are expected in the next years. In parallel to the main ISA100.11a working group, different working groups address complementary issues including the compatibility of the ISA100.11a standard with existing wired and wireless standards.

Network Elements

ISA100.11a WSNs can be organized according to different topology schemes and are composed of field devices, gateway(s), and handheld device(s) as depicted in Fig. 6. Some of the *field devices* responsible for sensor data collection and actuator management can also provide routing functionalities. According to the standard, there is no limitation of the amount of subnets that form the network, and therefore the total amount of devices is not limited. However, the amount of devices per subnet is restricted by the addressing space to 30 000 devices. One or several *gateways* ensure the connection between the WSN and the user application. The gateways also support the interoperability with different standards such as WirelessHART by translating and tunneling information between the networks and could act as security and network managers. Moreover, *handheld devices* support device installation, configuration and maintenance.

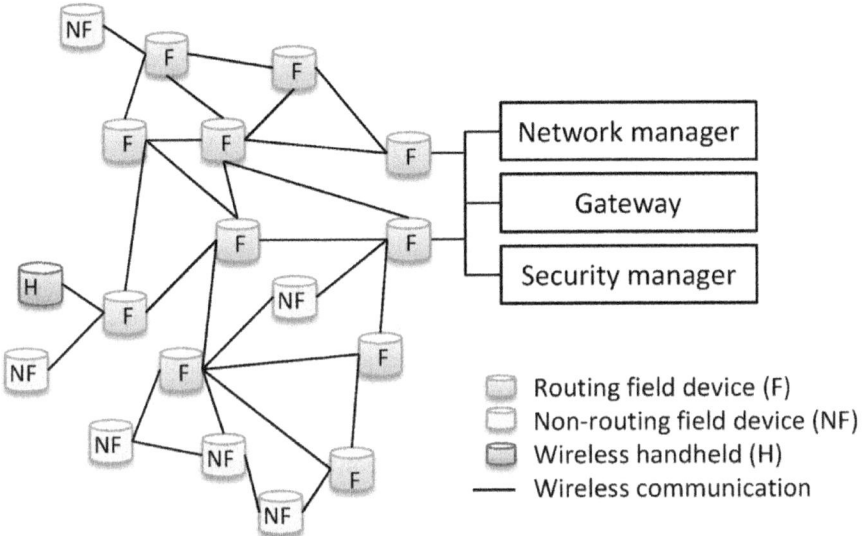

Figure 6. ISA100.11a network elements [25].

Architecture

The ISA100.11a standard addresses all OSI layers: physical, data link, network, transport and application layers. The*physical layer* is based on the IEEE 802.15.4 standard. However, additional requirements were defined. First of all, the ISA100.11a-2009 supports only the frequency band at 2.4 GHz; the lower sub-bands are not supported. In addition to the IEEE 802.15.4 standard, the ISA100.11a standard supports frequency hopping and also blacklisting that eliminates faulty frequency bands in order to improve the robustness against interferers. Additionally, the carrier sensing scheme can be disabled to reduce possible delay transmission.

The ISA100.11a *data link layer* is responsible for the management of the employed TDMA schemes by configuring the timeslot durations and managing the superframes. The configuration of timeslots can be done according to two different patterns: slotted channel hopping and slow channel hopping. The former scheme optimizes the bandwidth utilization and is adapted to energy-constrained routers, while the latter smooths the time synchronization requirements between neighbors by powering their receivers continuously during well-defined periods. The last scheme is therefore only adapted to routers offering unlimited energy budget. Both patterns can however be combined in a hybrid fashion by mixing superframes of both types. In addition to manage TDMA, the network manager assigns paths and links between the devices composing the WSNs. Each link is associated to one or multiple timeslots of a superframe and its type can be *transmit* and/or *receive* (Fig. 7). Information about the neighbors, the channel offset from the superframe hopping scheme as well as possible alternatives for the transmission and reception are also included. The ISA100.11a data link layer supports the previously described graph routing as well as source routing.

The *network layer* provides schemes for routing and QoS. It allows energy and bandwidth savings by mapping and translating the 128-bit addresses of the application endpoints to 16-bit short addresses used within subnets and vice versa. These savings can be increased by adapting the packet formats in function of the desired addressing, routing or QoS. Moreover, the network layer headers are compatible with the headers specifications conceived by the IETF 6LoWPAN working group and described in [26] in order to support future compatibility. Packet fragmentation and reassembly are also ensured at this layer. The packets can be routed at the backbone and the mesh levels, as defined in the standard. The first routing level is ensured by the data link layer, while the second is performed at the network layer by the end devices with routing capabilities. QoS is addressed in detail in the next section.

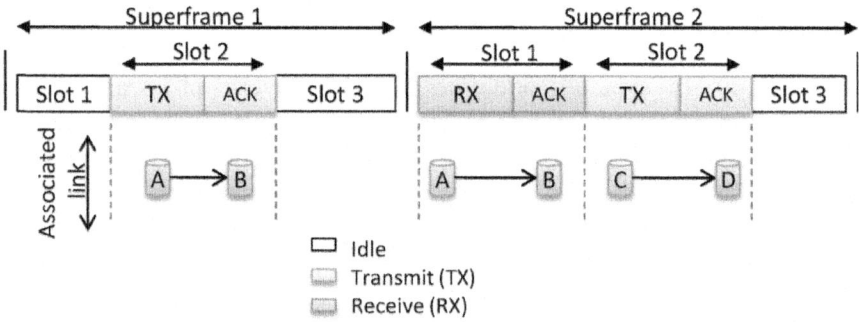

Figure 7. ISA100.11a superframes [24].

Depending on the level of reliability required by the application, the ISA100.11a *transport layer* can support end-to-end acknowledgements as well as unacknowledged communication. Additionally, flow control, segmentation and reassembly as well as security (see Section 4) are supported at transport layer, whereas the application layer ensures standard interoperability by using tunneling and native protocols at the gateways. The former carry protocols used in existing standards such as HART or FOUNDATION Fieldbus, while the latter provide efficient bandwidth utilization and therefore increase the battery lifetime.

ZigBee and ZigBee PRO

The ZigBee®standard, described in [27], was developed by the ZigBee Alliance and was originally designed for home automation. A new ZigBee PRO variant was released in 2007 to fulfill the industrial requirements. The ZigBee PRO standard is still based on the IEEE 802.15.4 physical and MAC layers and provides network and application layers with enhanced security features. However, the ZigBee PRO standard supports only frequency agility that consists of scanning available channels to determine the channel with the least interference, which is then selected and used by all ZigBee devices. Within the scope of this survey, we refer to both ZigBee and ZigBee PRO variants as ZigBee, except for the explicitly mentioned specificities.

Network Elements

ZigBee networks support hundreds of devices and should thus be suitable even for large deployments. They can be organized into star, tree or mesh topologies. The ZigBee standard is based on the two defined IEEE 802.15.4 device classes including *Full-Function Device* (FFD) and *Reduced-Function Device* (RFD) and proposes three different types of devices: ZigBee coordinator, ZigBee router and ZigBee end devices (Fig. 8). A unique FFD *ZigBee coordinator* manages the network by supervising the network formation as well as information storage, and bridges it with others ZigBee networks. The *ZigBee routers* are complementary to the network manager and also FFD devices with additional routing capabilities, responsible for linking group of devices and supporting multi-hop communications. *ZigBee end devices* are either RFD or FFD. They transmit the collected sensor or actuator data to a unique FFD including router or coordinator functionality. Consequently, a FFD becomes the master of RFDs organized according to a star topology. Furthermore, the ZigBee specifications introduce a *trust center* to manage the keys and the end-to-end configuration. Only one center trusted by all devices should be active and be associated with all network devices.

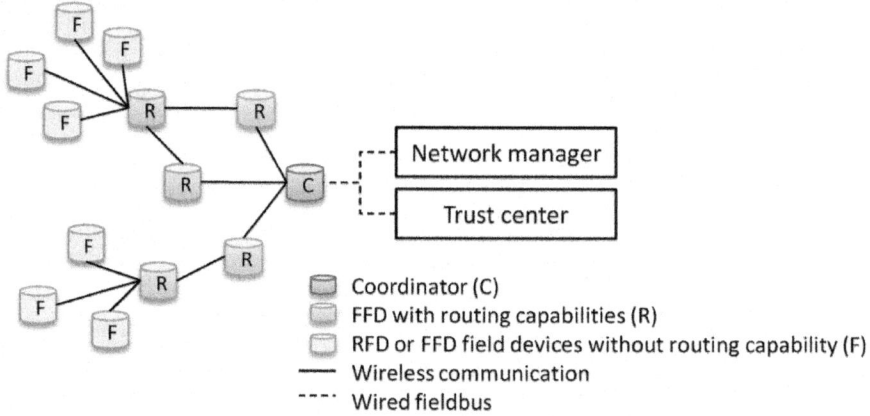

Network manager

Trust center

Coordinator (C)
FFD with routing capabilities (R)
RFD or FFD field devices without routing capability (F)
—— Wireless communication
- - - - Wired fieldbus

Figure 8. ZigBee network elements [28].

Architecture

The ZigBee stack is composed of the IEEE 802.15.4 physical and MAC layers as lower layers, and of the network and application layers specified by the ZigBee standard. After having set the selected common frequency for all devices, data transfers between ZigBee devices are possible. Two data transmission mechanisms are possible in ZigBee networks: with or without beacon. In the mode *with beacon*, the FFD sends a first beacon to synchronize all RFD sleeping phases and announces the superframe structure to manage the communication from end devices to the FFD. The first part of the superframe is slotted and CSMA/CA is used as channel access mechanism, while the second is composed of slots reserved for particular nodes by the network coordinator (Fig. 9). The FFD announces first the data transfer in the beacon to transfer data from the FFD to the RFD. Then, the concerned RFD must send a data request to the FFD to begin the data transmission. In case of FFD to FFD communication, the mechanism is similar, as one FFD acts as end device and is synchronized by the beacon originating from the second FFD. In the mode *without beacon*, no beacon and superframe are transmitted . The channel access is based on unslotted CSMA/CA. Each FFD coordinator remains continuously active to receive data coming from end devices during their limited active phase. RFDs send data requests to the FFD to receive data from the FFD. FFDs are permanently active and can thus communicate easily. In addition to transmission management, the MAC layer partially supports the admission of new devices in the network. The admission process starts by the scan procedure, during which the RFDs listen for beacon requests sent by a

FFD. Request and acceptance notification are then exchanged at the MAC layer to complete the admission process. However, the decision to accept or reject a device is left to the security mechanisms supported by the upper layers and in case of acceptance, a 16-bit short address is assigned to the new device.

The *network layer* is specified by the ZigBee standard and is responsible for network formation, address assignment as well as routing over the ZigBee network. The network layer is complementary to the MAC layer and takes part in the join procedure by initiating a network discovery mechanism to detect surrounding ZigBee networks. After the selection of the network by the application layer, the network layer chooses a parent to attach the joining device and requests the MAC layer to begin an association procedure, where the network layer assigns the 16-bit address to the joining device. The ZigBee network layer employs the *Ad hoc On Demand Distance Vector* routing algorithm (AODV) as route discovery mechanism to manage routing in mesh networks.

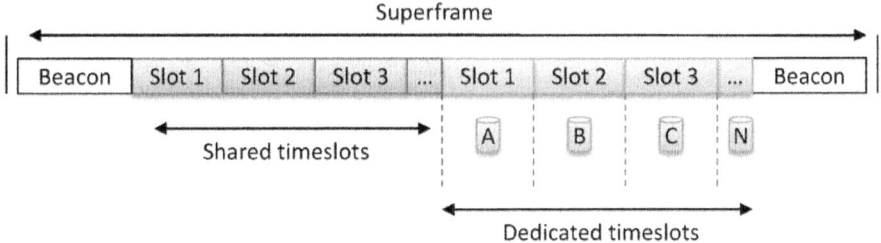

Figure 9: ZigBee superframe.

The ZigBee *application layer* proposes a framework for distributed application development and communication [27]. This application framework is composed of up to 240 *Application Objects* (APO). They consist of software units controlling dedicated device hardware and are disseminated over network devices. Each APO manages a set of variables and offers the possibility to set and read its values as well as report value changes. These functions are accessible by using the APO local number, which extends the device address. Additionally, the *Application Sub Layer* (APS) provides an interface to ensure security and data services between APO and *ZigBee Device Objects* (ZDO), which manage APO discovery services. Finally, application profiles described in the ZigBee specifications define formats and protocols for intra APO communication allowing the interoperability of ZigBee devices with the same application profile.

802.15.4e Factory Automation MAC Layer

The IEEE 802.15 Task Group 4e is currently developing a MAC layer [29] dedicated to factory automation and based on the IEEE 802.15.4 standard. The 802.15.4e Factory Automation MAC layer defines a deterministic TDMA communication scheme to fulfill the real-time requirement.

Network Elements

The network is composed of *sensors* and *actuators* organized in star topology around a *gateway* (Fig. 10). The *network manager* configures each end device via the gateway and allocates the dedicated time slots. After the configuration phase, sensor to gateway communication is unidirectional, whereas actuator/gateway communication is bidirectional.

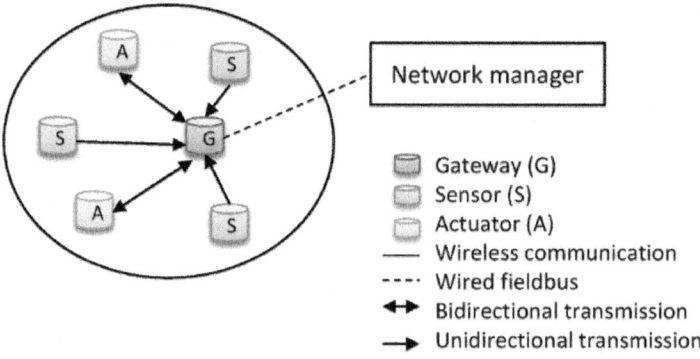

Figure 10. 802.15.4e Factory Automation MAC Layer network elements.

Architecture

The 802.15.4e Factory Automation MAC layer is based on the IEEE 802.15.4 physical layer and develops particular superframe formats as well as transmission modes to support deterministic TDMA [29]. The gateway supports three main transmission modes: discovery mode, configuration mode and online mode.

The *discovery mode* takes place during either network setup or joining procedure. The gateway sends superframes with beacons to indicate the discovery mode. When a device wanting to join the network receives such a beacon, it tries to access the transmission medium to send a *Discover Response* frame to the gateway with its current configuration parameters. The frame will be retransmitted by the device until the gateway receives it or changes its transmission mode.

During network setup or reconfiguration, the gateway is in a *configuration mode* and indicates this status in the superframe beacon. When the device receives the beacon and gets access to the transmission medium, it sends a *Configuration Response*frame to the gateway with its current configuration until the gateway receives it or changes its mode. As soon as the gateway receives the *Configuration Response* frame, it sends a *Configuration Request* frame with the new device configuration parameters and the device sends an acknowledgement in the next superframe.

In the *online mode*, devices can send data to the gateway in the timeslots allocated during the configuration mode and the gateway acknowledges the received data in the following superframe.

The superframes are sent to the end devices by the gateway and their structures depend on the current gateway transmission mode. The first slot is designed as the beacon slot (Fig. 11) and is common to all superframe structures. The end devices can detect the start of a new superframe at the reception of this first slot and synchronize themselves with it. Additionally, the beacon specifies the current transmission

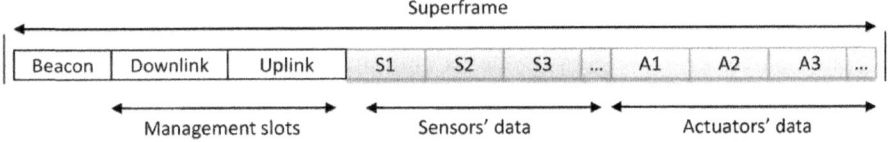

Figure 11: 802.15.4e FA MAC superframe structure [29].

mode and also acknowledgements for data transmitted in the previous superframe. In discovery, configuration and optionally online modes, the beacon is followed by up to two management time slots, which manage the bidirectional transmission between gateway and actuators. During online transmission mode, the next time slots are allocated to sensors. These timeslots can be either dedicated to a particular device or shared by a group of devices using CSMA/CA. In the first case, no addressing information is necessary, whereas the second case requires a simple addressing scheme. Then, actuator time slots are reserved in the superframe. The direction of the communication between actuators and gateway is indicated in the beacon, and each time slot can be either dedicated or shared.

QUALITY OF SERVICE

Quality of Service (QoS) refers to *"the collective effect of service performance which determine the degree of satisfaction of a user of the service"* [30]. More technically, QoS can be defined as the *"well-defined and controllable behavior of a system with respect to quantitative parameters"* [31]. Interesting parameters to monitor for computer networks can be delay, jitter, throughput, fairness, and packet losses for example. However, the requirements of industrial automation networks are different from usual computer networks. The harsh industrial operation environment may degrade the wireless communication performance due to path loss and shadowing, multi-path propagation, and interference [12]. Moreover, the traffic is mainly composed of short packets containing sensor measurements or actuator commands that need to be delivered timely, instead of large multimedia streams or interactive traffic that prevails in computer networks. We next identify the specific QoS requirements and evaluate the aforementioned selected specifications with respect to their compliance to these requirements. Open issues are identified and discussed based on the results of this evaluation.

Requirements

Industrial automation applications can be divided into two main categories of systems [32]: close-loop1 and open-loop systems2 as defined in [33]. The applications based on either close or open-loop systems have different QoS requirements, as close-loop systems monitor discrete operations, such as actuator control, and open-loop systems monitor continuous processes, such as cooking raw material.

Table 2: Industrial automation applications classified according to their real-time requirements [35]

	Real-time requirements			
	None	Soft	Hard	Isochronous
Cycle duration		greater than 10 ms	1 ms to 10 ms	250 μs to 1 ms
Jitter				less than 1ms
Application examples	Maintenance	Process automation	Control	Motion control
	Diagnosis	Data acquisition	Machine tools	

In close-loop applications, the sensors generate traffic that needs to be transmitted timely, reliably and accurately [33] to the control system, whereas in open-loop applications, real-time processing is typically not required, but the energy consumption is a crucial point. An additional classification [35] of the industrial applications refines these categories and is based on the degree of the required real-time guarantees. The classification is composed of four classes presenting increasing real-time requirements that are summarized in Table 2. We next focus on the support of real-time traffic and the reliability offered by the selected specifications.

Evaluation of the Selected Specifications

To evaluate the support of real-time transmission, the medium access control mechanisms including priority management schemes of the selected specifications are compared. Different diversity parameters including frequency and space diversity as well as protocol features such as acknowledgements are additionally considered in order to estimate the reliability provided by the different specifications.

Real-time Support

The selected specifications are mainly based on the IEEE 802.15.4 data link layer, except for the WISA specification relying on the IEEE 802.15.1 standard. As 802.15.4e FA MAC Layer and WISA only address the specifications of the physical and data link layers, the comparisons of the selected technologies are mainly based on the features of their two first layers. To complete this comparison, the 802.15.4 FA MAC needs to be evaluated in the context of a whole system. However, this aspect is outside the scope of this survey.

Medium Access Control Mechanisms

Each standard uses TDMA as medium access control mechanism (Table 3). However, TDMA is not the main medium access scheme in the ZigBee standard, which uses primarily CSMA/CA, but can also provide guaranteed timeslots in its beacon mode. Even if all specifications make use of TDMA access mechanisms, some differences between them can be observed. First of all, the timeslot length can be configured in the ISA100.11a, ZigBee and 802.15.4e FA MAC standards, while it is fixed in the others standards. Adapting the timeslot length to the needs of the applications allows taking into account the specific characteristics and optimizes the timeslot duration. Without such optimization, time can be lost between two successive slots if the slot is longer than the data to send. Configuring the length of the timeslots is thus an important feature to allow for optimizing the real-time support. However, all timeslots inside each superframe have the same length that limits a real adaptation to the requirements of individual applications, as only the overall traffic mix can be optimized. Another feature is the selection of the type of the timeslots, which can be either shared or dedicated. Most of the selected specifications offer both kinds of timeslots. Only the WISA specification uses exclusively dedicated timeslots. The choice between dedicated and shared timeslots is made difficult by the trade-off between real-time support and optimized utilization of the medium. Indeed, the dedicated timeslots are assigned to one particular device only. If this device wants to send commands or measurements, the data are transmitted immediately within the reserved timeslots. Otherwise the timeslot is lost and other devices wanting to transmit information may be constrained to wait longer. In case of shared slots, the devices have to compete to access the medium. The medium utilization is thus optimized because slots can be utilized by the stations in need of bandwidth, but the data transmission cannot always be immediate due to random back-off mechanisms for example.

Superframe Management

Furthermore, the WirelessHART, ISA100.11a and the ZigBee standards allows optimizing the transmission of superframes to fit the real-time communication constraints. The ISA100.11a standard supports insertion, removal and activation of superframes during the operating process, whereas additional superframes can be transmitted in parallel to the mandatory superframe in the WirelessHART standard. In the *without beacon* mode of the ZigBee standard, no superframe is transmitted and the devices do not have to wait for the next timeslot to transmit their data. However, this solution has also drawbacks. If one device has a great amount of data to send, the channel would be occupied for a long period of time and data requiring real-time transmission could not be send during this period. Moreover, the time to get access to the channel may be longer than in case of dedicated slots, if all the devices have data to be sent. The management of dedicated and shared timeslots has therefore to be tailored to application characteristics and the traffic patterns.

Priority Management

In addition to medium access control, flow control with assignment of priority to the packets can be introduced to support real-time at higher level. A priority flag indicates to each device on the path if the arriving packet has to be transmitted without delay or can be buffered. The WirelessHART standard defines four main priority levels according to classes of data [36]. Control and configuration information as well as network diagnostic messages have the highest priority, followed by a second category composed of process data measurements and network statistic messages. The lowest priority is assigned to the fourth category, which includes packets reporting information about events and alarms. Additional classes of data are classified into the third category. In comparison, the ISA100.11a standard allows to prioritize the QoS contracts established between the devices and the system manager in addition to

Table 3. Features relevant for real-time operation

	WISA	WirelessHART	ISA100.11a	ZigBee	802.15.4e MAC
TDMA	x	x	x	x^3	x
CSMA/CA				x	
FDD	x				
Superframe	x	x	x	x^4	x
Fixed timeslot duration	x	x			
Duration	64 and 128 μs	10 ms			
Shared timeslots		x	x	x^4	x
Dedicated timeslots	x	x	x	x^4	x
Superframe optimization		x	x		
Message-based priority		x	x		
Priority levels		4	2		
Contract-based priority			x		
Priority levels			4		

message priorities. The device wanting to transmit data first sends the required priorities in a contract proposal, as well as other additional parameters including reliability, periodicity and negotiability. The system manager replies to this request in a contract response and indicates the provided QoS for the requested communication. Depending on the traffic conditions, the system manager might not be able to provide the requested QoS. If the device indicates that the contract is negotiable, the system manager can propose another QoS level or postpone the contract to provide the requested QoS level. The contract priority is set during the contract establishment and concerns all messages exchanged during the contract duration. Depending on these priorities, the system manager manages the routing and the load balancing to provide the guarantees defined in the contract. Four levels of contract priority are available at network layer: network control, real time buffer, real time sequential and best effort queued. The first priority level can be used by the system manager to communicate critical information about the network management; whereas the second category is used for periodic data exchanges with buffer overwrite operations in case of fresher messages. The third class, real time sequential, is appropriate for applications requiring sequential and real-time data delivery like video or voice-based applications. The last level is adapted for client-server communications. Within the contract, message priority can also be assigned by setting one bit to either low or high. However, the contract priority takes precedence over the message priority.

Each standard supports real-time communication in a different way. While the WISA specification and the 802.15.4e FA MAC layer target applications with strong real-time requirements, the ZigBee and the WirelessHART standard are more adapted to applications with softer requirements. The recent ISA100.11a standard supports currently soft real-time requirements. More precisely, the evaluation of the selected technologies has shown that:

- Dedicated timeslots are supported by all standards. Their utilization is a key feature to fulfill strong real-time requirements and is particularly appropriate, if the complete set of the sensors have a continuous stream of information send. To optimize the transmission, dedicated as well as shared timeslots can be combined to support fluctuations of the traffic load. Such hybrid modes can be envisaged in most of the considered technologies except for the WISA standard, which only support dedicated timeslots.

- Tuning the timeslot length is a second key feature that allows to support application-specific traffic as well as optimize the real-time support. Except for the WISA and the WirelessHART standards, all standards offer this option. However, the timeslot length is common to all timeslots within a superframe, which limits the adaptation to variations of traffic as well as real-time requirements. Although this property may be useful for applications with various length of data to send, it would introduce additional overhead and complexity and may therefore unnecessary.

- Superframe management is an additional means to maintain real-time communication in case of very high traffic load. However, the superframe optimization proposed by WirelessHART and ISA100.11a remains a minor contribution to the real-time support in comparison to the aforementioned dedicated timeslots and adaptable timeslot lengths.

- Priority mechanisms are provided by the WirelessHART and the ISA100.11a standards. Both standards support message-based priority, while the ISA100.11a standard offers contract-based priority additionally. Although the proposed contracts are promising in terms of QoS guarantees, the overhead and the complexity introduced by their management as well as the additional resulting traffic may limit their practical feasibility. Even if message-based priorities may be less efficient than contracts, their utilization allows to reach a balance between overhead and efficiency and provides a complementary method to support real-time traffic.

Reliability Support

Industrial WSNs are located in spaces, where equipment moves, conditions change and interference perturbs the communication. Mechanisms, such as space and frequency diversity as well as acknowledgements (Table 4), are thus required to protect the wireless networks from these disturbances [37].

Space Diversity

Space diversity allows bypassing obstacles and interference by modifying the routing within networks organized in mesh topology. Such ability is however impossible with star topologies, as each device communicates exclusively and directly with a central coordinator. In case of obstacles that block the wireless communication, no alternative path is available and the communication cannot be established. Within the selected standards, the WISA specification and the 802.15.4e FA MAC standard are foreseen to be

Table 4. Features relevant for robust/reliable features

	WISA	WirelessHART	ISA100.11a	ZigBee	802.15.4e MAC
Mesh topology		x	x	x	
Channel agility				x^5	?⁶
Channel hopping	x	x	x		?⁶
Channel blacklisting		x	x		?⁶
DDL acknowledgements	x	x	x	x	x
TL acknowledgements		x	x		
Automatic repeat request	x	x	x	x	?⁶

deployed in star topology only. Their protection against wireless channel obstructions and fluctuations is therefore reduced in comparison with the other standards. However, the maximal distance between the central coordinator and the sensors is shorter than between devices deployed in mesh topology. The probability of an obstacle breaking the communication is therefore low, if careful network planning is performed.

Frequency Diversity

In addition to space diversity, frequency diversity reduces the effects of the environment on the wireless communication by limiting the interference. Two frequency diversity schemes are possible: channel hopping and channel blacklisting. The channel hopping allows avoiding interference by changing the transmission frequency, while the devices maintain lists of frequencies to avoid due to their significant interference with the channel blacklisting scheme. The WirelessHART and the ISA100.11a standards use both of them and possess therefore an efficient response against interference. The WISA specification uses channel hopping, but does not blacklist faulty channels. In comparison, the ZigBee and 802.15.4e FA MAC standards do not provide any mechanism to avoid potential interference, which may lead to erroneous transmissions or even worse to a total transmission break down. However, the ZigBee PRO standard proposes an enhancement of the ZigBee standard by offering frequency agility. The available channels are scanned during the network setup phase to select a frequency without interference. Then, the frequency is shared by all ZigBee compliant devices and remains unchanged until the next network setup period. Even if the frequency agility may limit the effects of potential interference during the network formation, it can rapidly become inefficient in case of additional sources of interference.

Acknowledgement Management

Interferences or obstacles can also lead to packet loss. To ensure transmission reliability, *acknowledgements* (ACK) at *Data Link Layer* (DLL) are supported by all the selected standards, as well as end-to-end acknowledgements at *Transport Layer*(TL) for the WirelessHART and ISA100.11a standards. Each standard however manages its ACK mechanisms according to its transmission scheme. For example, the WISA specification transmits each ACK in the downlink channel, whereas each ACK is transmitted during the same timeslot as the received data in the WirelessHART standard. Once an ACK is missing, all standards7 support the automatic retransmission of the data.

The comparison of the aforementioned mechanisms summarized in Table 4 shows that:

- The technologies targeting *process automation* applications provide a good protection against potential obstacles and node failures, as mesh topology is supported.

- The WirelessHART and ISA100.11a standards offer the most complete set of mechanisms with channel blacklisting and frequency hopping to avoid perturbations caused by interference. While channel blacklisting allows saving time and energy by avoiding scanning channels previously identified as faulty, the obtained benefits are minor in comparison with the frequency hopping capability that is fundamental. Even if the ZigBee PRO version shows some improvements, the frequency diversity proposed by the ZigBee standards is insufficient to fulfill the strong reliability requirements of industrial applications.

- All selected technologies use acknowledgements at data link layer and automatic repeat request to ensure reliable transmissions and identify packet losses. At transport layer level, only the WirelessHART and ISA100.11a standards support acknowledgements.

Open Issues

The previous evaluation has highlighted that the selected specifications differ with respect to the supported QoS requirements for industrial applications, particularly concerning real-time support and reliability.

Open Standard for Factory Automation

The first conclusion that can be drawn is that no standard provides currently an open solution to *factory automation* with strict real-time requirements. The WISA specification is dedicated to this kind of deployments, but it is based on the IEEE 802.15.1 standard that consumes more energy than the IEEE 802.15.4-based standards. Moreover, the WISA specification is proprietary, thus locking the user into a single vendor as well as the design and maintenance of a proprietary set of interfaces. Even if WISA successfully fulfills the requirements of factory automation, it does not support openness and interoperability. The 802.15.4 Factory Automation MAC provides a promising perspective, but is still under development. Furthermore, the recently released ISA100.11a standard is expected to be enhanced by future addenda to make it fit to applications with stronger real-time requirements. As a consequence, additional research and development is necessary in order to obtain open and IEEE 802.15.4-based standards, which suit the needs of factory automation.

QoS Support in Heterogeneous Networks

In the *process automation* domain, the standards WirelessHART and ISA100.11a occupy the first positions and a potential convergence is analyzed by the ISA100.12 Working Group. WirelessHART devices are foreseen to be deployed within ISA100.11a-based WSNs, as illustrated in Fig. 12. In parallel to the ISA100.12 Working Group, the ISA100 Wireless Backhaul Backbone Network Working Group addresses potential interoperability between the ISA100.11a and the ZigBee-based standards. Additionally, the ISA100.11a standard is compatible with existing wired standards including HART, FOUNDATION Fieldbus, Modbus, and Profi bus. The ISA100.11a standard offers therefore a wide panel of compatible standards, which allows manufacturers to reuse existing devices. However, questions regarding the provided QoS capabilities among heterogeneous WSNs have to be raised. Indeed, such compatibility requires translation between the standards at the gateways, which may increase the end-to-end transmission delay of end systems belonging to different kinds of networks. Moreover, options implemented in one standard may not be supported or may have to be disabled to support interoperability. In such case, segments of the network that do not support the same priority scheme or the QoS contracts established in the ISA100.11a standard may ignore the indicated priority leading to a best-effort QoS. Thus, the efforts to support the QoS requirements would be wasted. A central network manager for both or more network segments based on different standards would be a possible solution because options, configuration settings and capabilities could be centrally managed. The network manager would be able to obtain a global view of the network and thus is able to take the appropriate decisions. However, a single central network manager remains a single point of failure that threatens the resilience of the network, if it is not deployed redundantly. In case of malfunctions, the whole network would be affected and could catastrophically fail. Therefore, the support of interoperability between the investigated standards remains an open issue. Although addressing the heterogeneity has a number of interesting research challenges, its practical feasibility may be limited by the technical complexity and financial costs incurred to adapt different technologies.

QoS Support in Multi-hop Networks

The surveyed standards dedicated to *process automation* can also be deployed in mesh topology, thus raising the question of QoS support over multi-hop routes. The WirelessHART and ISA100.11a standards propose solutions based on priority mechanisms, which are centrally managed by the network manager. While the standards provide the basic mechanisms to support differentiated QoS, the detailed specification to support operation in a multi-hop network is outside the scope of the standard.

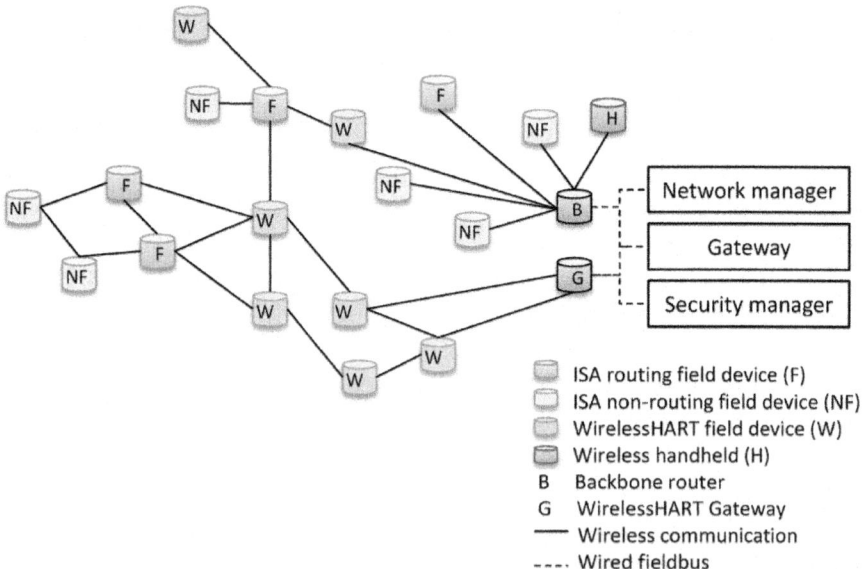

Figure 12: Interoperability of WirelessHART and ISA100.11.a networks [38].

Although mechanisms that support QoS in multi-hop networks were developed for other applications, e.g. the Internet, their reuse is limited as they do not provide sufficiently strong real-time guarantees and are mainly adapted to multimedia streams, which differ from data streams in sensor networks. Further research is therefore mandatory to provide an exhaustive description of the required mechanisms to support QoS in multi-hop networks.

SECURITY

In addition to the previously considered quantitative QoS parameters, security guarantees play an important role within industrial WSNs. Indeed, without any protection mechanism, the network could suffer from attacks or malfunctions that degrade the desired QoS by introducing additional delays, or not delivering correctly and timely the needed information. For example, these malfunctions can perturb the production chain, as one of the machines would move at an unexpected time or in the wrong direction. Such perturbations can have important consequences going from delayed and damaged production to broken equipment. Additional costs are not the only consequence; employee's lives can be endangered in the worst case; for example in case of explosions due to false temperature measurements in chemical industries. A threat analysis is conducted in this section followed by an evaluation of the selected standards in order to determine whether the WSNs are protected against the identified threats. Within the scope of this survey, only attackers located within the range of the WSNs and taking advantage of the wireless characteristics of the industrial networks are considered, as many methods like firewalls [12] are efficient to protect the networks against attacks coming from the outside. Moreover, attacks requiring physical capture of sensor platforms are excluded from this analysis.

Threat Analysis

To protect industrial WSNs efficiently against potential attackers, the following main security criteria have to be considered: confidentiality of information, integrity of information, authentication of communication peers and availability of information [35]. The first criterion ensures that the data access is restricted to authorized parties only, while the second ensures their protection against alteration and modifications by either malicious parties or the harsh surrounding environment. The authentication of communication peers allows guaranteeing that the exchanged data are coming from trusted devices. At last, the information availability ensures that data and services are accessible even in case of attacks. To perturb or even break down industrial

WSNs, the attackers can therefore target one or several of the aforementioned criteria and conduct the appropriate attack(s).

Confidentiality of Information

The wireless nature of the communication between the sensors and devices eases these attacks, as there is no strict physical boundary of the transmission medium. An attacker located close to the network can thus easily eavesdrop the communication and threaten the confidentiality of the transmitted information. The content of the packets can be revealed to the attacker, who can benefit from stolen information like network configuration data to conduct further attacks. Eavesdropping can also be coupled with network monitoring to perform traffic analysis. The aim of this attack is to determine the responsibility of each sensor and identify the data sink for example. An analysis of the packet content is not mandatory to success; the amount of exchanged packets can be a sufficient clue [39]. However, an attack directed against the data sink can be very efficient, as the entire data set may be damaged or lost.

Integrity of Information

In addition to the confidentiality of the exchanged information, its integrity can be threatened by attackers adding additional fragments to the packets or manipulating the data. However, malicious behavior is not the only source of packet manipulations; errors due to the harsh industrial environment are also possible. The modifications of the packet content may cause misbehavior of the equipment and thus have inconvenient effects on the production, or even worse.

Authenticity of Communication Peers

Packet manipulation can be one sign that one or several malicious nodes have succeeded in integrating itself with the network. Such intrusions widely open the doors to further attacks like Sybil [40] and node replication [41] attacks. Both attacks profit from weaknesses of the authentication mechanisms to insert malicious nodes. In the former case, these nodes take illegitimately multiple identifiers; while in the latter, they capture and use existing device IDs. The identifier manipulation allows the attacker to modify the content of the traffic exchanged between the devices as well as control messages such as routing messages. These attacks can therefore be the basis of further routing attacks like wormhole [42] or black hole [43] attacks, where the attackers are able to disconnect part of the network or make it totally inoperable.

Availability of Information

Such routing attacks also threaten the last criterion, as the data may not be delivered timely or even at all and the information are therefore not available. Additional attacks can be conducted at different layers to disturb the availability of information. At the physical layer, jamming may cause interference at different frequencies in an intermittent or constant manner that make the communication impossible. Jamming may be caused by malicious attacks or unintentionally by surrounding equipment. To fight against malicious jamming, the physical protection of the industrial sites is one of the first measures to adopt. However, most of the industrial sites still accept external visitors. Even if their visits may be strictly controlled, attackers might benefit from security weaknesses to introduce jammers within the factory. Additionally, uninterrupted transmission of data by the attacker can generate collisions and force retransmissions at data link layer. The energy budget of the node decreases rapidly due to the retransmissions and the sensor is made inoperable. Additional energy consumption can also be caused by flooding the network with many connection requests at transport layer for example.

Evaluation of the Selected Standards

The selected standards are evaluated to determine how the current industrial WSNs are protected against the aforementioned threats. The set of considered standards is restricted to the WirelessHART, ISA100.11a and the ZigBee standards8.

Confidentiality of Information

The evaluation begins with data confidentiality including protection against eavesdropping and traffic analysis. The most efficient way to protect the industrial WSNs against eavesdropping is to encrypt the exchanged data. The WirelessHART, the ISA100.11a as well as the ZigBee standards use the 128-bit AES encryption [44] coupled with different keys depending on the layer of encryption. For example, WirelessHART uses the session key to encrypt the message at transport layer, while link and network keys are used at data link layer and at network layer respectively in the ZigBee standard. As mentioned by [45], AES remains an efficient mechanism to keep the data secret. Moreover, its efficiency is increased by the utilization of keys with short lifetime and unique for each device such as the session key used in WirelessHART. Eavesdropping is consequently made difficult or even impossible in networks running the three considered standards. However, the confidentiality is not ensured at all layers. For example, even if the packets are encrypted at transport layer, header and payload of packets sent at network layer are transmitted unencrypted in the WirelessHART standard [45]. An eavesdropper can therefore discover the crucial information, such as source and destination addresses that are contained in the network header, and perform traffic analysis easily afterwards. Nonetheless, the traffic analysis attack can also be performed without any message decryption [39].

Integrity of Information

The selected industrial standards benefit from the security mechanisms included in the IEEE 802.15.4 standard that ensure data integrity at data link layer. An additional *Message Integrity Code* (MIC) is inserted at the queue of the data to protect. The data are signed and the receiver is able to determine whether the data have been tampered with or not. Data integrity protection can be provided in complement with encryption by using the *enhanced combined encryption and authentication block cipher mode* (CCM*). Depending on the desired security level, the length of the MIC can be set to 32, 64 or 128 bits. The longer the code is, the higher the integrity protection is, but also the greater the overhead is. The length should therefore be selected carefully.

Message and Device Authenticity

In addition to provide hop-to-hop data integrity, the MIC allows to authenticate the packets by using secret symmetric keys known by both sender and receiver. For example, the shared network key and the unique session key are used in WirelessHART to authenticate the messages at data link and network layers respectively [45]. The authenticated packets are thus recognized as originated by authorized members of the network. However, before message authentication can be performed, each device must be first authenticated during the join procedure. Even if the name of the keys may vary between the standards, their functions are similar. Devices willing to participate to the network exchange *join requests* and *join responses* with the network manager and use public key and asymmetric private key kept inside the joining devices. These keys are used for computing the data link and network MICs respectively and may be either preloaded in the devices at the factory or distributed by a unique trusted center, which maintains and updates the security keys. In WirelessHART, an additional join key is used to encrypt the *join request*. Once the device is recognized as authorized member, it can exchange authenticated data with the other members. As each standard is based on a central entity responsible for the network management and keeping tracks of the participating devices, the probability is very low that the attacks such as Sybil and node replication attacks may be performed successfully. For example, in WirelessHART, the network manager links each device with a unique identity [45]. The identification is completed by a list of unique IDs maintained at the gateways. The network manager identifier and the gateway ID are used conjointly with the session key to maintain sessions between the device and the network manager as well as the gateway respectively. Devices claiming the same identity as an existing one or sharing multiple identities would be immediately discovered, as these would already be listed.

Availability of Information

The last threat to be evaluated is the availability of information. First of all, the information availability can be threatened by jamming according to different patterns. In case of continuous jamming with one or several jammed frequencies, channel blacklisting provides an efficient solution, as the jammed channels are eliminated from the set of communication frequencies. In case of intermittent jamming, frequency hopping provides good results and allows keeping sufficient levels of information availability. With both frequency hopping and channel blacklisting features, the WirelessHART and ISA100.11a standards provide therefore a better protection against jamming than the ZigBee PRO standard that only offers frequency agility. At network layer, attacks modifying the routing scheme can be avoided by the authentication mechanisms, as devices would only be able to route packets, if they have been previously identified as reliable and authorized to take part to the WSN. Nonetheless, the selected standards do not provide dedicated mechanisms to avoid the generation of collisions by a malicious source transmitting continuously data, as well as solutions against flooding of connection requests at transport layer.

To summarize, we have shown that:

- Eavesdropping is made difficult or even impossible, but confidentiality is not addressed at all layers.
- Traffic analysis is still possible.
- The information integrity is sufficiently ensured.
- The probability of successful Sybil and node replication attacks are limited.
- The frequency diversity and agility is sufficient to protect the network against intermittent jamming.
- The current mechanisms do not provide protection means against malicious sources transmitting continuously or performing higher layer attacks such as flooding of connection requests at transport layer.

Open Issues

The selected standards are resistant against most of the considered attacks except for continuous jamming at all frequencies, collision and flooding attacks as well as traffic analysis. However, solutions against such particular kind of jamming and collision attacks are particularly difficult to find because the data transmission is made impossible in both cases. The only solution would necessitate human interventions to eliminate the interference source(s), as soon as a long-term communication breakdown is detected. Flooding of *connection request* is more delicate to solve, as regular nodes must still be able to send such request in order to join the network and authenticate themselves. Filtering the connection requests by a list of devices susceptible to join the network would not solve the problem, as the network manager would have to receive the requests in order to determine the sender identity. A solution remains therefore to be found. Even if payload encryption provides a partial solution to traffic analysis, it is not sufficient. Indeed, an analysis of the routing paths can be sufficient to determine the traffic scheme. Countermeasures to this analysis could be to insert fake packets and/or randomly distribute the traffic. However, the routing would not be optimized and the network performances including end-to-end delay would be degraded. Moreover, insertion of additional packets would drag the energy budget of the nodes down.

The existing security mechanisms are theoretically sufficient to cover the main attacks. However, the standards only provide specifications and leave design and implementation to the users. For example, the roles of each security key are described, but their management scheme has to be developed by the users, according to their requirements and respecting the standard specifications. To the best of our knowledge, Raza *et al.* are the first to have proposed a design and an implementation of a security manager adapted to the WirelessHART standard in [47]. The investigation of similar proof of concept for additional standards e.g. the promising ISA100.11a may be an interesting track to follow in order to provide already evaluated implementations that the users can tune to meet their requirements.

CONCLUSIONS

Within the scope of this article, we have provided a detailed survey on WSN standards dedicated to industrial automation networks. The standardization efforts are ongoing and targeting different application areas such as factory automation or process automation. We have focused on the IEEE 802.15.1 and IEEE 802.15.4 standard families, which have been adapted to industrial applications in need of short-range communication with high data rate, and energy-aware applications requiring larger coverage, respectively. We have selected the WirelessHART, ISA100.11a, ZigBee and 802.15.4e Factory Automation MAC standards among the IEEE 802.15.4 standard families and the WISA specification among the IEEE 802.15.1-based standards. Except for the WISA and 802.15.4e Factory Automation MAC, all the selected standards target mainly process automation applications. An overview of each standard has been provided with particular focus on the network elements and the features of the protocol stack.

In the next step, we have identified several QoS requirements posed by industrial applications, such as support of real-time communications as well as highly reliable communications. The selected standards have been evaluated to determine how real-time and reliability requirements are supported. The results have revealed that no officially released and open standard is currently able to fulfill the strong real-time requirements of the factory automation domain. Moreover, the questions of QoS provisioning over heterogeneous networks as well as over multi-hop routes in homogeneous networks have been raised.

We have then focused on security issues of the surveyed standards by identifying potential attacks that could threaten the industrial WSNs and affect their operation. The standards have also been evaluated to determine if the proposed security mechanisms are sufficient to protect the WSNs against the derived threats. The evaluation has shown that the standards are resistant against most of the investigated threats, except for continuous jamming at all frequencies, collision attacks and flooding of connection requests. Moreover, we have pointed out that the design and the implementation of the security managers are left to the users or implementers of the standard. Here, the detailed operation of such security and network managers and the corresponding protocol mechanisms are an interesting area for further research.

We conclude that the selected standards fulfill almost completely the identified QoS and security requirements as long as they operate in single-hop mode. However, some aspects that are of high interest for the domain of industrial automation, including multi-hop operation and support of QoS and security over heterogeneous network segments, need further research.

ACKNOWLEDGEMENTS

This work was supported by CASED (www.cased.de) and Siemens AG Corporate Technology. The authors would like to thank the anonymous reviewers for their valuable comments.

REFERENCES

1. Geng, H. *Manufacturing Engineering Handbook*; McGraw-Hill Professional: New York, NY, USA, 2004.

2. Shell, R.L.; Hall, E.L. *Handbook of Industrial Automation*; Marcel Dekker: New York, NY, USA, 2000.

3. Low, K.S.; Win, W.N.N.; Er, M.J. Wireless Sensor Networks for Industrial Environments. In Proceedings of the International Conference on Computational Intelligence for Modelling, Control and Automation and International Conference on Intelligent Agents, Web Technologies and Internet Commerce (CIMCA-IAWTIC), Vienna, Austria, November 2005.

4. Mukhopadhyay, S.C.; Huang, Y.M. *Sensors: Advancements in Modeling, Design Issues, Fabrication and Practical Applications*; Springer-Verlag: Heidelberg, Germany, 2008.

5. IEEE Computer Society. IEEE Standard for Information Technology, Telecommunications and Information Exchange between Systems, Local and Metropolitan Area Networks, Specific Requirements, Part 11: Wireless LAN Medium Access Control (MAC) and Physical Layer (PHY) Specifications. 2007.

6. Willig, A.; Matheus, K.; Wolisz, A. Wireless Technology in Industrial Networks. *Proceedings of the IEEE* 2005, *93*, 1130–1151.]

7. IEEE Computer Society. IEEE Standard for Information Technology, Telecommunications and Information Exchange between Systems, Local and Metropolitan Area Networks, Specific Requirements, Part 15.1: Wireless Medium Access Control (MAC) and Physical Layer (PHY) Specifications for Wireless Personal Area Networks (WPANS). 2002.

8. IEEE Computer Society. IEEE Standard for Information Technology, Telecommunications and Information Exchange between Systems, Local and Metropolitan Area Networks, Specific Requirements, Part 15.4: Wireless Medium Access Control (MAC) and Physical Layer (PHY) Specifications for Low Rate Wireless Personal Area Networks (LR-WPANs). 2007.

9. Scheible, G.; Dzung, D.; Endresen, J.; Frey, J.E. Unplugged but Connected - Design and Implementation of a Truly Wireless Real-time Sensor/Actuator Interface. *IEEE Industrial Electronics Magazine* 2007, *1*, 25–34.]

10. ABB. Reliable Factory Automation: Wireless Automation with WISA. http://www.abb.com (accessed on November 2009).

11. Steigmann, R.; Endresen, J. Introduction to WISA and WPS. http://www.abb.com (accessed on November 2009).

12. Willig, A. Recent and Emerging Topics in Wireless Industrial Communications: A Selection. *IEEE Transactions on Industrial Informatics* 2008, *4*, 102–124.]

13. ODVA. The DeviceNet Specification, Common Industrial Protocol (CIP) Specification. http://www.odva.org (accessed on November 2009).

14. Modbus Organization. Modbus Application Protocol Specification. http://www.modbus.org (accessed on November 2009).

15. Dzung, D.; Apneseth, C.; Endresen, J.; Frey, J.E. Design and Implementation of a Real-time Wireless Sensor/Actuator Communication System. In Proceedings of the IEEE Conference on Emerging Technologies and Factory Automation (ETFA), Catania, Italy, September 2005.

16. IEEE 802.15.1 Standard. http://www.ieee802.org/15/pub/TG1.html (accessed on December 2009).

17. De Biasi, M.; Snickars, C.; Landernäs, K.; Isaksson, A.J. Simulation of Process Control with WirelessHART Networks Subject to Packet Losses. In Proceedings of the Conference on Automation Science and Engineering (CASE), Washington, DC, USA, August 2008.

18. Lennvall, T.; Svensson, S.; Hekland, F. A Comparison of WirelessHART and ZigBee for Industrial Applications. In Proceedings of the IEEE International Workshop on Factory Communication Systems (WFCS), Dresden, Germany, May 2008.

19. Griessmann, J.L. WirelessHART, an Overview. http://www.hartcomm.org (accessed on December 2009).

20. HART Communication Foundation. WirelessHART☐ Technical Data Sheet. http://www.hartcomm.org (accessed on December 2009).

21. HART Communication Foundation. TDMA Data Link Layer Specification. http://www.hartcomm.org (accessed on December 2009).

22. Song, J.; Han, S.; Mok, A.K.; Chen, D.; Lucas, M.; Nixon, M. WirelessHART: Applying Wireless Technology in Real-Time Industrial Process Control. In Proceedings of the IEEE Real-Time and Embedded Technology and Applications Symposium (RTAS), St. Louis, MO, USA, April 2008.

23. Tanenbaum, A.S. *Computer Networks*; Prentice-Hall: Upper Saddle River, NJ, USA, 2002.

24. International Society of Automation. ISA-100.11a-2009, Wireless Systems for Industrial Automation: Process Control and Related

Applications. http://www.isa.org (accessed on November 2009).

25. International Society of Automation. ISA100.11a Status. http://www.isa.org (accessed on December 2009).

26. Montenegro, G.; Kushalnagar, N.; Hui, J.; Culler, D. RFC 4944: Transmission of IPv6 Packets over IEEE 802.15.4 Networks. http://www.ietf.org/rfc/rfc4944.txt (accessed on November 2009).

27. Baronti, P.; Pillai, P.; Chook, V.W.; Chessa, S.; Gotta, A.; Hu, Y.F. Wireless Sensor Networks: A Survey on the State of the Art and the 802.15.4 and ZigBee Standards. *Computer Communications* 2007, *30*, 1655–1695.]

28. Galeev, M. Home Networking with ZigBee. http://www.media.mit.edu (accessed on December 2009).

29. Winkel, L.; Bahr, M.; Vicari, N. 15-08-0572-00-004e Proposal for Factory Automation. http://www.ieee802.org/15/pub/TG4e.html (accessed on December 2009).

30. International Telecommunication Union. E.800 - Terms and Definition Related to Quality of Service and Network Performance Including Dependability. http://www.itu.int (accessed on November 2009).

31. Schmitt, J. *Heterogeneous Network Quality of Service Systems*; Kluwer Academic Publishers: Norwell, MA, USA, 2001.

32. Mathiesen, M.; Thonet, G.; Aakwaag, N. Wireless Ad-hoc Networks for Industrial Automation: Current Trends and Future Prospects. In Proceedings of the IFAC World Congress, Prague, Czech Republic, July 2005.

33. DiStefano, J.J.; Stubberud, A.R.; Williams, I.J. *Schaum's Outline of Feedback and Control Systems*; McGraw-Hill Professional: New York, NY, USA, 1994.

34. Liu, X.; Goldsmith, A. Wireless Communication Tradeoffs in Distributed Control. In Proceedings of the IEEE Conference on Decision and Control (CDC), Maui, HI, USA, December 2003.

35. Neumann, P. Communication in Industrial Automation: What is going on? *Control Engineering Practice* 2007, *15*, 1332–1347.]

36. Kim, A.N.; Hekland, F.; Petersen, S.; Doyle, P. When HART goes Wireless: Understanding and Implementing the WirelessHART Standard. In Proceedings of the IEEE International Conference on Emerging Technologies and Factory Automation (ETFA), Hamburg, Germany, September 2008.

37. HART Communication Foundation. Why WirelessHART? The Right Standard at the Right Time. http://www.hartcomm.org (accessed on

December 2009).

38. Sereiko, P. The ISA100 Standard: Characteristics and Benefits of the Standard, Latest Developments and Progress. http://www.isa. org (accessed on December 2009).

39. Walters, J.; Liang, Z.; Shi, W.; Chaudhary, V. Wireless Sensor Network Security: A Survey. *Security in Distributed, Grid, Mobile, and Pervasive Computing* 2007, 367–405.

40. Newsome, J.; Shi, E.; Song, D.; Perrig, A. The Sybil Attack in Sensor Networks: Analysis & Defenses. In Proceedings of the International Symposium on Information Processing in Sensor Networks (IPSN), Berkeley, CA, USA, April 2004.

41. Parno, B.; Perrig, A.; Gligor, V. Distributed Detection of Node Replication Attacks in Sensor Networks. In Proceedings of the IEEE Symposium on Security and Privacy (S&P), Oakland, CA, USA, May 2005.

42. Hu, Y.; Perrig, A.; Johnson, D. Packet Leashes: a Defense against Wormhole Attacks in Wireless Networks. In Proceedings of the IEEE Conference on Computer Communications (INFOCOM), San Francisco, CA, April 2003.

43. Karlof, C.; Wagner, D. Secure Routing in Wireless Sensor Networks: Attacks and Countermeasures. *Ad Hoc Networks* 2003, *1*, 293–315.]

44. Schneier, B. *Applied Cryptography*, 2nd ed.; John Wiley & Sons: Hoboken, NJ, USA, 1996.

45. Raza, S.; Slabbert, A.; Voigt, T.; Landernäs, K. Security Considerations for the WirelessHART Protocol. In Proceedings of the IEEE International Conference on Emerging Technologies and Factory Automation (ETFA), Mallorca, Spain, September, 2009.

46. Phan, R. Impossible Differential Cryptanalysis of 7-round Advanced Encryption Standard (AES). *Information Processing Letters* 2004, *91*, 33–38.]

47. Raza, S.; Voigt, T.; Slabbert, A.; Landernäs, K. Design and Implementation of a Security Manager for WirelessHART Networks. In Proceedings of the IEEE International Conference on Mobile Adhoc and Sensor Systems (MASS), Macau, China, October 2009.

Chapter 10

RELIABILITY AND AVAILABILITY EVALUATION OF WIRELESS SENSOR NETWORKS FOR INDUSTRIAL APPLICATIONS

Ivanovitch Silva [1], Luiz Affonso Guedes [1], Paulo Portugal [2] and Francisco Vasques [3]

[1]Department of Computer Engineering and Automation, Federal University of Rio Grande do Norte, Campus Universitário 59078-900, Natal, Brazil

[2]ISR, Department of Electrical and Computer Engineering, University of Porto, Porto 4200-465, Portugal

[3]IDMEC, Department of Mechanical Engineering, University of Porto, Porto 4200-465, Portugal

ABSTRACT

Wireless Sensor Networks (WSN) currently represent the best candidate to be adopted as the communication solution for the last mile connection in process control and monitoring applications in industrial environments. Most of these applications have stringent dependability (reliability and availability) requirements, as a system failure may result in economic losses, put people in danger or lead to environmental damages. Among the different type of faults that can lead to a system failure, permanent faults on network devices have a major impact. They can hamper communications over long periods of time and consequently disturb, or even disable, control algorithms. The lack of a structured approach enabling the evaluation of permanent faults, prevents system designers to optimize decisions that minimize these occurrences. In this work we propose a methodology based on an automatic generation of a fault tree to evaluate the reliability and availability of Wireless Sensor Networks, when permanent faults occur on network devices. The proposal supports any topology, different levels of redundancy, network reconfigurations, criticality of devices and arbitrary failure conditions. The proposed methodology is particularly suitable for the design and validation of Wireless Sensor Networks when trying to optimize its reliability and availability requirements.

INTRODUCTION

Traditionally, applications in industrial environments are based on wired communication solutions [1]. However, recently, the industry has shown interest in moving part of the communication infrastructure from a wired to a wireless environment, in order to reduce costs related with installation, maintenance and scalability of the applications. In this context, Wireless Sensor Networks (WSN) actually represent the best candidate to be adopted as the communication solution for the last mile connection in process monitoring and control applications in industrial environments [2]. Among many advantages, the absence of a wired infrastructure enables WSN to extract information in a simpler way than traditional monitoring and instrumentation techniques [3].

Industrial applications have usually stringent dependability (reliability and availability) requirements, since faults may lead to system failures which can result in economic losses, environmental damage or hurting people [4,5]. In this context, we can classify faults as transient or permanent [6]. Transient faults usually affect communication links between devices and are caused by noise or electromagnetic interferences. Permanent faults affect network devices and have their origin in hardware malfunctions. After a permanent fault a device is considered (permanently) failed, and to become operational again a repair activity is necessary. In this paper we focus on permanent faults that affect network devices leading to its failure (note: the failure modes of a failed element become the fault types for the elements interacting with it [6]). Permanent faults have, typically, a major impact on the system operation [7]. Their immediate consequence is that communications with the affected device are no longer possible. However, in worst case situations various network devices can become isolated, as when a network device that acts as a router fails. As a result, the control algorithm is disturbed which may lead to a system failure with serious consequences.

The use of a methodology to evaluate the dependability requirements of a WSN can anticipate decisions regarding the topology, criticality of the devices, levels of redundancy and network robustness, that can be used to take decisions during the system life-cycle, and particulary, on early planning and design phases. For example, depending on the topology, alternative paths to the sink can be created improving the overall reliability of the network. In the same way, if a sensitivity analysis is supported, critical devices can be identified and decisions about different redundancy approaches can be taken.

The main contribution of this paper is to propose a methodology to evaluate the reliability and availability of Wireless Sensor Networks in industrial environments that are subject to permanent faults on network devices. The approach is based on Fault Tree Analysis (FTA), which is a technique used to obtain the probability of occurrence of an undesired state or event [8]. In the addressed case, the undesired event is related to the failure of a specific device or group of devices. A device is considered to be faulty if it suffers a permanent failure or if there is not any route to sink that includes the device.

The proposal addresses several aspects, being very flexible and able to be easily adapted to different kinds of scenarios. When compared with the available approaches, the main advantages are:

- Support of all possible topologies: line, star, cluster and mesh;
- Network failure conditions can be specified in a very flexible way, ranging from a single device to groups of devices;
- Failure and repair processes can be characterized using different types of time distribution functions;
- Network devices can have redundant (internal) architectures;
- Topology reconfigurations due to device failures are considered (e.g., self-healing routing protocols);
- Different types of dependability measures can be obtained from the same model (e.g., reliability, availability, MTTF) as well as the criticality of the network devices (Birnbaum's measure).

To complement the proposed approach we have also developed a software tool that automatically evaluates the reliability and availability of a WSN. The tool automatically generates a fault tree with the minimal set of events that leads to the network failure condition. After that, the fault tree is translated into a language understandable by the sharpe (Symbolic Hierarchical Automated Reliability and Performance Evaluator) tool [9], which is used to compute the desired dependability measures.

The remainder of this paper is organized as follows: Section 2 surveys some of the most relevant research works on reliability and availability evaluation for Wireless Sensor Networks. In Section 3, we give an overview about Wireless Sensor Networks with a special attention to wireless industrial networks standards, such as WirelessHART and ISA 100.11a. Next, in the Section 4, is held a brief introduction to Fault Tree Analysis (FTA) and basic concepts used in the proposal. Section 5 describes the proposed methodology for the reliability and availability evaluation of Wireless Sensor Networks. In Section 6 several scenarios are evaluated using network topologies commonly adopted in industrial applications. Finally, Section 7 concludes the paper and presents directions for future studies.

RELATED WORKS

The network reliability problem is a classical reliability analysis problem [10] that can be classified as: *k-terminal, 2-terminal* or *all-terminal*. Suppose a network with N devices and a set of K devices ($K \subset N$ and $|K| < |N|$). K is a set composed by a sink node and $|K| - 1$ field devices. Defining a sink device $s \in K$, the *k-terminal* problem is expressed as the probability that there is at least one path from s to each field device in K. The *2-terminal* problem is the case where $|K| = 2$, whereas the*all-terminal* problem is the case where $|K| = |N|$. These cases are known to be NP-hard problems, however several algorithms can be found for networks with limited size [11].

The network reliability problem has been widely studied for wired networks. For example, in [12] the author deals with the problem of measuring the reliability and availability of a wired network assuming hardware and software failures. The author gives an important insight about the state-space enumeration and the topology adaptation strategy when failures occur. The main difference between the reliability analysis of wired and wireless networks is related to the dynamics of the network. In a wireless networks, the dynamics of the network is greater since links fail more often and also due to the mobility of some of the devices. An early work about the reliability evaluation for a radio-broadcast network was conducted by [10]. In that work, the authors considered unreliable devices and reliable links and showed that the two-terminal reliability problem for radio broadcast networks is computationally difficult.

In [13], the authors analyzed the reliability and the expected maximum delay for a distributed sensor network. The network is assumed to be dense and organized into clusters. The reliability was measured as the probability that there was at least one path between the sink device and a sensor node within a cluster. The authors assumed unreliable devices and reliable links. It was proved that the problem was, in general, NP-hard. However for a topology up to 40 devices the problem is still tractable. In [14], the network reliability was evaluated for mobile ad-hoc networks based on the 2-terminal problem. The authors assumed unreliable devices and dynamic network connectivity. The proposed algorithm, although not finding the minimal cut set for the network, can be extended for the type of static networks typically found in industrial applications. In [15], the authors analyzed the influence of adding redundant devices, in what concerns the reliability and availability of multi-hop wireless networks. This work provides an interesting discussion about the reliability and availability of a WSN, particularly if it is considered that a router node can be a redundant device.

A tentative effort to create a methodology to evaluate the reliability of a WSN infrastructure was performed in [16]. The authors created a scheme based on reduced ordered binary decision diagrams (ROBDD) to model a cluster topology, where a reliability evaluation was also conducted. The authors do not considered multiple paths connecting a device to the sink. Thus, it is no longer possible to use self-healing routing protocols. Common-cause failures were considered, but the technique was focused in a single cluster. The methodology was applied only for a cluster topology with non-flexible failure conditions, and the criticality of devices was not determined. By introducing the concept of coverage-oriented reliability, the same authors extended the previous work [16] creating other mechanisms to evaluate the reliability of a WSN [17]. They assumed that the network fails if a specific point in the cluster is not covered by at least K devices. This give a more flexible way to configure failure conditions. However it is not possible to create two or more coverage subsets for the same cluster.

Another coverage-oriented reliability mechanism was proposed in [18]. The authors propose a framework to evaluate the reliability of a WSN based on coverage requirements. Given an area A, the network fails if there is no subset of fully operating nodes whose own generated traffic can reach the sink and the total area covered by this subset is greater than A. The authors used a 3-state node reliability model to represent random failures in the devices. This model has been shown to work better over the conventional 2-state (operate/fail), but it neither supports the inclusion of spare devices nor indicates the criticality of the devices. Finally, the inability to create several coverage areas makes it difficult to specify flexible failure conditions.

Another methodology for the reliability evaluation of a WSN was proposed in [19]. The authors propose a new topology control mechanism and they used a methodology for evaluating the reliability of the network operating with this mechanism. The basic idea is to represent the network as a graph and to measure the reliability based on the number of functional spanning trees. If there is at least one functional spanning tree, then the network is considered reliable. The proposal is simple and works very well for the analysis of the topology control mechanism. However it is not suitable to evaluate arbitrary WSN. It is not possible to use and validate physical redundancy, neither to compute the criticality of the devices. Flexible failure conditions are also very difficult to represent due the failure dependences for a spanning tree condition.

As an alternative to the aforementioned approaches, Fault Tree Analysis (FTA) techniques can be used to evaluate the reliability and availability of the network. The main advantage of FTA is related to the intuitive procedure used to describe events that lead to network failures. However, for complex topologies the construction of the fault tree is a time-consuming task demanding much effort. The usual solution is to adopt an approach that automatically generates the fault tree based on the network specification. In [20], the authors developed a modeling methodology for automatic generation of fault trees. The idea is to split a system in different components that are represented by function tables and state transition tables. These components are connected to each other in order to describe the behavior of the whole system. After the modeling phase, a trace-back algorithm is used to create the fault tree. In [21], an automatic generation mechanism for the fault tree was described within the context of an automation system. The basic idea is to model the system using a timed automata and then perform a model checking to verify which situations may lead to a system failure. After that, the results are summarized and the fault tree is generated. Another way to automatically generate the fault tree is to use digraphs (directed graphs) [22]. A digraph is composed by nodes and edges. Nodes represent component failure whereas edges represent relationships between nodes. In [23], the authors developed an automatic generator for fault tree based on digraphs. This work was an improvement upon the algorithm previously proposed in [22]. In both approaches they create a digraph to model the behavior of the system. All aforementioned works use dependency relations between system components to generate a fault tree.

Recently, an interesting contribution to the dependability evaluation of Wireless Sensor Networks was proposed in [24,25]. The main idea is to compute a new dependability parameter called *producibility*, that measures the probability of a sensor node is in a active state and it is able to communicate with the sink at time t. This new measure combines the reliability of a sensor node with their battery level. Network failure conditions are related with the existence of a minimum number of sensor nodes (*k-out-n*) able to send data to the sink. Metrics are computed using analytic techniques based on Continuous Time Markov Chains (CTMC) and reward functions. In [26] the same authors propose an alternative approach based on Non-Markovian Stochastic Petri Nets (NMSPN). This numerical based approach was selected to relax some of the assumptions related to the analytical technique. In the same work, they also propose to use Fault Trees to compute the network failure condition. Although these are interesting works, they are too much focused on energy consumption problems, which makes it difficult to extend the proposed methodology for generic scenarios. The same applies for the metrics. Moreover, network failure conditions are defined in a very restrictive way (*k-out-n* devices) which are not suitable for industrial scenarios, where it is important to identify the failed device and not only the number of failed devices.

It becomes clear from the previous discussion that these works only provide partial solutions for the problem. Since most of them are focused on specific scenarios, they are very restrictive with regard to the definition of network failure conditions, dependability metrics, topology, network reconfiguration and redundancy aspects, as well as applicability to industrial scenarios. The present work aims to remove most of these limitations by proposing a methodology that considers the most important aspects of the network operation through a flexible approach.

WIRELESS SENSOR NETWORKS

Wireless Sensor Networks are a pervasive technology that targets the connectivity between sensor nodes in multiple environments. Its infrastructure is usually composed of a large number of sensor nodes, with small physical size, which runs upon relatively inexpensive computational processes. Sensor nodes measure local environmental conditions and forward sensed values to a set of central points, referred as sink nodes, for appropriate processing. Sensor nodes can sense the environment, communicate with neighbor nodes, and perform basic computations on collected data. Installation flexibility and easy configuration enable better usability and maintenance than traditional communication technologies [1]. These characteristics allow the use of WSN over a wide range of useful applications [3,27–29].

Currently, WSN solutions are based on standardized or proprietary protocols. There are many different protocols for the upper layers, but the IEEE 802.15.4 [30] is a *de facto* standard for the lower layers. Recently the IEEE 802.15.5 standard [31] has been released to provide multi-hop mesh functions. Both standards are compatible while maintaining simplicity. On the other hand, Zigbee (2004) and Zigbee Pro (2007) were the first standards to implement the upper layers. Both standards do not have the support for channel hopping and are still not scalable enough to support large topologies [32]. Channel hopping is an important feature when industrial applications are considered, due to its robustness against external interferences and persistent multi-path fading. A new standard, IEEE 802.15.4e, is being developed to support additional industrial requirements and it is expected to be approved by the end of 2011. Currently, only the WirelessHART and ISA 100.11a standards are suitable to be used in industrial applications.

The methodology proposed in this paper can be easily implemented to evaluate the reliability and availability of Zigbee, WirelessHART and

ISA 100.11a networks. However, as the application focus is for industrial environments, only the WirelessHart and ISA 100.11a standards will be described in the following sections.

WirelessHART

WirelessHART is an extension of the HART protocol to support wireless communication. The concept behind WirelessHART was first discussed in 2004 at the HART Communication Foundation (HCF) meeting. The main question was how to interoperate legacy devices with wireless devices, in order to take advantage of the amount of installed HART devices. It is estimated that more than 24 million HART devices are installed around the world and its shipping expected are around over 2 million per year [33]. In September 2008, the WirelessHART specification (HART 7.1) was approved by the International Electrotechnical Comission (IEC) as a publicly available specification (IEC 62591) [34]. WirelessHART was the first industrial wireless communication technology to attain this level of international recognition [35].

WirelessHART defines eight types of devices, as presented in Figure 1: network manager, network security, gateway, access point, field device, adapter, router and handheld device. All devices that are connected to the wireless network implement basic mechanisms to support network formation, maintenance, routing, security and reliability.

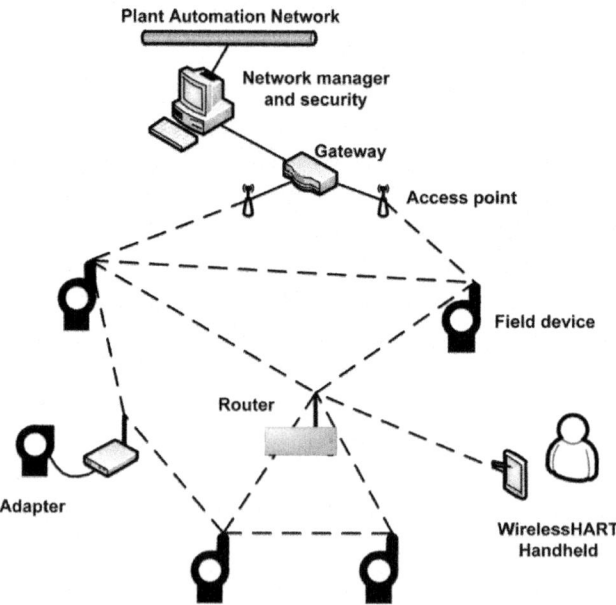

Figure 1: WirelessHART devices.

Field devices are the most basic WirelessHART devices. They are directly connected to the process and plant equipments. Field devices can transmit measurement data, receive and forward packets from/to any device. Usually they may be line, loop or battery powered. All field devices have a physical maintenance port, which is used for offline configuration and diagnostics. Compatibility with legacy HART devices is guaranteed through the use of adapter devices. The adapter devices are not directly connected to the plant equipments, however, they have to support the same functionalities of field devices. On the other hand, handheld devices are used during the installation, configuration and maintenance phases of the network. They do not have to support routing mechanisms.

Router devices are used for routing purposes, *i.e.*, forward packets from one device to another device. They are not directly connected to the industrial process, thus they can be installed anywhere in the plant. Their use is not really necessary since field devices have internal routing capabilities. However, router devices can provide redundant paths to the gateway, and they can also minimize energy consumption in field devices. The connection between the plant automation network and the wireless network is provided by the gateway. The gateway works as a sink point for all wireless traffic. The logical communication with the wireless network occurs through access points installed in the gateway. The amount of access points can be configured to increase redundancy and to improve the effective network throughput.

The security manager is the entity responsible for ensuring the security over the network. It provides join, network and session keys for all devices. These keys are used to authenticate and to encrypt data. The storage and management of keys is also under the responsibility of the security manager. The core of the WirelessHART is the network manager. It is logically connected to the gateway and manages the entire network. The communication with network devices occurs through the application layer protocol. The main duties of the network manager are related with scheduling, management of the device list, routing (redundant paths), collect information about performance, failure detection, and network formation.

WirelessHART has a physical layer based on IEEE 802.15.4, but implements its own medium access control (MAC) sublayer. The MAC is based on a TDMA (Time Division Multiple Access) communication mechanism that uses *superframes*. Superframes are composed by slots, and the amount of slots indicates its periodicity. To support multiple schedule requirements, a WirelessHART network can use multiple superframes with different number of slots. Each slot has a fixed duration of 10 ms, which is enough time to transmit a packet and receive an acknowledgment (the maximum packet size is 133 bytes including headers). Slots can be dedicated or shared. The use of dedicated slots is more common. Shared slots are used for transmission retries and advertising indication during the join procedure. A slot supports until 15 channels, thus, theoretically 15 devices can simultaneously transmit in the same slot time. The standard uses a mechanism of frequency hopping and a channel blacklist to minimize the influence of noise/interference in the network operation and consequently to increase the communication reliability.

An important procedure defined in the WirelessHART is the path failure indication [36]. The communication between two devices can fail due to hardware failures or due to interferences from the external environment. Therefore, it is essential that failure events are reported to the application. The WirelessHART defines the variable *path-fail-time* to control the path failure indication. If a device identifies that no packet was received from a specific neighbor within the *path-fail-time*, an alarm indicating that the path is no longer available is sent to the application.

ISA 100.11a

The International Society of Automation (ISA) has developed a wireless mesh networking standard known as ISA 100.11a [37] that guarantees a deterministic communication latency, while increasing the communication reliability. It focuses on process control and monitoring applications, with latency requirements around 100 ms. ISA 100.11a can coexists with other wireless technologies such as cell phones, IEEE 802.11×, IEEE 802.15×, IEEE 802.16×, and can provide tunneling for legacy protocols (HART, Foundation Fieldbus, Profibus, Modbus).

A typical ISA 100.11a network is presented in Figure 2. It may be composed of seven types of devices: gateway, system manager, security manager, router, backbone router, input/output (IO) devices and portable devices. Each device has a specific role definition that control its functions. The IO device is responsible for monitoring the environment. If minimization of the energy consumption is configured, the IO device only transmits messages. Otherwise, the IO device can also route messages. In addition to the routing functionality, a router device shares the function of provisioning devices to join the network. A router device can use slow slotted hopping to send advertising messages about the network for joining devices. On the other hand, backbone router devices are used to encapsulate external networks in order to carry native protocols over ISA 100.11a. The gateway device provides a connection between the wireless sensor network and the plant automation network. There is support for multiple gateways and backbone routers [38]. The most important tasks are performed by the security manager and the system manager. The system security management function is controlled by the security manager whereas the system manager governs all the network, devices and communications.

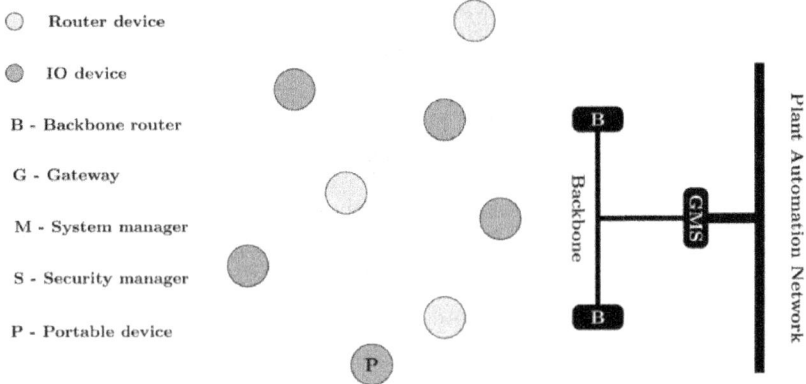

Figure 2: Typical ISA 100.11a network.

Similarly to WirelessHART, ISA 100.11a has a physical layer based on IEEE 802.15.4. On the other hand, the data link layer is slightly different from the one used on WirelessHART. The slot time has duration of 10 ms or 12 ms. The schedule mechanism was designed in a more flexible way than the WirelessHART schedule. There is support for slotted channel hopping (TDMA), slow channel hopping (CSMA) and a hybrid combinations of both. The TDMA approach is similar to the WirelessHART schedule. In the CSMA (Carrier Sense Multiple Access) approach, contiguous slot times are grouped into a single radio channel with a period ranging from 100 ms to 400 ms. During this period the radio of devices are always activated. This approach is indicated for neighborhood discovery procedures and frequency hopping in the case of overlapping with 802.11× networks, for example. On the other hand, the hybrid approach is more suitable for a flexible retry procedure. Other improvements when compared with the WirelessHART standard are related with the frequency hopping pattern. ISA 100.11a has defined five default hopping patterns to mitigate the influence of external communication interference. For example, the pattern 1 is configured to eliminate the overlap with the same channels of IEEE 802.11×.

The network and transport layers support mesh networks, similarly to WirelessHART. However, addressing in ISA 100.11a is compatible with the 6LoWPAN [39] (IPv6 over IEEE 802.15.4). ISA 100.11a also introduces a new mechanism to detect failures in the network based on the transmission of alert messages.

RELIABILITY, AVAILABILITY AND FAULT TREE ANALYSIS

In this section, we provide a brief introduction to reliability, availability and fault tree analysis concepts that are closely related with the proposed methodology.

Reliability

Reliability is a measure used to characterize if a component/system, is properly working according to its specifications during a specific period of the time [6]. Formally, it is defined as the probability that a component does not fail in the time interval $(0, t]$. Considering that the *time to failure* of a component, T, is a random variable defined by a cumulative distribution function $F(t)$ (CDF), the reliability $R(t)$ is given by:

$$R(t) = Pr(T>t) = 1-F(t) \qquad (1)$$

The reliability function is closely related with the *failure rate function* $\lambda(t)$. This function (also known as *hazard rate*) describes the instantaneous failure rate of a component. Formally, this function is defined as the probability that a component fails during the period of the time $[x, x + \Delta t]$, knowing that it is working at time instant $t = x$. The behavior of this function has been extensively discussed in the literature [40]. For many systems/components this function presents a characteristic shape which is similar to a *bathtub curve*. When the system is young, the failure rate is higher (infant mortality), and then quickly decreases until stabilizes (useful life). As the system/component gets older it increases again (wear out). For electrical/electronic systems it is common to consider that the failure rate is constant during the useful life period, *i.e.*, $\lambda(t) = \lambda$ [41]. It can be proved that $R(t)$ and $\lambda(t)$ are related according to the following expression [42]

$$R(t) = exp\left(-\int_0^t \lambda(u)du\right) \qquad (2)$$

Therefore $\lambda(t)$ establishes $R(t)$. Another metric related with $R(t)$ is the MTTF (*Mean Time to Failure*). Formally, it is defined as the expected (average) time during which a component is working properly, and is given by [42]

$$MTTF = E(T) = \int_0^\infty tf(t)dt \qquad (3)$$

Availability

Availability is a measure which is defined as the probability of a component/ system is functioning at time t. The availability at the instant t is referred as *instantaneous availability A* (*t*). The *steady-state availability* expresses the percentage of time that a component is working properly. Formally it is defined as $A = \lim_{t \to \infty} A(t)$ (note: this metric only makes sense in systems which have a stationary probabilistic condition [42]). Availability is closely related with repair actions. In fact, it is implied that the system is repaired after a failure, otherwise $\lim_{t \to \infty} A(t) = 0$. For a non-repairable system $A(t) = R(t)$.

Similarity to the failure rate $\lambda(t)$, it is possible to define a *repair rate* $\mu(t)$, as the rate at which a failed component is repaired. The MTTR (*Mean Time to Repair*) is defined as the expected (average) time that takes to repair a component. If failure and repair rates are assumed constant, respectively λ and μ, then it can be proved that $A(t)$ is given by [42]

$$A(t) = \frac{\mu}{\lambda + \mu} + \frac{\lambda}{\lambda + \mu} e^{-(\lambda + \mu)t}$$

(4)

Finally, if failure and repair actions are independent and described by *i.i.d* (independent and identically distributed) random variables, than the following relationship applies [42]

$$\lim_{t \to \infty} A(t) = A = \frac{MTTF}{MTTF + MTTR}$$

(5)

This expression is independent of the CDF that characterize failure and repair processes.

Fault Tree Analysis

Fault tree analysis (FTA) is a deductive technique commonly used to evaluate system's dependability [43]. It can be used to describe the root causes that lead to a system failure, in a qualitative or quantitative way. In the former case, it can be used during system development to identify potential problems that could lead to a system failure, or after commissioning, to identify events that caused a system failure. In the latter case, it is mainly used to obtain dependability measures, such as the system's reliability and availability.

Fault trees (FT) are a graphical model that represents the combination of events that lead to a system failure. The model uses a treelike structure composed by events and logic gates. Events represent either normal or faulty conditions, such as component failures, environmental conditions, human-made faults, *etc.* They are considered *boolean, i.e.*, they either occur or not occur. Logic gates are used to represent the cause-effect relationships among events. The inputs of these gates are either single events or combinations of events which result from the output of other gates. There are several types of gates available, such as *and, or* and *k-out-of-n* (Figure 3). The process of building a FT is performed deductively and starts by defining the top *event*, which represents the *system failure condition*. From this event, and by proceeding backwards, the possible root causes are identified. The events at the bottom of the tree are referred as *basic events*. If a basic event occurs two or more times in a FT it is called a *repeated event*.

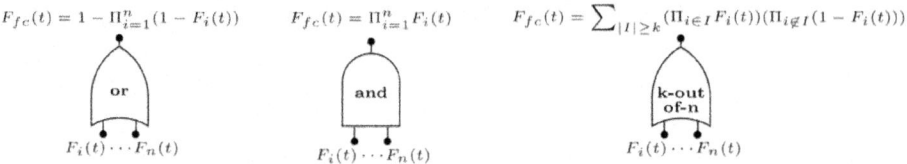

$$F_{fc}(t) = 1 - \Pi_{i=1}^{n}(1 - F_i(t)) \qquad F_{fc}(t) = \Pi_{i=1}^{n} F_i(t) \qquad F_{fc}(t) = \sum_{|I| \geq k}(\Pi_{i \in I} F_i(t))(\Pi_{i \notin I}(1 - F_i(t)))$$

Figure 3: Cumulative distribution function $(F_{fc}(t))$ for the gate output (*and, or, k-out-of-n*).

From a probabilistic point of view, the assessment of a FT consists of calculating the probability of the top event starting from the probabilities of the basic events. This calculation is performed differently for each type of gate. Assuming a gate with *n* independent inputs (events), where the occurrence of event *i* is described by means of a cumulative distribution function $F_i(t)$, then the gate output CDF $F_{fc}(t)$ is given according to the Figure 3 [42].

When an *and* gate is used, the failure condition occurs only if all input events have occurred. On the other hand, when an *or* gate is used, the failure condition occurs if at least one input event have occurred. Finally, if a *k-out-of-n* gate is used, the failure condition occurs if at least *k* input events have occurred.

When a FT does not contain any repeated event, the probability of the top event can be obtained through direct calculation using the formulas presented in Figure 3. However, if there are repeated events these equations are no longer valid. Therefore, in these situations it is necessary to employ different approaches. In the literature we can find several techniques to accomplish this task, such as inclusion-exclusion principle, sum of disjoint products, factorization and direct/indirect recursive methods [43]. In the context of this work, we will focus on the *sum of disjoint products* (SDP) [44].

The SDP method can be efficiently employed in fault trees with repeated events and it is easily automated. The basic idea of SDP is to find a boolean function $\varphi(x)$ that describes the failure condition (*i.e.*, the top event), and to transform this function into another function where the individual terms are mutually exclusive.

Consider a system with *n* components. The first step starts by obtaining the *structure function* of the system, $\varphi(x)$, which is given by

$$\phi(\mathbf{x}) = \begin{cases} 1 & \text{if the system has failed} \\ 0 & \text{if the system has not failed} \end{cases}$$

$$(6)$$

where $\mathbf{x}$ is referred as the *state vector*, $\mathbf{x} = (x_1, x_2, \ldots, x_n)$. Each element x_i is a boolean variable that represents the state of component i (e.g., $i = 1 \Leftrightarrow$ the component has failed). The function $\varphi(\mathbf{x})$ can also be expressed as the union of *minimal cut sets*.

$$\varphi(\mathbf{x}) = K_1 \cup K_2 \cup \cdots \cup K_n \qquad (7)$$

A cut set K_i is a subset of events whose simultaneous occurrence leads to the occurrence of the top event. A cut set is said to be *minimal* if does not contain other cut set. There are several algorithms available to automate the process of obtaining the minimal cut sets from a fault tree [43]. After obtaining the cut sets, $\varphi(\mathbf{x})$ can be transformed in a sum of disjoint products, as follows

$$\phi(\mathbf{x}) = K_1 \cup K_2 \cup \cdots \cup K_n = K_1 \cup \overline{K}_1 K_2 \cup \cdots \cup \overline{K}_1 \ldots \overline{K}_{n-1} K_n \qquad (8)$$

where K_i the i-cutset and K_i its complement. Since the terms are pairwise disjoint, the probability of the top event can be obtained as the sum of the probabilities of the individual terms.

It is possible to compute several dependability measures from a fault tree. In the context of this work we will focus on reliability and availability. Assume that the top event represents the failure of the system. Thus, the probability of this event occurring during a period of time t is the complement of the reliability $R(t)$. If the top event is expressed by its *minimal cut sets*, then to compute the reliability is only necessary to replace each event i, in the respective cut set, by its reliability function $R_i(t)$. After that, the reliability of the system $R(t)$ can be easily computed using simple probability laws (*i.e.*, probability of union and intersection of events).

Availability can be obtained in a similar way, by replacing each event by the availability function of each component $A_i(t)$. However, this computation in only valid if the repair processes are all independent and if the number of repairman (*i.e.*, number of repair actions) is not limited. Further details can be found in [43].

Component Importance

After computing the top event probability (or any other relevant metric, such as the reliability or availability), the user is able to foresee the system behavior from a dependability viewpoint. However, this does not highlight what is the contribution of each component to the final result. Such information is relevant because it allows the system designer to make decisions concerning the system structure, which can be used to optimize dependability metrics (e.g., availability), or other performance measures.

In this section we will review some importance measures that can be used to rank components in order of importance. We assume a system composed of n independent components, where each component i is characterized by a reliability function$R_i(t)$.

Birnbaum's Measure

Birnbaum's measure $I^B(i|t)$ is a metric that describes the reliability importance of a component [45]. This measure is defined as the partial differentiation of the system reliability with respect to the reliability of component i, as follows

$$I^B(i|t) = \frac{\partial R(t)}{\partial R_i(t)} \quad \text{for } i = 1, 2, \ldots, n$$

(9)

If $I^B(i|t)$ is large, a small variation in the reliability of component i will result in a major change in the reliability of the system. A component i is considered *critical* for the system, if when the component i fails, the system also fails. Thus, the Birnbaum's measure can also be interpreted as the probability of component i being critical for the system at time t [42].

Criticality Importance

The criticality importance $I^{CR}(i|t)$ is a measure particularly suitable for prioritizing maintenance actions [42]. This measure is defined as the probability that component i is critical at time t and is failed at time t, knowing that the system is failed at time t, being defined as follows

$$I^{CR}(i|t) = \frac{I^B(i|t)(1 - R_i(t))}{1 - R(t)}$$

(10)

In other words, the criticality importance is the probability that a component i has caused a system failure, knowing that system is failed at time t.

METHODOLOGY FOR RELIABILITY AND AVAILABILITY EVALUATION

The main objective of the proposed methodology is to provide a framework to support the evaluation of the dependability of a WSN, in order to provide valuable information to the system designer enabling it to develop robust and fault tolerant applications. The methodology can be applied on all stages of the network life cycle, allowing the identification of weaknesses (e.g., topology, devices, *etc.*) as well as helping to define a strategy to cope with these problems.

Introduction

As aforementioned in the Section 2, the reliability evaluation of a general network is a NP-hard problem. Nevertheless, as it will be discussed in the next section, this problem can be tractable for a low-medium number of field devices, as is the case of networks typically found in industrial applications.

Figure 4 overviews the proposed methodology. The process starts by providing information about the network topology, device types and redundancy, device's failure and repair process and network failure condition. The latter one is defined by a logical expression that combines the failure status of field devices. For attaining flexible failure conditions and to support self-healing routing protocols, it is necessary to find all paths between the gateway (sink) and the devices that encompass the failure condition. Next, a fault tree is generated using all the previous data. From that, the respective minimal cut sets are obtained using an inversion technique. This cut set is re-expressed as a (minimal) fault tree, which is used to produce input data for the tool that computes the results. For this task we use the sharpe tool [9], which is able to compute the metrics of interest, either symbolically or numerically. It is possible to evaluate the reliability, availability and mean time to failure (MTTF) of the WSN, and also the Birnbaum's and the criticality measures for all field devices. Finally, we also have developed a software tool that automates the previous steps.

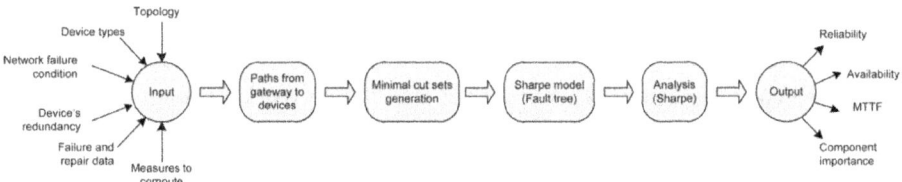

Figure 4: Overview of the methodology for reliability and availability evaluation.

Assumptions

The main assumptions considered in the methodology are the following:

- **Topology:** the network is composed of N field devices, which can belong to one of the following types: end device (e.g., sensor/actuator node), router, access point and gateway (*i.e.*, sink). Devices are arranged according to one of the following topologies: line, star, cluster and mesh. These elements are defined according to the WirelessHART, ISA 100.11a and Zigbee standards;

- **Faults:** only permanent faults are considered. The links, due to its wireless nature, are only affected by transient faults and thus are considered to be reliable (*i.e.*, they do not fail). Thus, only field devices can fail. After a permanent fault a device is considered failed (permanently). We assume that device failures are independent. In principle any type of distribution can be used to characterize the occurrence of device failures. However, the sharpe tool poses some restrictions. The tool imposes that CDFs must be expressed using *exponential polynomial* terms as following

$$F(t) = \sum_{j=1}^{n} a_j t^{k_j} e^{b_j t}$$

(11)

Many distributions can be expressed in this way (e.g., exponential, Erlang, hypoexponential, hyperexponential). Other distributions (e.g., Weibull, deterministic) can be approximated using exponential polynomial terms. Further details can be found in [46].

- **Repairs:** field devices can be repaired after failing, if necessary. After a repair the device is considered as new. We consider that repair processes are independent and that the number of repairman (*i.e.*, number of repair actions) is not limited. The time necessary to repair a device is characterized by a *repair distribution*. This distribution is defined in analogous way to the failure distribution discussed previously;

- **Redundancy:** field devices can have an internal redundant architecture with several available spares. We assume that when the main element fails, its replacement by a spare is always performed with success;

- **Reconfiguration:** when a device fails the network topology can change. We assume that the network manager (WirelessHART) or system manager (ISA 100.11a) is able to identify with success a device failure, and then update the network topology (communication paths). It is also assumed that the time required to perform this operation is negligible and that it is always successful (if alternative paths exist). Thus, the support of self-healing routing protocols is assured, since all paths between a field device and the gateway are considered;

- **Measures:** the following measures can be computed: reliability, unreliability, availability, unavailability, MTTF and component importance (Birnbaum and Criticality). Results can be presented both numerically and symbolically using exponential polynomial terms.

- **Inputs:** to compute the measures it is necessary to provide the following input data: network topology, type of devices, device's redundancy, network failure condition, characteristics of the device's failure and repair processes and measures to compute.

Topology

The first step of the proposed methodology is to define a structure through which the network can be modeled. In the proposed approach, the network is organized as a graph $G(V, E)$ with n vertices (V) and k edges (E). The vertices represent field devices whereas the edges represent the wireless links between devices. The network topology can be stored in the adjacency matrix $(A_{n\times n})$ of graph G. If a device N_i has a neighbor N_j, then the entries a_{ij} and a_{ji} of A will receive the value 1, otherwise they will receive the value 0. Thus, by using this structure we can represent any WSN topology. Figure 5 shows an example of a WSN represented using the aforementioned structure. In this example, the network is composed by 1 gateway (Gtw_0), 1 access point (Ap_0), 2 routers $(R_0$ and $R_1)$ and 3 field devices $(Fd_0, Fd_1$ and $Fd_2)$. The indexes located more close to the vertices are used to identify the devices in the adjacency matrix.

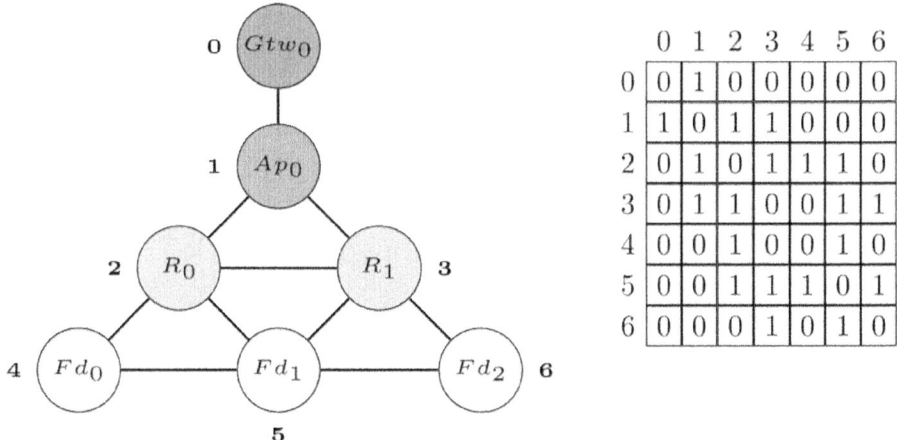

Figure 5: Example of a Wireless Sensor Network represented through a graph and its respective adjacency matrix.

Network Failure Condition

The network failure condition defines which combination of devices may lead to a network failure. In the proposed methodology we support any combination that can be expressed using boolean operators (*i.e.*, AND, OR). The failure condition associated to field device N_i is defined as fc_Fd_i. The case where the device failure condition is related with several failure events will be described in the next sections. A combination of devices that lead to a network failure is defined as nfc_and_j, where j is the identification of combination and is represented by the boolean *AND* of the failure condition of the devices (note: if the *AND* gate has only one input, the event nfc_and_j is replaced by the respective input). A device can belong to more than one failure condition. The network failure condition (*nfc*) is represented by the boolean *OR* of all combinations that lead to the network failure (Figure 6).

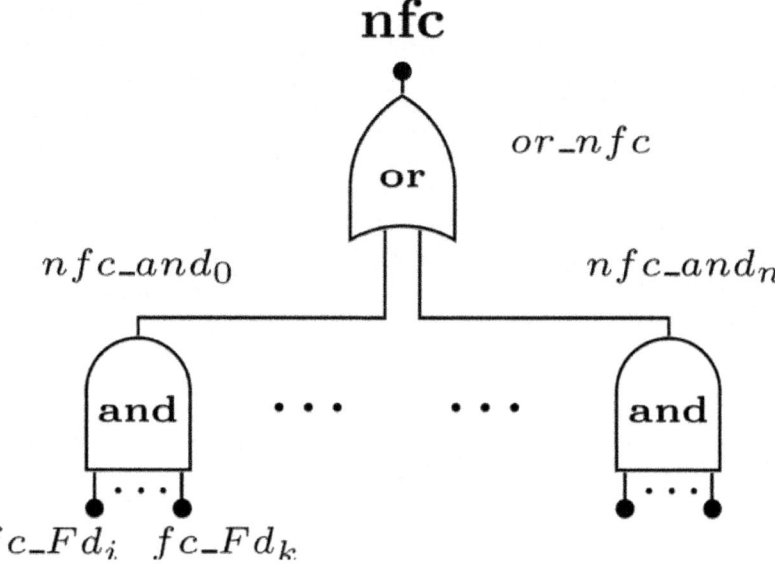

Figure 6. Network failure condition (nfc).

Device Redundancy

Regarding to redundancy issues, we consider that there are two types of field devices: with redundancy and without redundancy. The latter are simple field devices, while the former are composed by multiple devices arranged according to a fault-tolerant architecture based on a *hot standby sparing* approach. Hot standby sparing provides redundancy through the use of spare devices. A device is kept operational whereas the others devices (spares) are in standby. When the operational device fails, a spare module assumes the operation. We assume that this is an internal arrangement of the field device. That is, from the perspective of an external observer the behavior of a redundant device is indistinguishable from a device without redundancy. The number of spare devices available for each field device is an input of the model.

In the proposed model, devices are represented based on the failure event. The failure event for a redundant device is represented by the boolean *AND* of all spare devices whereas a device without redundancy is represented by a basic failure event.

Device Failure

After obtaining an expression for the network failure condition, it is necessary to define the conditions that may lead to the failure of a field device. Note that only devices that belong to the network failure condition are analyzed. We consider two possibilities for a device failure: (i) its hardware has failed; (ii) there is no path between the device and the gateway. This latter case corresponds to a *connectivity failure*, since the device itself did not failed in a strict sense (*i.e.*, it works), but it is considered non-operational from a network perspective because it is no longer possible to communicate with it. If a device along the path fails, the network may have the required mechanisms to reconfigure itself in order to use other paths. This type of reconfiguration is done by self-healing routing protocols. As aforementioned, the failure condition for a field device, fc_Fd_i, is split in two involving hardware and connectivity problems.

Regarding to hardware failures, we must consider the cases where the field devices are configured with or without redundancy. A redundant device fails (Figure 7(a)) if the current operating device suffers an hardware failure (Fd_{i_a}) and if all its spares have already failed (events Fd_{i_b} to Fd_{i_z}). This is represented by event r_Fd_i. On the other hand, for a device without redundancy (Figure 7(b)) the device fails when its hardware fails.

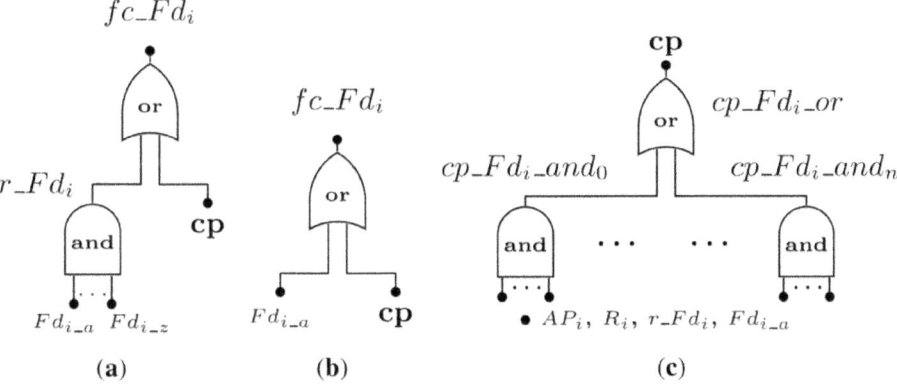

Figure 7: Device failure condition. **(a)** Redundant field device; **(b)** Simple field device;**(c)** Connectivity problem.

Regarding failures related with the connectivity problem (represented by the event *cp* in Figure 7(a,b), a device is considered to be faulty if there is no path from the device to the gateway (*i.e.*, sink). In other words, if a device has *j* paths connecting it to the gateway, at least one path must be working properly to consider that the device is operational. The event*cp* that represents this situation (Figure 7(c)) results from the combination of failures in access points (AP_i), routers (R_i), redundant devices (r_Fd_i) and devices without redundancy (Fd_{i_a}).

We will use an example to clarify the notation described in Figure 7. Consider a WSN composed by 4 field devices: Fd_1,Fd_2, Fd_3 and Fd_4. The addressed problem is to find the failure condition associated to device Fd_2. Assume that device Fd_2 is redundant and has one spare device, while devices Fd_1, Fd_3, and Fd_4 are not redundant. Regarding to the connectivity problem, if device Fd_3 fails or if devices Fd_1 and Fd_4 fail, then device Fd_2 will also fail since there is no path to the gateway. Based on this scenario, the failure condition for device Fd_2 (fc_Fd_2) is presented by Figure 8. Note that the event $cp\_Fd_2\_and_0$was replaced by a single event Fd_{3_a} because it makes no sense to build an *AND* gate with just one input.

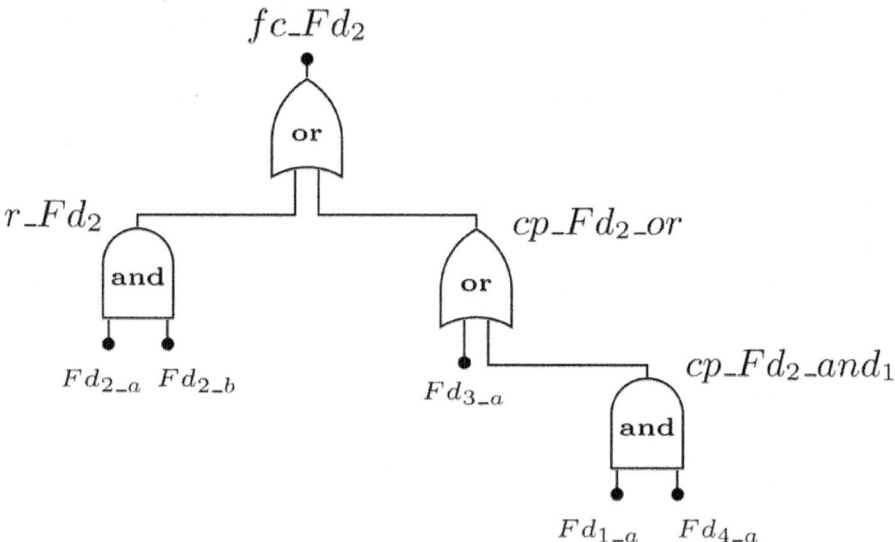

Figure 8: Example of a failure condition defined for the field device Fd_2.

Algorithm 1: Algorithm to generate all paths between a device and the gateway.

Algorithm: DFS(Paths, Device, Current_Path)

Output: All paths between a field device and the gateway.

```
1 for i ← 0; i < Number of Devices; i ++ do
      // If the device is the gateway then a path was found
2     if Device == Gateway then
3         Paths.add (Current_Path) ;
4         break;
      // Do not search for paths in the neighborhood of access points
5     if Device == Access Point and i ≠ Gateway then
6         break;
      // Searching for neighbor devices
7     if adjMatrix [Device][i] == 1 then
          // Eliminate cycles
8         if Current_Path.indexOf (i) < 0 then
9             Current_Path.add (i) ;
10            DFS (Paths,i,Current_Path) ;
11            Current_Path.remove;
```

Note that to find all combinations that lead to a connectivity failure necessarily requires some effort. To attain this, it is necessary to search all paths between the gateway and field devices that belong to the network failure condition. This procedure is described in Algorithm 1. The basic idea is to perform a depth-first search (DFS) in the adjacency matrix that represent the WSN. The procedure recursively traverse all devices on the path until the gateway is reached. Two restrictions were introduced to simplify the DFS. First, it makes no sense to search neighbor devices of the access point (line 5). Access points are directly connected to the gateway, and an access point does not communicate with another access point. The second restriction is related to the elimination of paths within a cycle. During the recursion, a device only joins to the current path if it is not already part of that path (line 8).

For the sake of understanding Algorithm 1, considers the example of the output flow produced when the paths from device Fd_0 to the gateway are being searched for the scenario presented in Figure 5. The output flow for this example is shown in Figure 9 (note: only two paths are described (dark gray circles) due to lack of space, however the others paths can be easily deduced.). The output flow starts at device Fd_0 and explores as far as possible each branch (neighbor) until reaching the gateway. After that, a backtracking procedure is conducted and another branch will be evaluated. Figure 9 is composed by 5 levels. The first level is composed only by the target device (Fd_0). The second level is composed by the neighbors of device Fd_0 (R_0 and Fd_1). The other levels follow the same procedure taking into account the restrictions imposed by Algorithm 1 (lines 5 and 8).

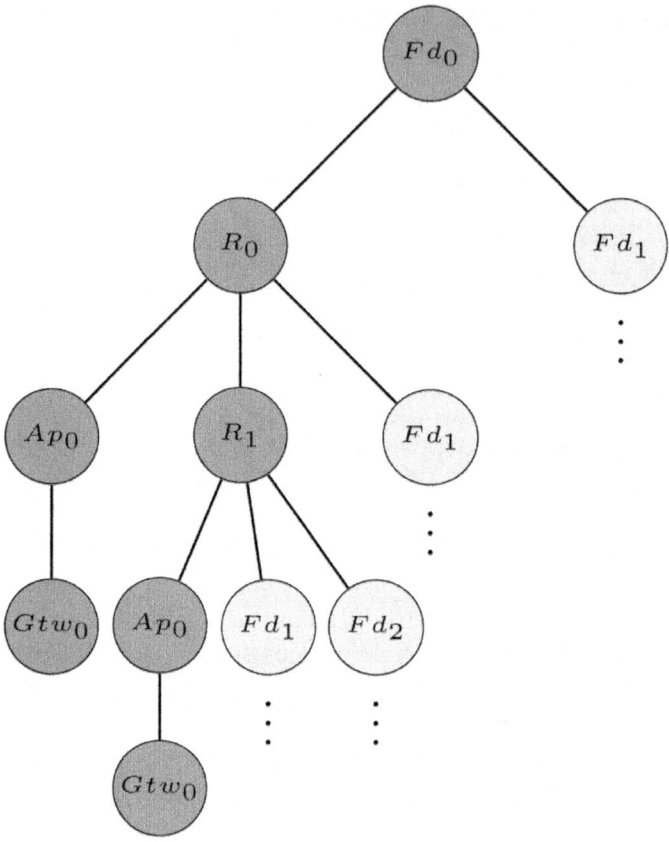

Figure 9: Output flow of Algorithm 1 for the device Fd_0 of Figure 5.

All paths generated by Algorithm 1 are stored in a data structure based on a fault tree. This data structure will be used to generate the minimum cut sets in the next section. Each path i belonging to device j is mapped for an *and* gate. The inputs of each *and* gate are composed by the devices of the respective path. An *or* gate is used to connect all the paths (*and* gates) from a device to the gateway.

Minimal Cut Set Generation

After defining both the network failure condition and finding all the paths between the gateway and field devices that belong to the network failure condition, it becomes possible to generate a fault tree describing the network failure process. In this faulty tree, the top event represents the network failure condition and the basic events represent the device failure events. It could have been possible to feed this model directly into an evaluation tool in order to compute the dependability measures. However we choose to reduce this tree to a *minimal* one. The goal is to obtain a less complex fault tree in order to enable a faster computation of the dependability measures. Therefore, we compute the minimal cut sets of the fault tree and use this data to build a *minimal* fault tree.

The algorithm used to compute the minimal cut sets is similar to the one proposed in [47]. The major difference is that in the original proposal the authors consider reliable devices and unreliable links, whereas in our proposal we consider the opposite (*i.e.*, unreliable devices and reliable links). An inversion technique to generate the minimal cut sets from the *minimal path sets* (MPS) was proposed in [48]. A MPS represents a set of components such that, if all components are properly working then the system is operational. However, if a component of the MPS fails, then the system also fails. It was proved in [48] that the application of DeMorgan's law to an appropriate boolean polynomial of the minimal path set is related to a boolean polynomial of the minimal cut set. An efficient algorithm for path inversion was extended in [47] based on the procedure proposed in [48].

To clarify the minimal cut set generation, consider the following example that assumes the topology described in Figure 10. In this scenario, the field device Fd_2 is the source whereas the gateway is the sink. L_i indicates if a link between two network devices is operational, whereas $\overline{Li}$ indicates the link is down. Based on [47], the minimal path set for this example is described by Equation (12).

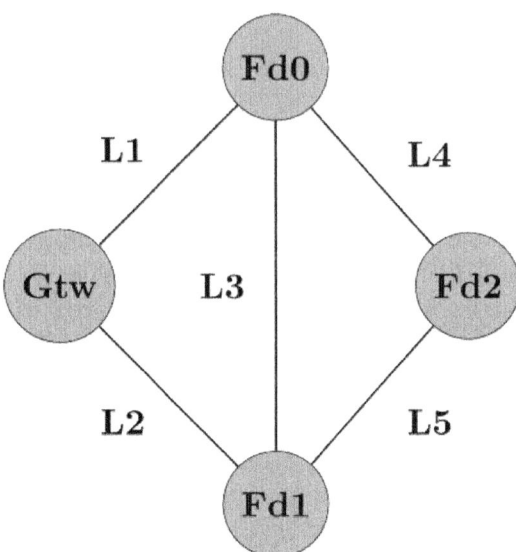

Figure 10: Topology used to exemplify the minimal cut set generation.

$$L_1L_4 + L_2L_5 + L_2L_3L_4 + L_1L_3L_5 \tag{12}$$

As described in [48], applying the DeMorgan's laws to the Equation (12) we can find a boolean polynomial related to the minimal cut set according to Equation (13). By applying the distributive law $((A + B)C = AC + BC)$ and the absorption law $(A + AB = A)$ to Equation (13), we obtain the result presented in Equation (14). In other words, the communication between Fd_2 and the gateway is broken if either the links L_1 and L_2 fail or if the links L_1, L_3 and L_5 fail or if the links L_2, L_3 and L_4 fail or if the links L_4 and L_5 fail.

$$(\overline{L_1} + \overline{L_4})(\overline{L_2} + \overline{L_5})(\overline{L_2} + \overline{L_3} + \overline{L_4})(\overline{L_1} + \overline{L_3} + \overline{L_5}) \tag{13}$$

$$\overline{L_1L_2} + \overline{L_1L_3L_5} + \overline{L_2L_3L_4} + \overline{L_4L_5} \tag{14}$$

On the other hand, if we assume unreliable devices and reliable links the minimal path sets of Figure 10 would be different from the represented in Equation (13). In this case, the minimal path sets are described by Equation (15). In the same way, if we apply the DeMorgan's law and use the distributive and the absorption laws, the minimal cut sets for the example of Figure 10 can be obtained according to Equation (16). In other words, the communication between Fd_2 and the gateway is interrupted if either the gateway fails or if Fd_2 fails or if Fd_0 and Fd_1 fail.

$$Fd_2Fd_0Gtw + Fd_2Fd_1Gtw + Fd_2Fd_0Fd_1Gtw + Fd_2Fd_1Fd_0Gtw \tag{15}$$

$$\overline{Gtw} + \overline{Fd_2} + \overline{Fd_0}\,\overline{Fd_1} \tag{16}$$

Sharpe Source Code Generation

After computing the minimal cut sets it becomes possible to generate a *minimal* fault tree that represents the network failure process. As discussed previously, we choose to proceed in this way to enable faster computations, but also due to the fact that the sharpe tool does not accept models expressed as cut sets, but accept models described as fault trees.

In this *minimal* fault tree the top event results from two events. The first one is related to the failure of the gateway, whereas the second one is related to the network failure condition that is application dependent. The top events are described in Figure 11. Note that event *nfc* is based on the Figures 6 and 7 (Sections 5.4 and 5.6).

Failure

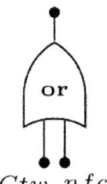

Gtw nfc

Figure 11: Events that conduct to a network failure.

For the sake of understanding of the top event definition, consider the example represented in Figure 10. We assume that the failure condition is defined as the following: $Fd_0 + Fd_1 \cdot Fd_2$. In other words, the network fails if the device Fd_0 fails or if devices Fd_1 and Fd_2 fail. We also assume that all devices are configured without redundancy. The first step is to find the device failure condition (fc_Fd_i) for each device present in the network failure condition. In this case, $fc\_Fd_0 = Fd_{0\_a}$, $fc\_Fd_1 = Fd_{1\_a}$ and $fc\_Fd_2 = Fd_{2\_a} + Fd_{0\_a} \cdot Fd_{1\_a}$. The top event for this example is presented in Figure 12.

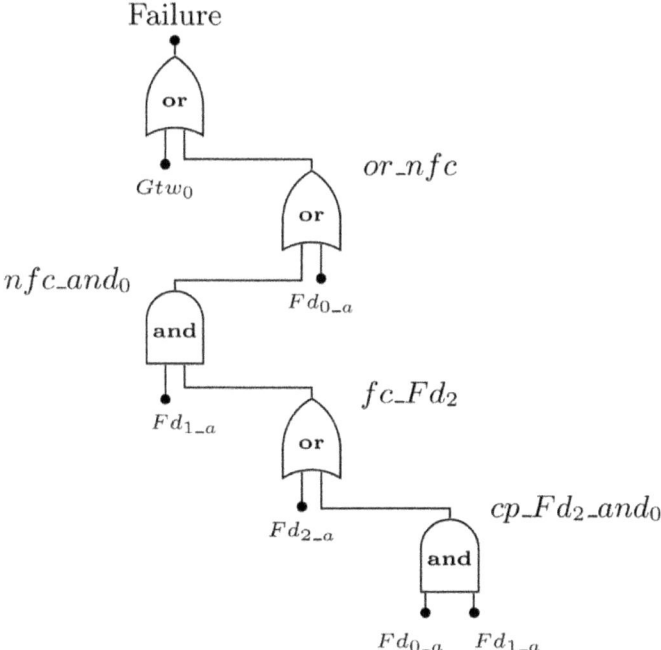

Figure 12. top event definition for the example of Figure 10, when the network failure condition is configured for $Fd_0 + Fd_1 \cdot Fd_2$.

The following step is to transform the generated fault tree into an input file for the sharpe tool. The source code exported to the sharpe is simple and intuitively understandable. The procedure includes: (a) define constants, functions and events; (b) decide if an event is basic or repeated; (c) eliminate inconsistencies; (d) build the fault tree and define the measures to be computed.

The functions used in the source code are related to failure and repair CDFs associated with each field device. If we intend to evaluate the network reliability $R(t)$ then it is only necessary to replace each event related with the failure of device i by the respective CDF (as described in Section 4.3), since sharpe computes the device reliability $R_i(t)$ from that. However, if we are interested in evaluating the network availability $A(t)$ then the process is slightly different. If failure and repair rates are constant (*i.e.*, defined by an exponential distribution) then we can use these parameters to compute the availability function of the device $A_i(t)$ as described in Section 4.3. Then, it is only necessary to replace each event related with the device i by its availability function $A_i(t)$ (note: sharpe has already a function that computes this function from the rates). When other CDFs are used to characterize the failure and repair processes (or at least one of them) it is necessary to use the concept of *model hierarchy* provided by sharpe. The reasoning behind this concept is to use the output of one model as the input of another model. Applied to this case, we can model the behavior of device as a semi-Markov chain with two states: *operational* and *failed*. In the former case the device is operational, and in the latter it is failed. Transitions between these two states are described by failure and repair CDFs. As discussed in Section 5.2, these functions must be expressed using exponential polynomial terms. This model can be solved by sharpe and the respective availability (*i.e.*, *operational* state probability) can be used as the input event of the device in the fault tree. Note that it is not necessary to solve this model in advance. The model's code can be placed in the same code that describes the fault tree, since sharpe analyzes the dependencies between models before computing the results. Details about the use of hierarchical models can be found in [49].

If a device is redundant, sub-events are created according to Figure 7(a); otherwise sub-events are created according to Figure 7(b). An important aspect is to decide if an event is basic or repeated. This aspect is easily analyzed through a search in the basic events of the fault tree. If an event occurs only once, it is considered to be basic; otherwise it is considered to be repeated.

The creation of the fault tree is completely based on discussions presented in Sections 5.4, 5.6 and 5.8. A few precautions should be taken in cases where the logic gates have only one input. This is solved by replacing the logic gate by the input event. Finally, the choice of evaluation function (reliability, unreliability, MTTF, availability, unavailability) or the importance measures (Birnbaum or criticality) is inserted in the source code.

An example of generated sharpe source code is presented in Figure 13. This example is based on the fault tree of Figure 12and assumes that device i failure and repair rates are constant, respectively λ_i and μ_i. Note that there are two repeated events, Fd_{0_a} and Fd_{1_a}. In this example the unavailability function (*inst_unavail*()) was used for the sake of illustration of the functions supported by sharpe. The notation adopted in the source code used to build the fault tree is identical to Figure 12.

$$
\begin{aligned}
&ftree \;\; FaultTreeModel \\
&\quad basic \qquad\; Gtw_0 \;\; inst\_unavail(\lambda_{Gtw_0}, \mu_{Gtw_0}) \\
&\quad repeat \qquad Fd_{0\_a} \;\; inst\_unavail(\lambda_{Fd_0}, \mu_{Fd_0}) \\
&\quad repeat \qquad Fd_{1\_a} \;\; inst\_unavail(\lambda_{Fd_1}, \mu_{Fd_1}) \\
&\quad basic \qquad\; Fd_{2\_a} \;\; inst\_unavail(\lambda_{Fd_2}, \mu_{Fd_2}) \\
\\
&\quad and \qquad\; cp\_Fd_2\_and0 \;\; Fd_{0\_a} \;\; Fd_{1\_a} \\
&\quad or \qquad\;\; fc\_Fd_2 \qquad\;\; Fd_{2\_a} \;\; cp\_Fd_2\_and0 \\
\\
&\quad and \qquad\; nfc\_and0 \qquad Fd_{1\_a} \qquad fc\_Fd_2 \\
&\quad or \qquad\;\; or\_nfc \qquad\;\; nfc\_and0 \;\; Fd_{0\_a} \\
&\quad or \qquad\;\; Failure \qquad\; Gtw_0 \quad\; or\_nfc \\
&end
\end{aligned}
$$

Figure 13: sharpe source code generated by the proposed methodology and based on the fault tree of Figure 12.

RESULTS

In this section we will present some results obtained when using the proposed methodology to evaluate some dependability metrics in WSN. Our main goal is to highlight some of the capabilities of the proposed methodology, regarding the identification of dependability bottlenecks in WSN, and the capability to evaluate reliability and availability in typical industrial application scenarios. The main assumptions considered in this section are listed below:

- **Scenarios:** we have used line, star and cluster (particular case of mesh topology) topologies.

- **Failure rate:** we assume that device failures occur with a constant rate (*i.e.*, exponential distribution). The gateway and the access point have typically a reliability higher than other network devices. Thus, the gateway and the access point have been configured to have a failure rate with one order of magnitude lower than the field devices and routers. The failure rate of the devices is unknown, however we can use different range of values to measure different behaviors. We assume a MTTF (hours) range to 1 year ($\lambda \cong = 1e–4$), 5 years ($\lambda \cong = 2e–5$) and 10 years ($\lambda \cong = 1e–5$).

- **Repair rate:** similarly, we assume a constant repair rate. Although this could be an unrealistic assumption, it can be proved that this approximation results in small errors if $\mu \gg \lambda$ (which is the case). We assume different range of values for the gateways and access points, when compared with the field devices and routers. The MTTR range for the first two devices was configured to 5 h ($\mu = 0.2$) and 10 h ($\mu = 0.1$), whereas the MTTR range for the other devices was configured to 1 day ($\mu \cong = 0.04$) and 2 days ($\mu \cong = 0.02$).

Star Topology

The first assessed scenario was the star topology. Consider an application that is monitoring the temperature of four boilers. A sensor node is installed in each boiler as described in Figure 14. Within this context, it is assumed that the application fails if at least one field device fails.

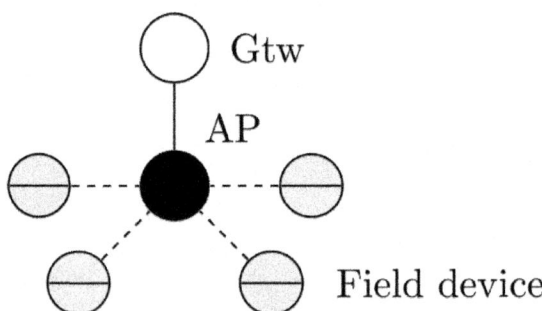

Figure 14: A star topology composed of four field devices.

It is intuitive to realize that the network reliability will increase as long as more reliable devices are used. This behavior is shown in Figure 15. If there is no redundancy, the scenario that uses sensor nodes with failure rate of 1E–4 (MTTF—1 year) presents a network reliability lower than all the other scenarios using more reliable field devices.

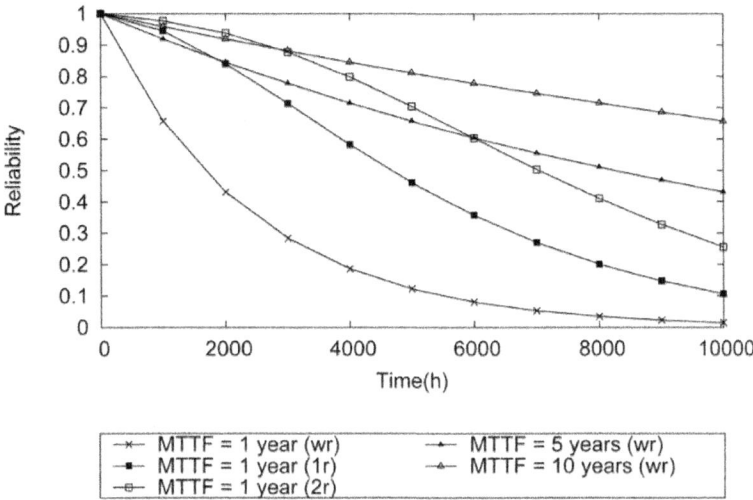

Figure 15: Reliability evaluation for a star topology.

In what concerns dependability requirements, design decisions are usually related to the selection of more reliable and more expensive devices *vs.* less reliable and inexpensive devices. In general, there is a global policy whose goal is to improve the reliability of the applications. The result presented in Figure 15 can be used for that purpose.

On the other hand, in some cases the application requires an increase of the network reliability, but the acquisition of more robust devices is not an option. A possible solution for this problem is the use of redundancy. For instance, consider the topology described in Figure 14, composed of sensor nodes with failure rate 1E–4. Depending on the number of spare devices that can be used, the network reliability can reach levels comparable with those achieved when more reliable devices are used. According to the results shown in Figure 15, when just one spare device is used (1r) in each field device, during 2000 h the WSN achieves reliability levels comparable to the case where field devices five times more reliable are used. If two spare devices are used (2r) in each field device, the network reliability during 3500 h presents levels comparable with a scenario where devices ten times more reliable are used. On the other hand, during 6000 h the network reliability has a performance level that is comparable to the case where devices five times more reliable have been used. This result is three times better than the result found when using just single redundancy (1r).

Another way to evaluate the influence of redundancy is through a MTTF analysis, which is dependent on the network failure conditions. We assume four types of failure conditions: case I, at least one field device fails; case II, at least two field devices fail; case III, at least three field devices fail; case IV, all field devices fail. The results are summarized in Figure 16. For example, if it is considered that the network fails if at least one field device fails, then the use of single redundancy (1r) may increase the network MTTF by 125%, whereas the double redundancy (2r) may increase it by 220%. This MTTF-based analysis can be useful to find the desired application requirements.

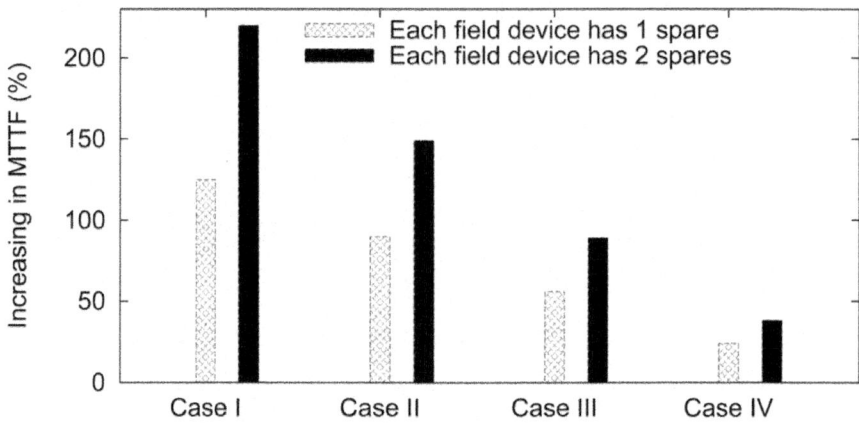

Figure 16: Influence of failure conditions and redundancy levels to the network MTTF.

Line Topology

Line topology is a typical solution used for monitoring pipeline applications. In this case, the information is relayed hop-by-hop until the gateway. Figure 17 illustrates an example of a line topology for a WSN. If a device along the line fails, the monitoring application will also fail.

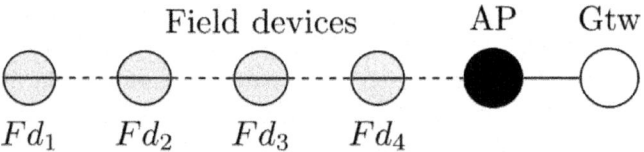

Figure 17: Typical line topology for a WSN.

One of the targets of the proposed methodology is to identify dependability bottlenecks in the network. These impairments can be identified through the use of Birnbaum's measure and criticality importance. A component importance analysis for this scenario is illustrated in Figures 18 and 19. Based on the Birnbaum's measure (Figure 18), the field deviceFd_1 is the device more susceptible to cause a network failure. This behavior is confirmed by the criticality importance, as illustrated in Figure 19. In other words, if the field device Fd_1 fails, there is a high probability that there is no path to the gateway due to a failure in an intermediate device.

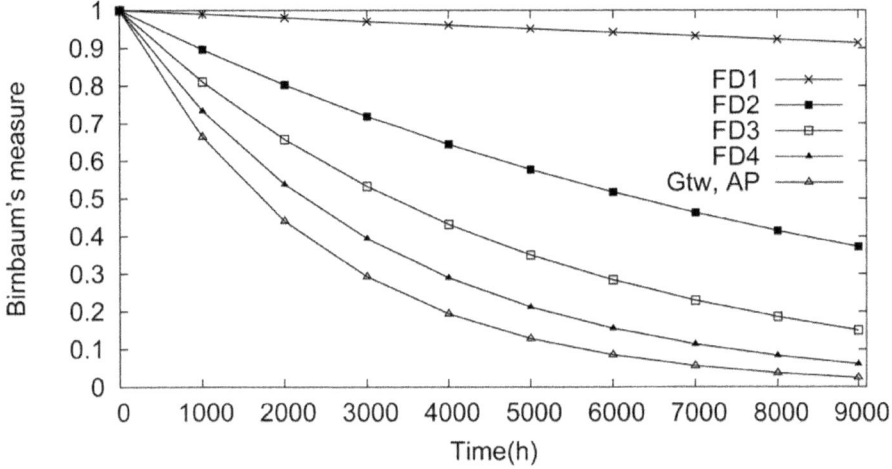

Figure 18: Analysis of the component importance based on Birnbaum measure for the line topology.

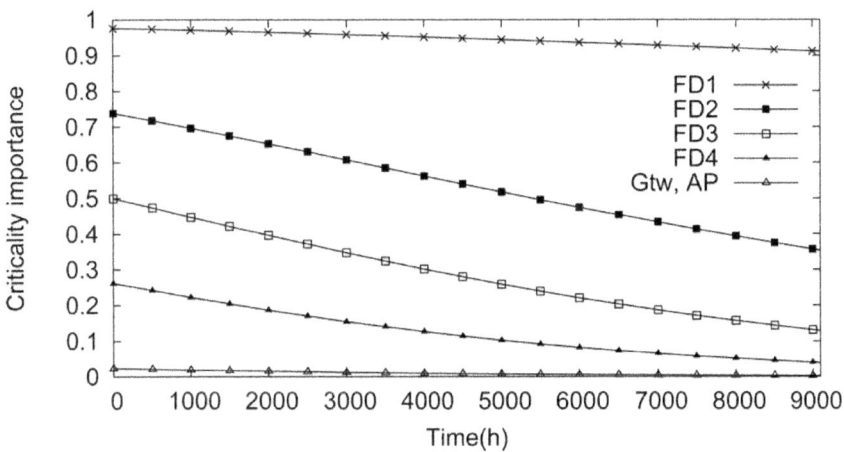

Figure 19: Analysis of the component importance based on Criticality measure for the line topology.

Consider a communication scenario where the application requires an improved reliability level and there is just one available spare device. The problem that must be addressed is to select which device should be equipped with the available spare device, so that the reliability level of the system is maximized. A sensitivity analysis for this problem is shown in Table 1. According to the results presented, if the spare device is configured to be at the gateway or access point, the MTTF of network will be increased by around 2%. Thus, there is no real advantage to configure two spare devices in the gateway or in the access points. The best result, as expected, is attained when the spare device is configured to be at device Fd_1. In this configuration, the network MTTF is increased by 19.23%. This configuration presents even better performance than the configuration with two spare devices in Fd_3 or Fd_4. Note that if the spare device have been configured in Fd_4, the network MTTF would have been decreased by 1.95%.

Table 1: Sensitivity analysis

Device	Increase in MTTF	
	1 spare device	2 spare devices
Gtw	2.32%	2.43%
AP	1.98%	2.40%
Fd_4	−1.95%	14.05%
Fd_3	3.60%	17.19%
Fd_2	10.44%	20.90%
Fd_1	19.23%	25.43%

Cluster Topology

In general, a cluster topology is used when there is the need to segregate partially a network. Each cluster may assume specific tasks, for example, monitoring a region, traffic prioritization, provide redundancy, *etc*. Figure 20 illustrates a typical cluster topology for a WSN, where the clusters communicate each other through router devices.

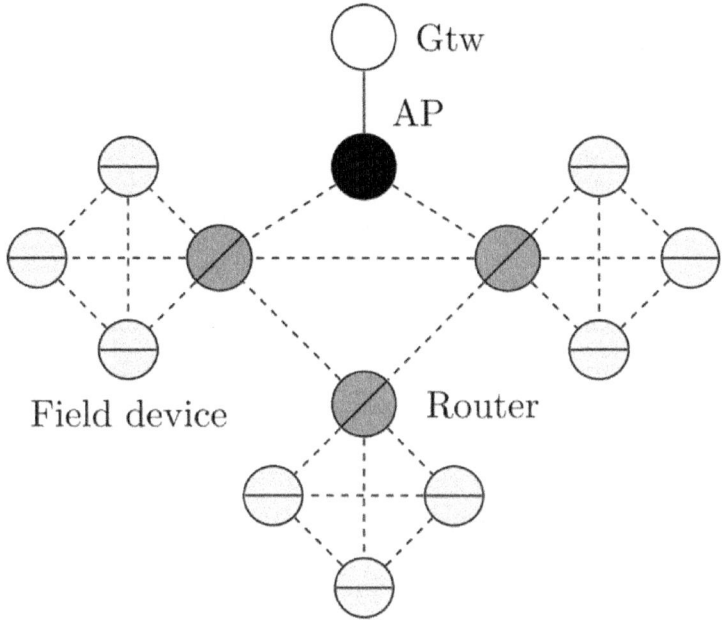

Figure 20: Typical cluster topology for a Wireless Sensor Network.

Consider for example an application where each cluster monitors an industrial control loop. The application will fail if at least one of the cluster fails. On the other hand, a cluster will fail if all field devices within the cluster fail. The dependability target is to maximize the availability of the application and consequently to minimize the application outage (measured in hours per year).

We have analyzed the influence upon the network unavailability of maintenance operations. The results of this analysis are illustrated in Figure 21. It can be observed that changes in the repair rate of the gateway or the access point do not cause significant changes in the network unavailability. On the other hand, the repair rates of the field devices and the routers have strong influence on the network unavailability. If, for example, the repair rate of one these devices is doubled, the application outage is decreased by around 50% (131 hours per year to 66 hours per year).

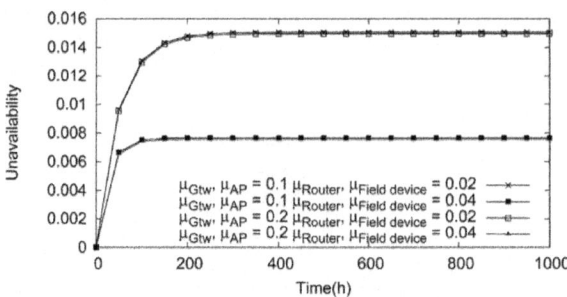

Figure 21: Influence of the maintenance of devices in the network unavailability.

According to the results presented in Figure 20, a critical device is the head of cluster. This function is executed by a router device. Assuming that there are two spare router devices available, the bottleneck of cluster can be minimized in two different ways: configuring a structural redundancy (scenario illustrated in Figure 20) for two router devices where each one receives a spare, or configuring the two spare devices as new router devices in the network as illustrated in Figure 22. The results from this analysis are presented in Table 2. In the first case, when redundancy has been used, the application outage time was around tree times more efficient than the scenario presented in Figure 20 (μ_{gtw}, μ_{AP} = 0.2 and μ_{router}, $\mu_{field\ device}$ = 0.02). On the other hand, the use of two new routers can significantly decrease the application outage by around 65 times. This result is due to the creation of new paths to the gateway, when new router devices are added to the network.

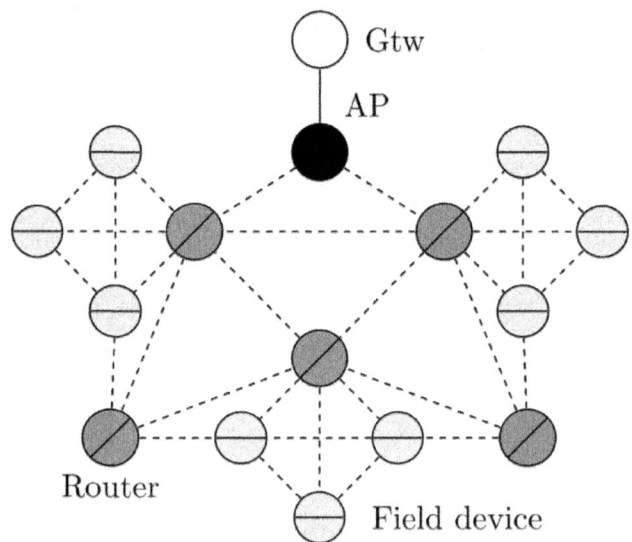

Figure 22. Adding router devices to improve reliability and availability.

Table 2: Means to improve the availability of network

Scenario	Unavailability	Outage (hours per year)
Normal	0.015	131
Repair 2× more fast	0.007	66
Redundancy	0.005	44
2 new routers	0.00025	2

CONCLUSIONS

In this paper we have proposed a methodology to evaluate the dependability of Wireless Sensor Networks in typical industrial environments. For this purpose, we have modeled a WSN using a fault tree-based formalism, considering permanent faults that occur in field devices due to hardware problems and the absence of routes to the gateway.

To validate the proposed methodology, we select several scenarios commonly found in industrial applications. The results obtained show that the proposal is useful to identify dependability bottlenecks, to estimate the required redundancy level and to aid the design throughout the life cycle of the network.

In future works we intend to support unreliable link, thus considering transient failures. Furthermore we also intend to consider the coverage factor related with the reconfiguration mechanisms, and also common-cause failures.

REFERENCES

1. Gungor, V.C.; Lambert, F.C. A survey on communication networks for electric system automation. *Comput. Netw* 2006,*50*, 877–897.

2. Zhao, G. Wireless sensor networks for industrial process monitoring and control: A survey. *Netw. Protoc. Algorithms*2011, *3*, 46–63.

3. Johnstone, I.; Nicholson, J.; Shehzad, B.; Slipp, J. Experiences from a wireless sensor network deployment in a petroleum environment. Proceedings of the 2007 International Conference on Wireless Communications and Mobile Computing (IWCMC '07), Honolulu, HI, USA, 12–16 August 2007; pp. 382–387.

4. Talevski, A.; Carlsen, S.; Petersen, S. Research challenges in applying intelligent wireless sensors in the oil, gas and resources industries. Proceedings of the 7th IEEE International Conference on Industrial Informatics (INDIN '09), Cardiff, UK, 23–26 June 2009; pp. 464–469.

5.   Kirrmann, H. Fault tolerance in process control: An overview and examples of european products. *IEEE Micro* 1987, *7*, 27–50.

6.   Avizienis, A.; Laprie, J.C.; Randell, B.; Landwehr, C. Basic concepts and taxonomy of dependable and secure computing. *IEEE Trans. Dependable Secur. Comput* 2004, *1*, 11–33.

7.   Wolf, F.G. Operationalizing and testing normal accident theory in petrochemical plants and refineries. *Prod. Oper. Manag* 2001, *10*, 292–305.

8.   Ericson, C.A. *Hazard Analysis Techniques for System Safety*; Wiley-Interscience: Hoboken, NJ, USA, 2005; p. 528.

9.   Trivedi, K.S.; Sahner, R. SHARPE at the age of twenty two. *SIGMETRICS Perform. Eval. Rev* 2009, *36*, 52–57.

10. AboElFotoh, H.; Colbourn, C. Computing 2-terminal reliability for radio-broadcast networks. *IEEE Trans. Reliab* 1989,*38*, 538–555.

11. Ball, M.O. Computational complexity of network reliability analysis: An overview. *IEEE Trans. Reliab* 1986, *35*, 230–239.

12. Hou, W. Integrated Reliability and Availability Analysis of Networks with Software Failures and Hardware Failures. Ph.D. Thesis,. University of South Florida, Fowler Avenue Tampa, FL, USA, 2003.

13. AboElFotoh, H.; Iyengar, S.; Chakrabarty, K. Computing reliability and message delay for cooperative wireless distributed sensor networks subject to random failures. *IEEE Trans. Reliab* 2005, *54*, 145–155.

14. Kharbash, S.; Wang, W. Computing two-terminal reliability in mbile *ad hoc* networks. Proceedings of the IEEE Wireless Communications and Networking Conference (WCNC '07), Kowloon, HongKong, 11–15 March 2007; pp. 2831–2836.

15. Egeland, G.; Engelstad, P. The availability and reliability of wireless multi-hop networks with stochastic link failures.*IEEE J. Sel. Areas Commun* 2009, *27*, 1132–1146.

16. Shrestha, A.; Xing, L.; Liu, H. Infrastructure communication reliability of wireless sensor networks. Proceedings of the 2nd IEEE International Symposium on Dependable, Autonomic and Secure Computing, Indianapolis, IN, USA, 29 September–1 October 2006; pp. 250–257.

17. Shrestha, A.; Xing, L.; Liu, H. Modeling and evaluating the reliability of wireless sensor networks. Proceedings of the Annual Reliability and Maintainability Symposium (RAMS '07), Orlando, FL, USA, 22–25 January 2007; pp. 186–191.

18. AboElFotoh, H.; Shazly, M.; Elmallah, E.; Harms, J. On area coverage

reliability of wireless sensor networks. Proceedings of the 36th Annual IEEE Conference on Local Computer Networks (LCN '11), Bonn, Germany, 4–7 October 2011; pp. 584–592.

19. Qureshi, H.K.; Rizvi, S.; Saleem, M.; Khayam, S.A.; Rakocevic, V.; Rajarajan, M. Poly: A reliable and energy efficient topology control protocol for wireless sensor networks. *Comput. Commun* 2011, *34*, 1235–1242.

20. Majdara, A.; Wakabayashi, T. Component-based modeling of systems for automated fault tree generation. *Reliab. Eng. Syst. Saf* 2009, *94*, 1076–1086.

21. Hussain, T.; Eschbach, R. Automated fault tree generation and risk-based testing of networked automation systems. Proceedings of the IEEE Conference on Emerging Technologies and Factory Automation (ETFA '10), Bilbao, Spain, 13–16 September 2010; pp. 1–8.

22. Lapp, S.A.; Powers, G.J. Computer-aided synthesis of fault-trees. *IEEE Trans. Reliab* 1977, *R-26*, 2–13.

23. Kim, J.; Kim, J.; Lee, Y.; Moon, I. Development of a new automatic system for fault tree analysis for chemical process industries. *Korean J. Chem. Eng* 2009, *26*, 1429–1440.

24. Bruneo, D.; Puliafito, A.; Scarpa, M. Dependability evaluation of wireless sensor networks: Redundancy and topological aspects. Proceedings of the 2010 IEEE Sensors, Kona, HI, USA, 1–4 November 2010; pp. 1827–1831.

25. Bruneo, D.; Puliafito, A.; Scarpa, M. Energy control in dependable wireless sensor networks: A modelling perspective.*Proc. Inst. Mech. Eng. O. J. Risk. Reliab* 2011. [CrossRef]

26. Bruneo, D.; Puliafito, A.; Scarpa, M. Dependability analysis of wireless sensor networks with active-sleep cycles and redundant nodes. Proceedings of the 1st Workshop on Dynamic Aspects in Dependability Models for Fault-Tolerant Systems (DYADEM-FTS '10), Valencia, Spain, 28–30 April 2010; pp. 25–30.

27. Silva, I.; Guedes, L.; Vasques, F. A new AODV-based routing protocol adequate for monitoring applications in oil & gas production environments. Proceedings of the 2010 8th IEEE International Workshop on Factory Communication Systems (WFCS '10), Nancy, France, 18–21 May 2010; pp. 283–292.

28. Petersen, S.; Carlsen, S. Performance evaluation of wirelessHART for gactory automation. Proceedings of the IEEE Conference on Emerging Technologies Factory Automation (ETFA '09), Mallorca, Spain, 22–25

September 2009; pp. 1–9.

29. Petersen, S.; Doyle, P.; Vatland, S.; Aasland, C.; Andersen, T.; Sjong, D. Requirements, drivers and analysis of wireless sensor network solutions for the oil gas industry. Proceedings of the IEEE Conference on Emerging Technologies and Factory Automation (ETFA '07), Patras, Greece, 25–28 September 2007; pp. 219–226.

30. *IEEE 802.15.4: Wireless LAN Medium Access Control (MAC) and Physical Layer (PHY) Specifications for Low-Rate Wireless Personal Area Networks (LR-WPANs)*, Available online: http://ecee.colorado.edu/liue/teaching/commstandards/2010F802.15/home.html (accessed on 9 December 2011).

31. *IEEE 802.15.5: Mesh Topology Capability in Wireless Personal Area Networks (WPANs)*, Available online: http://www.ieee802.org/15/ (accessed on 9 December 2011).

32. Lee, M.; Zhang, R.; Zheng, J.; Ahn, G.S.; Zhu, C.; Park, T.R.; Cho, S.R.; Shin, C.S.; Ryu, J.S. IEEE 802.15.5 WPAN mesh standard-low rate part: Meshing the wireless sensor networks. *IEEE J. Sel. Areas Commun* 2010, *28*, 973–983.

33. HCF. *Why WirelessHART? The Right Standard at the Right Time*; White paper; HART Communication Foundation: Austin, TX, USA, 2007.

34. *IEC 62591: Industrial Communication Networks—Wireless Communication Network and Communications Profiles—WirelessHART*, Available online: http://www2.emersonprocess.com/siteadmincenter/PMCentralWebDocuments/EMRWirelessHARTSysEngGuide.pdf (accessed on 9 December 2011).

35. Song, J.; Han, S.; Mok, A.; Chen, D.; Lucas, M.; Nixon, M. WirelessHART: Applying wireless technology in real-time industrial process control. Proceedings of the IEEE Real-Time and Embedded Technology and Applications Symposium (RTAS '08), St. Louis, MO, USA, 22–25 April 2008; pp. 377–386.

36. Chen, D.; Nixon, M.; Mok, A. *WirelessHART: Real-Time Mesh Network for Industrial Automation*; Springer: Berlin, Germany, 2010.

37. *ISA 100.11a-2009 Wireless Systems for Industrial Automation: Process Control and Related Applications*, Available online: http://www.isa.org/Template.cfm?Section=Standards2&template=Ecommerce/FileDisplay.cfm&ProductID=10766&file=ACF5AB8.pdf (accessed on 9 December 2011).

38. Ishii, Y. Exploiting backbone routing redundancy in industrial wireless systems. *IEEE Trans. Ind. Electron* 2009, *56*, 4288–4295.

39. Kushalnagar, N.; Montenegro, G.; Schumacher, C.P. *IPv6 over Low-Power Wireless Personal Area Networks (6LoWPANs): Overview, Assumptions, Problem Statement, and Goals*, RFC 4919. Available online: http://tools.ietf.org/html/rfc4919(accessed on 9 December 2011).

40. Shooman, M.L. *Probabilistic Reliability: An Engineering Approach*, 2nd ed; Krieger: Long Beach, CA, USA, 1990; p. 702.

41. *Military Handbook—Reliability Prediction of Electronic Equipment (MIL-HDBK-217F)*; Technical report; United States Department of Defense: Arlington County, VA, USA, 1991.

42. Rausand, M.; Hsyland, A. *System Reliability Theory: Models, Statistical Methods, and Applications*, 2nd ed; John Wiley & Sons, Inc: Hoboken, NJ, USA, 2004; p. 644.

43. Limnios, N. *Fault Trees—Control Systems, Robotics and Manufacturing Series*; John Wiley & Sons, Inc: Hoboken, NJ, USA, 2007.

44. Choi, J.S.; Cho, N.Z. A practical method for accurate quantification of large fault trees. *Reliab. Eng. Syst. Saf* 2007, *92*, 971–982.

45. Birnbaum, Z.W.; Saunders, S.C. A new family of life distributions. *J. Appl. Probab* 1969, *6*, 319–327.

46. Malhotra, M.; Reibman, A. Selecting and implementing phase approximations for semi-Markov models. *Commun. Stat. Stoch. Models* 1993, *9*, 473–506.

47. Shier, D.R.; Whited, D.E. Algorithms for generating minimal cutsets by inversion. *IEEE Trans. Reliab* 1985, *R-34*, 314–319.

48. Locks, M.O. Inverting and minimalizing path sets and cut sets. *IEEE Trans. Reliab* 1978, *R-27*, 107–109.

49. Sahner, R.A.; Trivedi, K.; Puliafito, A. *Performance and Reliability Analysis of Computer Systems: An Example-Based Approach Using the SHARPE Software Package*; Kluwer Academic Publishers: Dordrecht, The Netherlands, 1996.

CITATION

CHAPTER 1

Felipe Padilla, Aurora Torres, Julio Ponce, María Dolores Torres, Sylvie Ratté and Eunice Ponce-de-León (2011). Evolvable Metaheuristics on Circuit Design, Advances in Analog Circuits, Prof. Esteban Tlelo-Cuautle (Ed.), ISBN: 978-953-307-323-1, InTech, DOI: 10.5772/14688.

CHAPTER 2

David C. Potts (2011). Statistical Analog Circuit Simulation: Motivation and Implementation, Advances in Analog Circuits, Prof. Esteban Tlelo-Cuautle (Ed.), ISBN: 978-953-307-323-1, InTech, DOI: 10.5772/15680.

CHAPTER 3

Reza Hashemian (2011). A New Approach to Biasing Design of Analog Circuits, Advances in Analog Circuits, Prof. Esteban Tlelo-Cuautle (Ed.), ISBN: 978-953-307-323-1, InTech, DOI: 10.5772/14244.

CHAPTER 4

Bruno Apolloni, Simone Bassis, Angelo Ciccazzo, Angelo Marotta, Salvatore Rinaudo and Orazio Muscato (2011). Advanced Statistical Methodologies for Tolerance Analysis in Analog Circuit Design, Advances in Analog Circuits, Prof. Esteban Tlelo-Cuautle (Ed.), ISBN: 978-953-307-323-1, InTech, DOI: 10.5772/15164.

CHAPTER 5

Martin Holger Keding, Olaf Lüthje and Heinrich Meyr, Design and DSP Implementation of Fixed-Point Systems, DOI: 10.1155/S1110865702205065.

CHAPTER 6

G. Stojanovski and M. Stankovski, "A Hybrid System Approach for High Consumption Industrial Furnace Control," Intelligent Control and Automation, Vol. 3 No. 4, 2012, pp. 404-412. doi: 10.4236/ica.2012.34044.

CHAPTER 7

Kleanthis Thramboulidis Georg Frey, Towards a Model-Driven IEC 61131-Based Development Process in Industrial Automation, doi:10.4236/jsea.2011.44024 Published Online April 2011 (http://www.SciRP.org/journal/jsea).

CHAPTER 8

Aamir Shahzad, Malrey Lee Neal Naixue Xiong, Gisung Jeong, Young-Keun Lee, Jae-Young Choi, Abdul Wheed Mahesar, and Iftikhar Ahmad, A Secure, Intelligent, and Smart-Sensing Approach for Industrial System Automation and Transmission over Unsecured Wireless Networks, doi:10.3390/s16030322

CHAPTER 9

Delphine Christin, Parag S. Mogre ,and Matthias Hollick, Survey on Wireless Sensor Network Technologies for Industrial Automation: The Security and Quality of Service Perspectives, doi:10.3390/fi2020096

CHAPTER 10

Ivanovitch Silva, Luiz Affonso Guedes, Paulo Portugal, and Francisco Vasques, Reliability and Availability Evaluation of Wireless Sensor Networks for Industrial Applications, doi:10.3390/s120100806

INDEX